The Contaminants in the Nordic Ecosystem

ECOVISION WORLD MONOGRAPH SERIES

Series Editor

M. Munawar

Managing Editor

I.F. Munawar

Scientific Committee

M. Munawar
M. Luotola
J. Malm
E. Nikunen
E. Nurmi

Technical Commitee

J. Järvi
M.-L. Lindell
J. Lorimer
S. Munawar
N. Munawar
S. Rice

Degradation of the halogenated organic fraction of biologically treated bleached kraft
pulp mill effluents in Finnish lake water mesocosms
E.K. Saski, K. Salonen, A. Vähätalo & M. Salkinoja-Salonen — 109

Polychlorinated diphenyl ethers and chlorophenolic compounds in Salmon (*Salmo
salar*) from the Arctic Teno River compared to the Baltic Sea
P.J. Vuorinen, J. Koistinen, J. Paasivirta, M. Vuorinen & J. Hoikka — 125

Metals in cod (*Gadus morhua morhua* L.) from the Barents Sea
V.M. Savinov — 135

Biomobility of organic halogen compounds from contaminated soil – earthworms as
a tool
M. Laine, J. Jokela & M. Salkinoja-Salonen — 143

Influence of soil and climatic factors on the kinetics of transformation of herbicides
in soil
L. Torstensson & J. Stenström — 151

Variation of cyanobacterial hepatotoxins in Finland
*K. Sivonen, M. Namikoshi, R. Luukkainen, M. Färdig, L. Rouhiainen,
W.R. Evans, W.W. Carmichael, K.L. Rinehart & S.I. Niemelä* — 163

Human impacts on the Hudson Bay Region: Present and future environmental
concerns
P.G. Sly — 171

Workshop sessions — 265

List of Participants — 269

TABLE OF CONTENTS

Preface
    M. Munawar & M. Luotola — vii

Salutation
    L. Tarasti — ix

Opening Address
    K. Bärlund — xi

Chemicals control in the Nordic countries: Approaches to risk reduction
    B. Bucht, I.G. England, A. Lundgren, E. Nikunen & H. Tyle — 1

Modelled and observed fate of selected organochlorines in the Nordic environment
    J. Paasivirta, S. Sinkkonen, T. Rantio, D. Calamari, A. Di Guardo, M. Matthies & S. Trapp — 11

Cadmium and mercury in north European forest ecosystems
    M. Lodenius — 25

Swedish risk assessment of antifouling products: pleasure boats and commercial vessels
    C. Lyle, L. Törnquist, C. Debourg, A. Johnson & C. Unger — 33

Management of environmentally hazardous chemicals in Finland
    L. Ylä-Mononen — 39

Use and emission scenarios for methyl chloroform and carbon tetrachloride in Finland
    A.B. Mukherjee — 43

Ecological situation in the Leningrad Region: Emissions, transboundary air pollution, decline of the conifer forest (results of a three-year assessment and monitoring survey of the effects of air pollution on forests)
    N.I. Goltsova, T.V. Vasina & B.G. Popovichev — 49

Pollution-related environmental gradients around the "Severonikel" smelter complex at the Kola Peninsula, Northwestern Russia
    M.V. Kozlov & E. Haukioja — 59

Effects of environmental pollutants on the cold-hardiness of Arctic and boreal ectothermic animals
    K.E. Zachariassen & R. Lundheim — 71

Mechanisms of thermal acclimation in relation to ecotoxicology
    K.Y.H. Lagerspetz — 85

The role of natural organic material on the fate and toxicity of xenobiotics in aquatic environment
    J. Kukkonen — 95

CIP-DATA KONINKLIJKE BIBLIOTHEEK, DEN HAAG

Contaminants

The contaminants in the Nordic ecosystem: the dynamics processes and fate / ed. by M. Munawar and M. Luotola. - Amsterdam : SPB Academic Publishing. - Ill. - (Ecovision world monograph series)
With ref.
ISBN 90-5103-108-4 bound
Subject headings: Nordic ecosystems / aquatic ecology / aquatic ecotoxicology.

ISBN 90-5103-108-4

Distributors:

For the U.S.A. and Canada:
SPB Academic Publishing bv
c/o Demos Vermande, Order Department
386 Park Avenue South, Suite 201
New York, NY 10016
Telefax (+212) 683-0072

For all other countries:
SPB Academic Publishing bv
P.O. Box 11188
1001 GD Amsterdam, The Netherlands
Telefax (+31.20) 638 0524

© 1995 SPB Academic Publishing bv, P.O. Box 11188
  1001 GD Amsterdam, The Netherlands

All rights reserved. No part of this book may be translated or reproduced in any form by print, photoprint, microfilm, or any other means without prior written permission of the publisher.

THE CONTAMINANTS IN THE NORDIC ECOSYSTEM: THE DYNAMICS, PROCESSES AND FATE

edited by

M. Munawar and M. Luotola

SPB Academic Publishing bv/Amsterdam/The Netherlands

Preface

M. Munawar[1] & M. Luotola[2]

[1]*Canada Centre for Inland Waters, P.O. Box 5050,*
687 Lakeshore Rd., Burlington, Ontario, Canada, L7R 4A6
[2]*Finnish Environmental Agency, P.O. Box 140,*
FIN-00251 Helsinki, Finland

The First International Nordic Symposium on Chemicals in the Arctic-Boreal Environment was held in Helsinki from 13-15 May 1993. This book includes selected papers and posters presented at this meeting. In addition it contains summaries of the workshops to provide a holistic over view of the Symposium.

It is important to understand clearly the objectives and rationale of the Symposium. Is northerliness of special significance or relevance when harmful chemicals are considered? If so, why? In Finland the research on hazardous chemicals received special attention during 1992 at the Water and Environment Research Institute, Helsinki. A research program was established to emphasize the problem areas relevant to Finland. By far the most critical issues of the program dealt with the determination of special characteristics of northerliness which may have the strongest effect on the environmental fate and toxicity of harmful chemicals. Furthermore, considerable attention was given to the usage of such characteristics in the decision-making for chemical's management. This issue is of prime importance now as Finland and Sweden are members of the European Union. This would mean increased integration of data requirements and criteria applied in risk assessment.

When the above-mentioned project (research in Northern parts and periphery) was being planned, background information was extremely hard to find. Very little information is available about the ecotoxicology of Northern areas. Consequently the idea of convening an international symposium emerged (M. Luotola, Symposium Chair) to bring together active researchers, administrators, managers, and students who are deeply interested in exploring the complexity of Nordic ecosystems. The Symposium was to serve as a forum for reporting and exchanging research data, information, and for establishing contacts. Scientists, mainly from the Nordic countries (Finland, Sweden, Norway, Denmark), as well as from Canada, the United States and Russia were invited.

Approximately 125 participants worked hard and enthusiastically for two and one half days on a program consisting of five sessions (54 presentations; 19 oral and 35 posters) and two workshops. The program focused on the following themes:

* Northern Climate and the Decision-Making on Chemicals
* The Critical Features of the Northern Environment in Respect of Toxicity and the Environmental fate of Chemicals
* The Effects of Chemicals from the Ecological Point of View
* Environmental Modelling
* Individual Presentations with Relevance to the Symposium Theme

Workshops:

* The means of assessing the ecotoxicological effects and risks related to chemicals in northern conditions for the administrative decision-making procedures.
* Data gaps in research concerning chemicals in the Nordic environment and possibilities for cooperation in research.

Approximately twenty manuscripts were submitted to the Ecovision World Monograph Series (M. Munawar, editor) for publication. The peer review process for the manuscripts included in this book was strict, extensive, meticulous, and time consuming. Each manuscript was reviewed by at least two referees, and the Co-editors. The accepted manuscripts were then thoroughly reviewed and evaluated by the technical editor for their linguistic and stylistic requirements. The manuscripts were then sent to authors for revision. The revised manuscripts were once again evaluated by the co-editors and referees. Some of the manuscripts were subjected to more than one revision to facilitate further improvement and amendments. The resulting revised version was then assessed by the Desk editor for its completeness, format (according to Ecovision guidelines), and overall presentation. The final version following the desk editing was checked by the Co-editors and accepted for publication. Several people listed in the Scientific & Editorial/Technical Committees assisted significantly in various stages of processing which lead to the publication of this monograph, including the beautiful front and back cover design. We thank the referees (>50 in number) for their patient, meticulous, and constructive reviews of the manuscripts, the authors for their careful revisions and cheerful cooperation, the Technical and Desk editors for their hard work, which enhance the quality of this publication. We are also grateful to Wil Peters and P. Bakker of SPB Academic Publishing, The Netherlands, and Iftekhar Munawar, Managing editor of the Ecovision World Monograph Series, for their continued support, encouragement, and assistance throughout the publication process.

The Symposium was sponsored by the Nordic Council of Ministers, the Finnish Ministry of the Environment, and the Finnish Environment Agency (formerly known as: Finnish National Board of Waters and the Environment). The Symposium was organized by the Water and Environment Research Institute and the Chemicals Control Unit of the Finnish Environment Agency. The encouragement and assistance received from the Aquatic Ecosystem Health and Management Society in various aspects of this publication is greatly appreciated. On behalf of the Organizing Committee, we would like to offer our sincere thanks to the sponsors and all those who contributed to the success of the first Nordic Symposium. Many useful and essential research contacts have been established as a result of this maiden Symposium. We hope that this peer reviewed publication will provide the state-of-the art status of the research in the Nordic ecosystems and will be considered as a significant contribution to science by future generations to come.

Salutation

Lauri Tarasti

Secretary General,
Ministry of the Environment,
Finland

Until the present time, studies on the occurrence of chemicals and their effects on the environment have mainly been undertaken in warm and temperate climates. During the last few decades, however, man has extended his activities even further into remote areas of our earth, to the Antarctic and the Arctic. There are plans for exploiting the hitherto unobtainable natural resources in extreme Arctic surroundings. Meanwhile, production technologies developed in warm and temperate climates do not necessarily function as envisaged in a situation where the temperature is extremely low. One of the results may be environmental damage from chemicals. Thus, we have to intensify research on how chemicals behave in extreme climates.

The Finnish environmental administration has in recent years paid increasing attention to the state of the Arctic environment. At the initiative of Finland, Ministers from eight Arctic states convened in Rovaniemi in June 1991, to discuss and approve a common strategy and an action plan for the protection of the Arctic environment. The action plan also focussed on chemicals, and measures were proposed for further investigation of their harmful effects in Arctic environments. The Ministers convened again in Nuuk, Greenland, during September of 1993, to discuss an amended action plan dealing with environmental consideration of chemicals.

Transboundary pollution plays a major role in the deterioration of the Arctic and boreal environment. Airborne pollution has gained special attention, particularly pollution consisting of stable organic compounds and heavy metals. It is very important that the ECE Convention on Long-range Transboundary Air Pollution be supplemented in these respects.

The Nuuk meeting prepared proposals for extending the coverage of the protocols on sulphur and nitrogen emissions under the Air Pollution Convention to all of the Arctic area as well. One of the problems in Arctic environmental cooperation is the insufficiency of data on emissions in the area. The investigation of environmental effects of chemicals is also hampered by our scant knowledge of the basic structures of nature in the far North, which means that we cannot as yet fully understand the Arctic ecosystems.

In addition to their detrimental effects on the natural environment, chemicals in Arctic surroundings also affect the health of people. This very year is the UN Year of Indigenous Peoples. Most of these peoples inhabit Arctic and boreal areas. In these areas the daily food supply depends on what nature provides. The shortness of the food chain, especially as regards reindeer, seals and fish, becomes an added danger to human health when food sources are contaminated by dangerous and toxic chemicals. One-sided nutrition and the rapid transfer of chemicals in the food chain to human food are serious health problems. The investigation of the occurrence of chemicals and the forestalling of detrimental effects is a matter not only of culture, business and livelihood, including daily food of the Arctic peoples, but even a matter of life and death.

An additional problem in cold climate is the slowness of biological processes. In the natural cycles the biological decomposition of any superfluous substances, such as waste, is extremely limited. For this reason waste management, particularly as refers to chemicals and hazardous wastes, is a matter of prime importance. Waste recovery and transportation away from the Arctic assumes the same importance as in the Antarctic, where internationally agreed waste transportation has been found to be a necessity. The organization of waste management calls for more information on the effects of chemicals on northerly ecosystems.

In introducing the concept of Arctic environmental cooperation, Finland has proven her concern for the Arctic environment and the wellbeing of the indigenous peoples. We have to promote research especially on chemicals, ecology and toxicology, in order to establish a base for required environmental protection and conservation measures. Researchers from different countries should take part in this work in accordance with the international character of this field, sharing and learning from the difficulties ensuing from the harsh local conditions.

Today we are lucky to be able to discuss research on chemicals in an Arctic environment under pleasant local conditions. I would like to present my best wishes for a successful meeting and internationally useful results of this important symposium on chemicals in the Arctic-boreal environment.

Symposium opening and theme introduction

Kaj Bärlund

Environmental Director,
United Nations Economic Commission for Europe,
Geneve, Switzerland.
(Formerly: Director General,
Finnish National Board of Waters and the Environment)

On behalf of the National Board of Waters and the Environment I take this opportunity to welcome all of you: experts, officials and students in the environmental sciences. Welcome to Finland and to this International Nordic symposium on Chemicals in the Arctic-Boreal Environment. I am pleased to see that so many experts from abroad have had the opportunity to come to our spring-time capital.

The location of this meeting, the Marina Congress Centre, is actually an old warehouse that was converted for the purpose of hosting the fourth CSCE Follow-up Meeting in Helsinki in July 1992 and continues to serve as a conference centre and banquet room.

The central topic of our symposium is northerliness and how this northerliness may aggravate the impacts of chemicals and pollutants on our ecosystems, which are so vulnerable to the harsh natural conditions of the north.

As the representative of the Ministry of the Environment stated earlier, the environmental authorities in Finland, together with their counterparts form other Arctic nations, have in recent years paid a lot of attention to the state of the Arctic environment and its protection. In fact, some progress has already been achieved, but a great deal remains to be done.

Located on the southern border of the Arctic and Subarctic zone is the boreal forest zone, a vast forest biome which covers inland Canada, Fennoscandia and a wide slice of Russia. This area may be less barren than the Arctic and Subarctic zones, but it has many features in common with those zones: snow cover for several months during the year; long, sometimes very cold, winters; a widely variable climate, and so on.

For Finland, the boreal zone is very important. It is the area in which we carry on farming and forestry. Finland is the northern-most agricultural country in the world. Our cultivated area actually begins at a latitude where it ends in the rest of the world. The warm Atlantic Gulf Stream makes this possible.

International and national legislation on chemicals has advanced in recent years. It seems that some environmental authorities are even sighing with deep relief, believing that their problems are practically solved now that laws have been enforced.

The matter might well be seen form another perspective, however. Indeed, researchers in ecotoxicology fear that many of these environmental authorities have been creating a "virtual reality".

This "virtual reality" is the idea that the effects of chemicals can be assessed by standardized laboratory tests and by monitoring the environment. This "virtual reality" is that once a chemical has been tested and the specific requirements for it have been decided upon, the problem is considered to be solved. It is as if the chemical had ceased to exist – even though it has only just begun its migration into soil, air, water and organisms.

However, many researchers rightly fear that this "virtual reality" can never simulate reality and the effects of chemicals in ecosystems.

True, a lot of progress has been made since the late seventies. Chemicals administrators around the world have been working systematically to predict and minimize the risks of chemicals.

This work will hopefully result in integrated controls which can prevent those situations which have repeatedly surprised and confused researchers and environmental authorities.

Rather, I refer to those situations in which a chemical is first tested and then approved and finally verified as not harmful. It is verified not to migrate in the ecosystem but to degrade quickly and to be in every way acceptable for use. Then, when already in the field, it is identified as being the cause of unforeseen damage and is found in all sorts of places where it should not be.

The conclusion is: tests can be standardized – the ecosystems cannot.

The variability of biological factors is unlimited. Abiotic factors, too, may take observers by surprise. Some winters are almost without snow, some summers are excessively rainy, other summers are dry and hot.

What I have said is by no means meant to lessen the accomplishments of environmental authorities. Indeed, great progress has been made especially in the chemicals sector of environmental protection.

But I want to warn that a very real problem must not be hidden in red tape – the problem of the effects of chemicals on wildlife populations, biotic communities and ecosystems. I want to warn that any false sense of security must be discouraged.

In the assessment of environmental and health risks, the continued presence of "old" chemicals represents a real problem area. As no advance inspections or registration were required in the past, authorities may not have even the most basic information on chemicals that have been or still are in common use.

The existing chemicals programme which has been launched by the OECD and the EC is a very positive development. The programme is made up of a number of projects which provide risk assessment information on "old" chemicals which continue to be in use.

The way ahead is through cooperation among researcher, authorities and chemical manufacturers – within a forum that is both national and international. Financial support for basic research should be increased and never be out-weighed by administrative costs.

There are two great challenges before us. The first challenge is to create a more firm scientific basis for the development of new and more relevant test procedures – test procedures that will enable us to predict the actual effects of chemicals on wildlife populations, biota and ecosystems. The second challenge before us is to perceive and to predict serious ecological change.

The ecotoxicology of northern areas is a relatively unknown field. Authorities in the Nordic countries have a very challenging task in administrative decision-making when the effects of chemicals are largely assessed on the basis of research results obtained from climate zones which are temperate and warm.

Under these circumstances, it is not unusual that chemicals are admitted into the northern environment on the basis of what has been concluded from *ad hoc* standardized tests carried out at room temperature and carried out on unfamiliar species.

Another issue of current interest is the effect of European integration and the EEA agreement on our legislation. Should our decisions be made in accordance with the European standards, or may allowance be made for the special conditions of our environment.

In northern conditions, the variability of the behaviour and the effects of chemicals is induced by a number of factors such as coldness, shallow humic waters, the quality of organic matter in soil as well as the unique biotic species and ecosystems.

Coldness slows down the degradation of chemicals. Thus they remain present for long periods, thereby extending the time for organisms to accumulate them.

Northern organisms have developed different means of adapting to the cold. The survival stress involved under such conditions is considerable. Anything extraordinary, exposure to chemical contaminants, for example, markedly contributes to the stress. In a land that remains under snow and ice for most of the year, the physiology of organisms is heavily taxed in the process of adapting to the changing conditions. As the impacts of pollutants reflect this seasonal variation, they can hardly be assessed by standard condition tests of short duration.

Northern organisms typically accumulate fat for use as an energy source during the cold period. Many chemicals are fat-soluble and as such they are richly accumulated in fat. When organisms in certain energy-consuming processes use large amounts of the accumulated fat, chemicals may be mobilised in the metabolism in amounts large enough to bring about a serious exposure. This can happen, for example, during lactation. It can happen in the nest building season and also in winter under the ice cover. Many fish use up their entire stored energy, and so on.

Coldness and contaminants make a fatal combination for vegetation and especially for boreal forests. Examples of this are abundant. Although boreal forests are particularly adapted to growing in a cold climate, they have weaker mechanisms for fighting pollutants than the more southern forests.

Several speakers at this symposium will deal with these specific features of northern wildlife and we hope they will shed light on the most important aspects.

Chemicals and pollution control policies are international in outlook. Chemicals policy and pollution policy are often separated. One reason for this is that for practical reasons they are mostly handled by different people. Chemicals policy is usually understood to refer to the control over the production and use of chemicals before they become wastes. Pollution policy is understood to refer to the control over wastes discharged into the environment: atmosphere, sea or ground. This distinction is quite artificial and hard to maintain, especially as the two fields obviously are more or less overlapping.

At present, a fruitful integration of these two fields is an issue much discussed at many levels, for example the OECD and also the European Community.

In matters related to chemicals, the Nordic countries often act as a group using their joint influence in international organizations for the common purpose of having their specific needs taken into consideration in international decisions and agreements. The most important organizations in this respect are the European Community and the OECD. There are several important projects in progress to lessen the harm done by chemicals.

We shall hear more on the chemicals control policies in the Nordic countries during the first session of the day.

Although the name of this symposium includes the word chemical we shall deal with the matter rather flexibly, ignoring the clear difference that the language of law makes between chemicals and pollutants, which I mentioned earlier. In this respect, we prefer to be a little ahead of the development.

As the time reserved for the symposium is limited, it is not possible to discuss all the topics to their full extent. Following the general idea of the symposium, it is our aim to provide an occasion where northerliness is viewed as a key factor that more or less contributes to the effects and fate of chemical pollutants. For our discussions, we naturally need information about the occurrence and migration of chemicals. The most important aim we have is that of putting the available data within a framework. A framework that can provide an overall view and a setting for creative thinking. We hope to reach an understanding that will help us see how we can protect the delicate northern environment in line with sustainable development; retaining the values of nature and yet providing for the continuance of industries vital to human welfare.

It is not the policy of this meeting to produce communiques. But we hope the workshops will produce motions for starting concrete cooperation.

Once again, welcome to this symposium. I hope you enjoy the meeting and return home with a wealth of new ideas for your work and many new contacts both at the scientific as well as the personal level.

Chemicals control in the Nordic countries: approaches to risk reduction

B. Bucht,[1] I.G. England,[2] A. Lundgren,[1] E. Nikunen,[3] & H. Tyle[4]

[1]National Chemicals Inspectorate, P.O. Box 1384, S-171 27 Solna, Sweden.
[2]State Pollution Control Authority, P.O. Box 8100 Dep, N-0032 Oslo 1, Norway.
[3]Finnish Environment Agency, P.O. Box 140,
FIN-00251 Helsinki, Finland.
[4]Danish Environment Protection Agency, Strandgade 29,
DK-1401 Copenhagen K, Denmark.

Keywords: legislation, hazards, idustrial, risk, safey, international

1. Introduction

Chemicals play an important role in practically all sectors of modern societies and the chemical industry has expanded dramatically during recent decades. New chemicals are continuously introduced to the market and the existing chemicals find new uses. The use of chemicals has contributed to our high standard of living and our modern lifestyle. However, this development has also increased the chemical load on the environment. Therefore, the management of chemicals, with respect to their effects on the environment, needs keen attention and improvement.

National legislation constitutes the basis for comprehensive chemicals control with respect to hazards to the environment as well as to human health. The Nordic countries have modern legislation in this area, which originates as described: the Swedish Chemicals Act 1986, the Finnish Chemicals Act 1989, the Norwegian Product Control Act 1991 and the Danish Chemicals Act 1993. The acts are supplemented by several governmental ordinances and by regulations issued by governmental agencies.

The acts cover all chemicals for industrial, consumer or other use with the exception of pharmaceuticals, foodstuffs and animal feedstuffs. Pesticides are
included in the Swedish and Danish Chemicals Act, but are regulated separately in Finland and Norway.

The emphasis in the chemical laws is put on the marketing of chemicals but the laws cover all stages of the 'life-time' of a chemical from production, through processing, treatment, packaging, storage, sale, to use and, finally to waste disposal (with the exception of the Norwegian Act). According to the regulations, the manufacturers and importers have the main responsibility in most cases. However, the use of chemicals is mainly regulated by other laws, *e.g.* acts on work environment and on environmental protection.

2. Strategies for chemicals control

2.1 *The objectives*

The legislation on chemicals control deals mainly with the overall aim that the use of chemicals should not harm humans or the environment.

OBJECTIVES FOR CHEMICALS CONTROL

1. Better knowledge of the kind of risks the use of chemicals give rise to.	- Increased basic knowledge of the toxicological and the ecotoxicological properties of chemicals and of the risks caused by them.
	- Improved information on the users of chemicals
2. Better choices of and a safe handling of chemicals.	- Exchange of hazardous chemicals for less hazardous ones. (*The principle of substituion*).
	- Safe handling of the chemicals with respect to workers health, consumer safety and environment

---

In order to reach the risk reduction needed, each one of the objectives has to be accomplished to a satisfactory degree; the precautionary principle being the leading principle.

2.2 *Transparency*

A systematic stepwise approach in decision making is essential for transparency and for more effective risk reduction. This stepwise approach is applied for classification and labelling as well as for decisions to ban or restrict chemicals or other precautionary measures.

The first four steps are purely fact oriented, giving background information for decision making. The final step, the risk-benefit assessment, includes value-based assessments. The results of the assessments may therefore vary between countries and, with time, even within a country.

Classification and labelling of chemicals is based on hazard analysis whereas in all other risk management measures, risks and benefits are to be accounted for as well.

STEPWISE DECISION MAKING

1. Hazard analysis	- Assessment of the intrinsic properties of chemicals and their capacity to harm humans or environment
2. Risk analysis	- Estimation of the probability of any harm occurring and its likely extent
3. Benefit analysis	- Assessment of possible benefits with a certain use of a chemical
4. Consequence analysis	- Prediction of the consequences of choosing a certain decision alternative
5. Risk-benefit assessment	- Assessment based on an acceptable risk as perceived at the time of decision
6. Decision	- Also influenced by value and political, economical etc. judgements.

2.3 International cooperation

Chemicals are, to a high degree, subject to international trade. Consequently, international cooperation is needed in order to effectively cope with the hazards caused by chemicals. This applies for testing and hazard assessment, for classification, labelling, and for bans or restrictions. Another driving force for international attempts to coordinate or even harmonize chemical legislation, is the general objective to remove the technical barriers to free trade with chemicals.

International cooperation is an important part of the strategy for chemicals control in the Nordic countries. Cooperation between Nordic countries has been a long tradition in this field and as such was formalized a decade ago, by the establishment of the 'Chemicals Group' under the Nordic Council of Ministers. The task of the 'Chemicals Group' is to coordinate, supervise, make priorities, and allocate funding of joint Nordic projects in the field of chemicals control and thereby promote cooperation between the Nordic countries. The 'Chemicals Group' has sponsored a large range of research and development projects in this context, many of which have been used directly by the administrations of various Nordic countries; some of which have significantly influenced decision making dealing with chemicals control at an international level.

Each Nordic country, however, devotes also its own resources for participation with organizations such as various UN bodies and OECD, as well as for cooperation with the Council of Europe, European and Mediterranean Plant Protection Organization (EPPO), and different countries. Since the 70's, cooperation with EC has been of utmost importance because Denmark has been a member of EEC since 1972. The reason being, that most of the Danish chemicals legislation is totally harmonized with EC decisions, directives and regulations. In addition, close cooperation with the EC has increased for the other Nordic countries as a result of the agreement of the European Economic Area (EEA). Finland and Sweden recently became members in the European Union and consequently, their chemicals legislation is being harmonized with EC directives.

2.4 *Division of responsibility*

In the Nordic Acts on chemicals, it is stated that the main responsibility for accomplishing the goals for the chemicals control lie with the manufacturers and the importers. They have to investigate and assess their chemicals with respect to hazardous properties and decide whether or not to market them with those properties in mind. The chemicals marketed have to be classified and labelled accordingly and accompanied with Safety Data Sheets when sold for occupational use.

The users of chemicals have also their responsibilities. They are urged to choose chemicals with the lowest possible degree of hazard and to organize their use in order to minimize or avoid negative effects on humans or to the environment.

The main role of the authorities is to issue the regulations needed to implement the legislation, to give guidance, and to supervise the chemicals management practiced by industry and others handling chemicals. The authorities are expected to anticipate risk to humans or to the environment and take action based on scientific reasons. If a company does not agree, then it is the company's duty to demonstrate that there are no grounds for any action.

Each Nordic country has appointed a competent authority (in Finland and Norway, several authorities), which bears the main responsibility for the enforcement of the legislation especially concerning the marketing of chemicals. This authority has the power to issue the regulations needed to enforce the law and to supervise the importers and producers of chemicals (as substances in preparations and in various materials and products).

The use of chemicals can be supervised by other authorities whose main responsibility is for workers' safety and the environment. There are also supervising bodies at the regional and local community levels.

3. Knowledge of the intrinsic properties of chemicals

The pressing need for better understanding of the toxicological and ecotoxicological properties of old (existing) chemical substances can only be remedied by means of international cooperation. Therefore, Sweden was one of the initiators of the OECD-programme on existing chemicals in which all Nordic countries participate actively. This includes data gathering, testing and assessment of data on certain High Production Volume Chemicals (HPVC). Further, the Nordic countries also actively participate in the OECD Test Guidelines Programme under which the development of the prescribed or recommended testing methods for obtaining the data on the intrinsic properties of chemicals is carried out.

According to the notification system of new chemical substances in the EC, a standard package on various physicalchemical, toxicological, and ecotoxicological test data on new chemicals ('the base set of data') have to be given to the competent authority in the member state before the substance may be marketed for the first time within the

EC. Thus, this notification system for new chemicals has been in force in Denmark – like the other EC countries – since 1981. The notification system has now been updated in Denmark and Finland according to the '7th Amendment' (dir. 92/32/EEC), effective November, 1993. Sweden and Norway have recently introduced a notification system for new substances in their legislation.

The risk assessment procedures agreed upon for new substances marked the beginning of the development of risk assessment principles under the EC Council Regulation on Existing Substances. The aim of the work on existing chemicals, as well as on new chemicals, is to increase our knowledge of the toxicological and ecotoxicological properties of chemicals to a level sufficient to assess risks and to regulate those of greatest risk in order to avoid negative effects on humans and the environment.

4. Classification, labelling and Safety Data Sheets

The Nordic authorities have issued regulations for the classification and labelling of chemicals and for Safety Data Sheets which are identical (Denmark and Finland) with or very similar to the EC regulations. Classification criteria for health hazards have been in operation for a long time, but the criteria for classification of environmentally dangerous chemicals are relatively new. A proposal for such criteria was developed in a joint Nordic project in 1988. The criteria were further elaborated by the EC and formed the basis for the common environmental classification system now in operation in most European countries.

The fundamental strategy behind the classification and labelling system is that the user should have ready access to relevant information on significant intrinsic, harmful properties of the chemicals they use. Presently, one of the problems of the implementation of the EC classification system is, (according to the Nordic viewpoint), that it does not take epidemiological evidence on chronic neurotoxicity of chemicals (chemically induced dementia) appropriately into account. Depression of the function of the central nervous system caused by long term exposure to a range of organic solvents is, however, recognized both as a scientific fact (WHO 1985) and in the context of labour compensation legislation in all of the Nordic countries – in contrast to many EC countries. The Nordic countries are therefore continuously engaged in toxicological and epidemiological research in this area and are taking various initiatives to influence decision making regarding classification, labelling and risk assessment of neurotoxic organic solvents.

The Norwegian and Swedish systems for classification, labelling of carcinogenic substances and preparations, include a differentiation in labelling requirements according to the carcinogenic potency of the substance. According to this potency ranking system, carcinogenic substances are divided into three categories: 'high', 'medium', and 'low' potency carcinogens. Labelling requirements are then differentiated according to this potency ranking. The EC classification criteria for carcinogens do not include a potency ranking system, and therefore do not systematically classify the most potent carcinogenic preparations strictly.

Work is continuing, to improve and update existing criteria for classification and to elaborate on new ones. This work is also done on a Nordic basis and in close cooperation with the EC, where the contribution of Denmark and the other Nordic countries are significant. Also, the Nordic countries play an important role in the work done in OECD and in UN organizations on harmonization of regulations for classification and labelling.

5. Increasing knowledge on the use of chemicals

In order to assess chemical risks, it is necessary to have information – not only on the inherent properties of chemicals but also on their use. The use patterns and the volumes used, determine, to a very large extent, the exposure for humans and/or the environment. To address this problem, the Nordic countries have established and maintained Product Registers to identify end-uses of chemical substances. These registers contain computerized information on the chemical composition of a vast number of preparations (products). The importers and manufacturers are obliged to give the notice of chemicals they market (*e.g.* trade and chemical names, composition of substances in preparations, volumes, type of chemical, intended use, classification etc). In fact, the significance of the unique information of the Nordic Product Registers has also been recognized and used in various contexts on an international scale.

One of the main conclusions of the OECD Workshops on consumers, workplace and environmental exposure assessment was, in fact, to recommend that the OECD countries 'take advantage of mandatory release reporting requirements or other data sources of several member countries. Product Registers are examples of this kind of relevant data' (OECD 1993). Further, the 'use categorization' methodology developed and employed by the Nordic Product Registers have had a major influence on the development of use categorization methodologies which are now in force in various OECD and EC programs on new and existing chemical substances.

The information from Product Registers, especially the composition of preparations, is also being used in the present expansion of the existing EC classification and labelling system. This system, which covers human health hazards of chemical preparations, is being expanded to also cover environmentally dangerous preparations, . The information in the registers gives a basis for the risk identification needed for planning and priority setting, as well as for the choice of the most adequate methods for risk reduction. When needed, the information from the registers is supplemented by investigations of the use and exposure to chemicals in various selected sectors of industry, etc.

6. Very hazardous chemicals are to be avoided

The Swedish and Danish Chemicals Acts state that everyone who handles chemicals should avoid chemicals which can be substituted with less hazardous ones, and that consideration should be taken in this context to promote use of cleaner technology and safe disposal of chemicals. Recently, this principle of substitution has become a common, internationally applied principle. It is, for example, stated in the UN document Agenda 21 from the UN Conference in Rio 1992, as well as in some EC documents. It is also the rationale for bans or restriction of chemicals in many countries, as well as for frequent decisions in industry to substitute hazardous chemicals with less hazardous ones.

Normally, it is and must be the producers and users who, by means of responsible marketing and use, ensure us a safe use of chemicals. To a high degree, the crucial decisions needed for an effective chemicals control are to be taken in the interfaces between the producers and the users of chemicals. Certainly the pressure for safer chemicals from the users, (both professional and private consumers), is one of the most effective ways to influence the producers.

7. Recent work on risk reduction

7.1 *Pesticides*

In all Nordic countries, pesticides have to be approved before import, marketing or use. An application for approval of a pesticide must be accompanied by comprehensive scientific documentation, with test results and other data necessary to assess the hazardous properties, risks, and efficacy. The need for a pesticide is assessed as well. Approval is necessary not only for plant protection products but, with some exceptions, for biocides in general, such as: wood preservatives, antifouling agents, and industrially used biocides.

In 1986, the Swedish and Danish governments decided on a 5-year plan for risk reduction with the aim to reduce the use of pesticides in agriculture by 50 percent. After that, Finland and Norway (1992) also published comparable reduction programs. Lowering of the environmental impact of pesticides as expressed *e.g.* by total dosage and frequency of application, was the most important goal of the programs. Field trial data supported a cut of the dosages for herbicides by approximately 50 percent without loss in production.

Other parts of the plan included, for example, voluntary testing of old sprayers, increased control of the performance of existing spray equipment, restrictions for spraying in order to prevent contamination of adjacent areas or surface waters, and extended advisory service to training of farmers. The measures taken were for the most part voluntary. The plan included a more rigorous assessment of pesticides. Use of less hazardous pesticides is promoted. For example, in Sweden, pesticides are not approved if it can be shown that they give rise to increased risk in comparison with existing ones. Old approvals are withdrawn when new pesticides implying significantly lower risk become available (*the principle of substitution*). The period for approval is limited in all Nordic countries to 4 to 10 years.

According to the EC directive on marketing of plant protection agents (dir. 91/414/EEC), approval should be granted according to the 'Uniform Principles' (EC 1994).

RISK REDUCTION PROGRAMME FOR PESTICIDES
- Swedish example

The 5-year programme resulted in a 47% reduction of the use of pesticides, calculated on the volume of active substance. After 6 years the reduction was 56% compared with the mean use in 1981-85. This reduction has been possible without any negative effects on agricultural production indicating that there has been a significant overuse of pesticides. The government has recently decided on a second plan for risk reduction with the aim of reducing use by a further 50%.

The number of approved pesticides has decreased by 50%, and a considerable number of old pesticides of a high degree of hazard have been withdrawn from use. It is anticipated that in 1995 all pesticides in use will have modern, comprehensive documentation on their toxicological and ecotoxicological properties.

The Nordic countries, and Denmark in particular, are presently engaged in these discussions; trying to influence the interpretation of the directive so that 'the Nordic protection level' for approvals will not be undermined, according to the present evaluation procedures.

7.2 Other chemicals

Limiting the marketing or use of chemicals is, (with some exceptions), not a very common method for risk reduction. This is also true for chemicals which are known to be very harmful to humans or to the environment. Apart from certain active ingredients in pesticides, very few substances have been banned or severely restricted on a more general worldwide basis. PCB is an early example, CFC's are late ones. In the beginning, control of existing chemicals consisted of restricting the use of single hazardous chemicals (mercury, cadmium, DDT, etc). Today, the Nordic countries are striving towards more comprehensive risk management programs in cooperation with other countries.

The Swedish and Danish governments have decided upon bans or severe restrictions on mercury and several chlorinated organic solvents. For some other chemicals, voluntary measures to reduce or phase out the use should be undertaken by the industry and followed up by government agencies.

General criteria for chemicals to be phased out or restricted are to be elaborated. Such criteria should include data on *e.g.* toxicity/ecotoxicity, volumes in use, use patterns and exposure estimates, in order to identify chemicals of concern, and to give a basis for risk assessment and risk management. The Nordic Chemicals Authorities are dedicating increasing resources to this area.

International cooperation is required to make phasing out/limitations effective. This is of particular importance when taking action against chemicals which are dispersed via products such as electronic equipment, instruments, and cars. Heavy metals, phthalates and flame retardants are typical examples of such chemicals. Sweden has therefore initiated a risk reduction programme in the OECD in which all the Nordic countries have actively taken part.

Although close international cooperation on chemicals is essential, regional environmental characteristics should also be taken into account. Such Nordic characteristics are, for example: a long, cold winter; acid forest soils, cold, acid waters rich in humus; brackish Baltic waters, etc. Nordic cooperation will continue to have an important role in this respect in the future.

The development of legislation on chemicals is only possible with sufficient knowledge of their effects on human health and on the environment. Much of the research in the near future, aims at assessing the effects of the northern climate on the toxicity and fate of chemicals.

8. Summary

The main objective of the modern chemicals legislation of the Nordic countries is to enhance knowledge of the toxicological and ecotoxicological properties of chemicals to a level sufficient to assess risks. Furthermore, it is designed to regulate those of greatest risk, in order to minimize or avoid negative effects on humans or the environment.

Chemicals are, to a high degree, subject to international trade. Consequently, international cooperation is an important part of the Nordic strategy for chemicals control. Work is in progress to improve and update existing criteria for classification. This work is done on a Nordic basis and in close cooperation with the European Community (EC). The Nordic countries also actively participate in the OECD Chemicals Programme.

CHEMICALS TO BE PHASED OUT
- Swedish programme

Heavy metals:	Lead, mercury, cadmium, organotin compounds
Chlorinated solvents:	Methylene chloride, trichloroethylene, tetrachloroethylene
Wood preservatives:	Creosote, arsenic
Others:	Brominated flame retardants, phthalates, nonylphenoletoxylates, chlorinated paraffins

---

The Nordic countries have established and maintained Product Registers which currently contain computerized information on the chemical composition of a vast number of preparations. The significance of this information has also been recognized and used in various contexts on an international scale.

All the Nordic governments have established risk reduction programmes for pesticides. As far as the other chemicals are concerned, the Nordic countries are striving towards more comprehensive risk management programmes in cooperation with other countries. Although close international cooperation on chemicals is essential, inclusion of regional environmental characteristics is also imperative.

References:

EC 1994. Council Directive 94/43/EC of 27 July 1994 establishing Annex VI to Directive 91/414/EEC concerning the placing of plant protection products on the market. Official Journal of the European Communities, 1.9.1994, No L 227/31 – L 227/55.

OECD 1993. Report of the OECD Workshop on the Application of Simple Models for Environmental Exposure Assessment. OECD Environment Monographs No 69, Paris, p. 26.

WHO 1985. Organic Solvents and the Central Nervous System, Environmental Health Series Vol. 5, WHO, Copenhagen. Nordic Council of Ministers, Oslo, 273 p.

Modelled and observed fate of selected organochlorines in the Nordic environment

J. Paasivirta,[1] S. Sinkkonen,[1] T. Rantio,[1] D. Calamari,[2] A. Di Guardo,[2]
M. Matthies[3] & S. Trapp[3]

[1]*Department of Chemistry, University, P.O. Box 35, 40351 Jyväskylä, Finland*
[2]*Institute of Agricultural Entomology, University of Milan,
via Celoria 2, 20133 Milan, Italy*
[3]*Applied System Science, University, Albrechtstrasse 28,
D-4500 Osnabrück, Germany*

Keywords: Exposure modelling, fugacity models, chlorocymenes, chlorohydrocarbons, chlorophenols, chloroguaiacols

1. Introduction

There is an increasing need for reliable predictive systems to assess the potential exposure of humans and the environment to chemical substances. In response to this need, an international working group on 'Chemical Exposure Prediction,' supported by the European Science Foundation (ESF), concluded that there existed useful methods and data for preliminary environmental hazard assessment for anthropogenic chemical substances, but they needed further extension and improvement (Calamari, 1993). A large number of models exist for the assessment of the fate of chemicals in the environment, including bioaccumulation, but they have not been properly validated by using field data. A number of field observations on relevant systems were available but had not been compared with the models.

Leaching, run-off and plant uptake are major pathways of chemicals in soil/plant systems thus leading to contamination of food and drinking water. In aquatic systems, bioaccumulation is an important exposure route. Comparing calculated concentrations in soil, plant, water and animals with field observations would enhance the confidence into model predictions as well as the validity of laboratory results for exposure assessments. In Finland particularly, long-term field observations of organochlorine compounds (Paasivirta, 1989, 1990; Paasivirta & Rantio 1991; Paasivirta *et al.*, 1993; Vuorinen *et al.*, 1985, 1992, 1993) can be used for studies on the applicability of environmental modelling systems in Nordic conditions.

2. EDA programme

EDA (Environmental Damage and its Assessment) is an international cooperative effort supported by ESF. The program on chemical exposure prediction is coordinated by Professor Davide Calamari from the University of Milan, Italy. There are participating research institutes from Finland, Germany, Great Britain, Greece, Italy, Netherlands and Switzerland. The programme's purpose is to evaluate the models used to predict the exposure as result of the environmental behaviour and fate of hazardous chemicals. In particular, field data obtained in different European countries is used to assess the influence of the environmental variability and the capability of predictive models to cope with different environmental situations and scales (local, regional, global). The expected outcome of the EDA three-year programme, 1992–1994, is validated models to evaluate the exposure of humans and the environment to both existing and new chemicals, in order to ensure proper management of chemical substances. The validity of model predictions on the behaviour and fate of chemicals in terrestrial and aquatic systems is tested by comparing estimations against measurements in various field studies.

The influence of the variability of different (European) environmental situations will be quantified. The tool-box of submodels will be made useful in microcomputers through a user-friendly way, thus enhancing the general use for chemical exposure predictions.

3. Finnish chemicals and environments in EDA

Large number of field observation data has been accumulated in the University of Jyväskylä since 1970, regarding research on organochlorine compounds (OCls) in various Nordic aquatic environments. The first chemicals modelled are persistent chlorohydrocarbons which were assumed to be mainly air-transported in Nordic environments (Set I; Fig. 1). The second group of chemicals consists of key components of bleaching pulp mill effluents (Set II; Fig. 1).

The modelling area chosen for the Set I compounds is the Bay of Bothnia (1 in Fig. 2), where these chlorohydrocarbons have been measured intensively since the 1960s. For pulp mill effluent compounds (Set II), three recipient areas (2–4 in Fig. 2) in Finland are modelled, at present. Area 2, the Äänekoski watercourse, is a chain of shallow lakes. Area 3 consists of the Kymijoki River with a strong flow. Area 4 is a brackish water system in the Pietarsaari Archipelago, on the coast of the Bay of Bothnia. For these areas, for example, the fate of selected OCls is modelled in constant inflow conditions and compared with measured concentrations in water, sediment and biota. In this report, examples of modelling at areas 1 and 2 only are presented.

4. Modelling parameters for the chemicals

Examples of the parameters needed in modelling for the substances are presented in Table 1. Most water solubilities (S) and octanol/water partition coefficients (Kow) are taken from literature (Suntio et al., 1988a, 1988b, Xie et al. 1984). The remaining S and Kow values are measured (HPLC) or estimated from the regression equation for log S and log Kow (Lyman, 1990). For S and Kow of chlorophenolics, pH of 6.5 for water is assumed. This value corresponds the pure and oligotrophic conditions at the areas studied. Vapour pressures (P) are estimated from literature values of boiling points at reduced or normal pressure using the method of K. M. Watson, modified by Grain (1990). These estimates of P are consistent with literature data (Suntio et al., 1988a, 1988b).

The vapour pressures of chemicals at environmental temperatures (T K) can be represented by the equation (Grain, 1990):

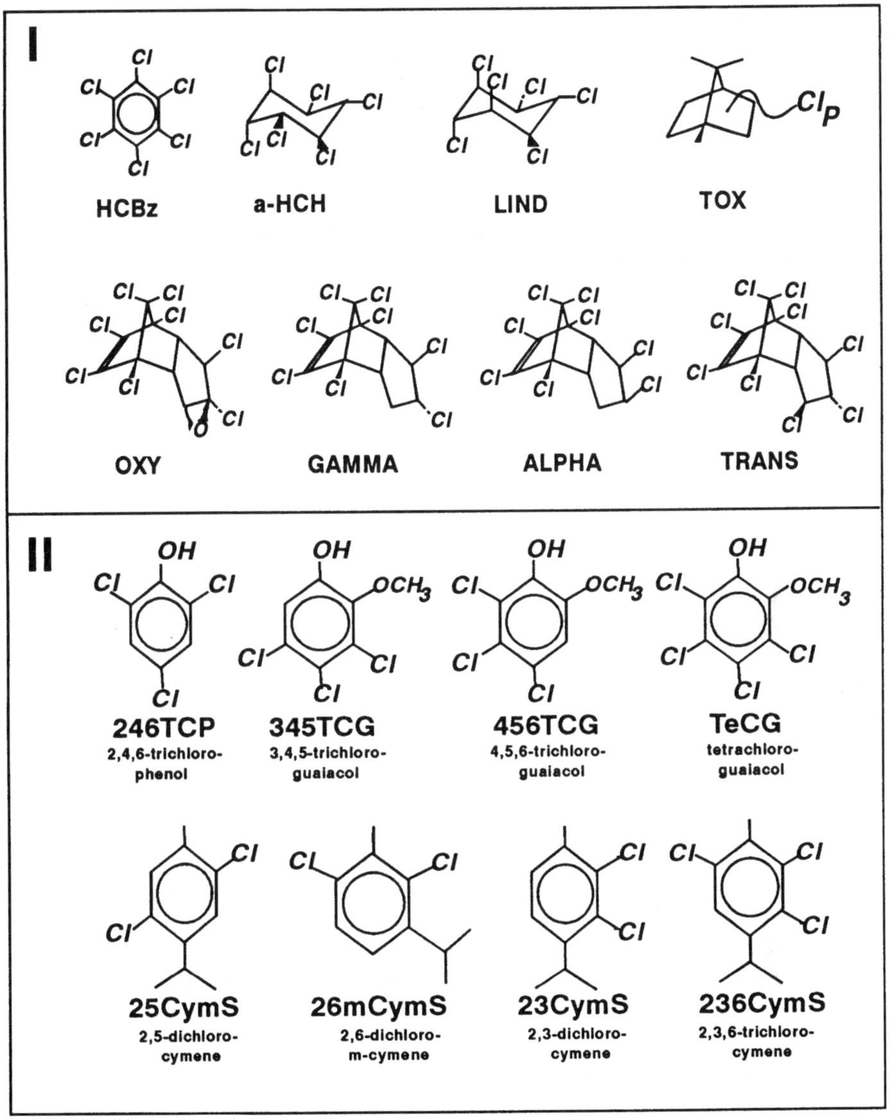

Fig. 1. Formula and name abbreviations of the chemicals modelled.
Set I: hexachlorobenzene (HCBz), α-hexachlorocyclohexane (a-HCH) γ-hexachlorocyclohexane or lindane (LIND), sum of chlordanes (SCHLOR) = oxychlordane + transchlordane + cis-chlordane + transnonachlor, and toxaphene (TOX).
Set II: 2,4,6-trichlorophenol (246TCP), 3,4,5-trichloroguaiacol (345TCG), 4,5,6-trichloroguaiacol (456TCG), tetrachloroguaiacol (TeCG), 2,5-dichloro-p-cymene (25CymS), 2,6-dichloro-m-cymene (26mCymS), 2,3-dichloro-p-cymene (23CymS), and 2,3,6-trichloro-p-cymene (236CymS).

$$\log P = A - B/T$$

The factors A and B are determined from the estimated or measured P values by linear regression. P corresponds to the vapour pressure of pure chemicals. If the environmental temperature (T) is lower than the melting point (Tm), the chemical behaves in the

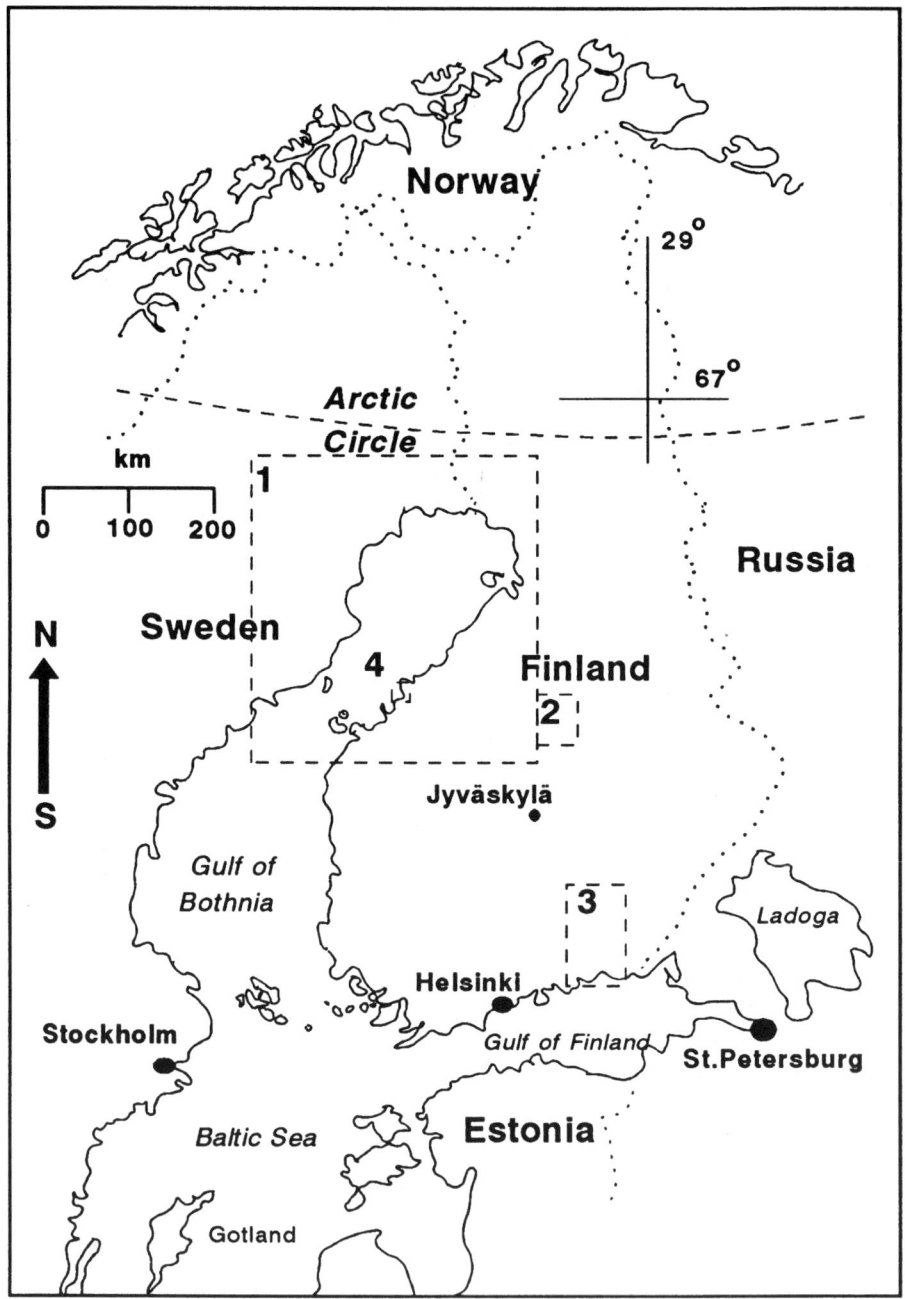

Fig. 2. Location of the model areas: 1. Bay of Bothnia, 2. Äänekoski Watercourse, 3. Kymijoki River, and 4. Pietarsaari Archipelago.

Table 1. Modelling parameters: M = molecular mass; S = water solubility; Tm = melting point; pKa = – Log acid dissociation constant; A and B factors for the vapour pressure; Kow = octanol/water partition coefficient; t1/2 = half-lifetime of degradation.

Parameter	LIND	SCHLOR	TOX	246TCP	TeCG	236CymS
M g/mol	290.8	409.8	413.8	197.4	261.9	237.6
S mg/L	10	1.85	0.55	794	89.4	0.5
Tm K	386.1	378.2	355.2	342.7	394.2	¤
pKa	-	-	-	6.5 *	6.2*	-
A	16.99	11.45	11.06	11.05	12.41	9.96
B	5566	4267	4231	3247	3950	2952
Log Kow	3.95	4.90	5.20	3.69	4.61	5.5*
t 1/2 d (25°C)						
Hydrolysis	130	300	500	∞	∞	∞
Photolysis	22	50	100	20	15	30
Biodeg./soil	260	240	1000	200	50	140
Biodeg./water				160	30	1400

¤ 236CymS is liquid at environmental temperatures studied (1 to 25°C)
* Measured by HPLC

environment like a supercooled liquid. The estimated P is then = P(s), which is converted in modelling to P(l) by the equation (Grain, 1990):

$$\ln P(s) = \ln P(l) + 6.8(1-Tm/T)$$

where factor 6.8 is an estimate of the entropy of fusion.

Degradation rates (as lifetimes t1/2) of LIND, SCHLOR (cis-chlordane as representative molecule) and TOX are from literature (Jury *et al.*, 1983; Suntio *et al.*, 1988a). Photodegradation lifetimes of 246TCP, TeCG and 236CymS in air are partly estimates from literature data (Yoshida *et al.*, 1987; Svenson & Björndahl, 1988). Their total (photo- and bio-) degradation rates in water-sediment system were measured by microcosm experiments of substances mixed with water, suspended solids and sediment taken from the model area (Mikkelson *et al.*, 1993; and our unpublished results). The lifetime of hydrolysis and biodegradation is assumed to double for every 10°C decrease of temperature. The lifetime of photolysis is assumed to be constant in the temperatures studied.

5. Modelling values, results for environments, and discussion

5.1 Simple fugacity model FATEMOD

As a first estimation of the environmental fate of a chemical, simple multi-media models can be used, which take into account partition between and transformation within the various environmental compartments under steady-state conditions. Mackay (1991) level I, II, and III fugacity models (all assuming steady-state) are widely adopted for screening purposes, because they take into account most processes involved in the distribution and fate of a substance.

In the level I model, equilibrium is assumed, and transformation of the chemical is

Table 2. Properties of the Bay of Bothnia and 'Southern Sea' environments.

Value	Bay of Bothnia	Southern Sea
Temp. °C (annual average)	2	25
Org. Carbon Fraction:		
Soil	0.04	0.01
Sediment	0.05	0.02
Susp. Sediment	0.15	0.04
Air volume m^3	3.7E+13	3.7E+13
Water volume m^3	7.4E+11	7.4E+11
Soil volume m^3	1E+07	1E+07
Sediment volume m^3	7.4E+12	7.4E+12
Biota (fish) volume m^3	7.4E+5	7.4E+5
Air depth m	1000	1000
Water depth	20	20
Soil depth	0.1	0.1
Sediment depth	0.02	0.02
Air flow m^3/h	3.7E+11	3.7E+11
Water flow m^3/h	7.4E+8	7.4E+8

excluded. Output consists of the relative concentrations (equilibrium distribution) of the substance in compartments of an environment. The compartments are Air, Water, Soil, Sediment, Suspended Sediment and Fish (biota).

In the level II model, equilibrium is also assumed, but transformation and advection are taken into account. In addition to the steady-state concentrations, reaction and advection rates and residence times are obtained. The obtained concentrations are arbitrary, calculated from the assumed total emission rate, but their ratios are characteristic of the environment and compound chosen.

The level III model gives an output similar to level II, but with greater precision and in non-equilibrium conditions. Estimates of chemical quantities, concentrations and lifetimes in four compartments (Air, Water, Soil and Sediment) are obtained. Concentration in Fish is given (as in Level II) based only on the partition between biota (lipid) and water.

In this study, FATEMOD model program (GWBASIC) was used. It is a modification of Mackay GENERIC program, which contains estimations at levels I, II and III. Fates of chlorohydrocarbons LIND, SCHLOR (cis-chlordane as a representative molecule) and TOX (octachlorobornane as a representative molecule) were modelled in the boreal Bay of Bothnia environment and in a fictitious 'Southern Sea' environment which has the same sizes and fluxes but the average temperature and organic carbon fractions differ compared to the Bay of Bothnia. Properties used are presented in Table 2. The values of emissions for the level III estimation were derived by trial modelling, to give approximately the same concentrations as observed in air, water or fish at the Bay of Bothnia (Bidleman et al., 1987; Gaul, 1992). Then, the same emissions were used in modelling the 'Southern Sea' case. Some modelling results (partial outputs from FATEMOD) are shown in Table 3.

There are very few reports on the chlorohydrocarbon levels in air, water and sediment at the Bay of Bothnia. Concentrations in fish have been studied more extensively. However, the levels obtained from the FATEMOD can be considered to be in accordance with the present knowledge. Consequently, a comparison to the 'Southern Sea'

Table 3. FATEMOD results for the Bay of Bothnia compared with similar more southern area and observed average levels in the Bay of Bothnia. Ca, Cw, Csed and Cb are the concentrations in air, water, sediment and biota (fish), respectively

Area Compound	Bay of Bothnia			Southern Sea		
	LIND	SCHLOR	TOX	LIND	SCHLOR	TOX
Level I						
% in Air	1.75	.003	.0005	1.75	8.64	1.60
% in Water	87.6	6.15	2.03	87.6	14.13	4.95
% in Sed.	10.4	91.6	95.7	10.4	84.3	93.4
Level II						
Res.time h	509	1274	4669	152	338	1264
Level III						
Emission kg/h						
to air	.055	0.10	0.005	.055	0.10	0.005
to water	4.00	0.36	1.355	4.00	0.36	1.335
Ca pg/m^3	29.4	64	3.2	66.7	93.0	12.5
Cw ng/L	2.3	0.25	0.65	0.84	.15	0.63
Csed ng/g fw &	.010	.033	0.78	.00043	.015	.073
Cb ng/g fw	0.72	3.4	21.6	0.27	2.2	26.7
Observed						
Ca #	30		3			
Cw ¤	1.5-2.3					
Cb Salmon *	1.72	34.3	134.8			
Cb Trout *	0.76	3.12	21.5			

& Conc. in fresh sediment; Csed dw is approximated by division with 0.37
# Extrapolated from values of Bidleman et al. (1987)
¤ From Gaul (1992)
* From Paasivirta & Rantio (1991) and Paasivirta et al.(1993)

modelling gives an idea of how these three chlorohydrocarbons behave in a boreal, compared to a temperate sea, environment. Lower concentrations in air (level I distributions) and higher persistency (longer level II residence times) are obvious. Low water soluble SCHLOR and TOX pollutants are taken into sediment to a greater extent in boreal rather than in temperate conditions (level I), but the final concentrations there are unknown. This is the case because the real degradation rates in suspended solids and sediments are not known. Much more in the way of experiments and environmental level measurements must be done in order to fill in these gaps. According to the level III model, bioaccumulation of LIND and SCHLOR was slightly higher and that of TOX about the same in the Bay of Bothnia compared to the 'Southern Sea'.

Another FATEMOD application was done for Lake Kuhnamo, the first pulp mill recipient area in the Äänekoski watercourse (Fig. 3). Trial modelling parameters for this environment at two periods are presented in Table 4. Results for the three compounds are presented in Table 5. These substances were components in effluents from the bleaching kraft pulp mill and intensively monitored by analyses of emissions, water and biota during the modelled time periods, August 1986 and March 1987 (Priha & Välttilä, 1987; Paasivirta et al., 1988a, 1988b, 1990; and unpublished results from Paasivirta et al.).

Fugacity level I, II and III modelling in the pulp mill effluent case (Tables 4 and 5) gave realistic values of water concentrations, obviously due to approximately correct

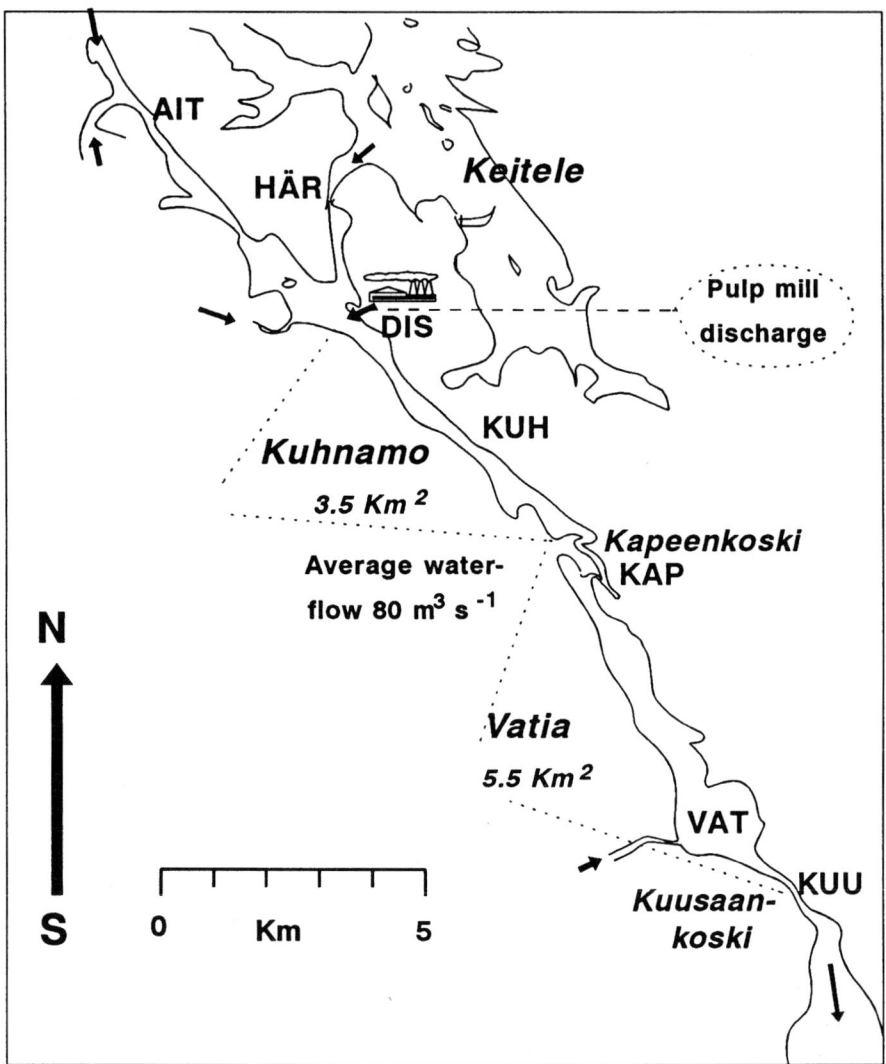

Fig. 3. The Äänekoski watercourse.

values of emissions and the (bio)degradation rate in the water compartment (fed into the model as a lifetime in water). The model gave too low values for concentrations in sediment. This is due to limited knowledge on degradation rates in sediment. Also, at least in the case of 246TCP and TeCG, processes other than partitioning, (*e.g.* binding), are important sources to higher levels in environment than in the model. The estimated and observed concentrations in biota are of the same order of magnitude for 246TCP and 236CymS, but deviate very much for TeCG. Obviously, bioconcentration processes other than partitioning, which were omitted in the model, are important.

Table 4. Properties of the pulp mill recipient Lake Kuhnamo.

Value	August 1986	March 1987
Temp. °C	20	1
Org. Carbon Fraction:		
Soil	0.22	0.22
Sediment	0.22	0.22
Susp. Sediment	0.30	0.30
Air volume m^3	3.5E+9	3.5E+9
Water volume m^3	2.065E+7	2.065E+7
Soil volume m^3	10	10
Sediment volume m^3	70000	70000
Susp.sed. volume m^3	123.9	123.9
Biota (fish) volume m^3	20.65	20.65
Air depth m	1000	1000
Water depth	5.9	5.9
Soil depth	0.1	0.1
Sediment depth	0.02	0.02
Air flow m^3/h	3.5E+7	3.5E+7
Water flow m^3/h	2.1623E+5	1.3685E+5

Table 5. FATEMOD results for the recipient Lake Kuhnamo. For notations see Table 3.

Period	AUGUST 1986			MARCH 1987		
Compound	246TCP	TeCG	236CymS	246TCP	TeCG	236CymS
Level I						
% in Air	.363	.059	.934	.066	.0073	.202
% in Water	22.39	3.36	.026	22.45	3.37	.026
% in Sed.	77.11	96.4	98.9	77.34	96.4	99.6
Level II						
Res.time h	90.5	94.9	266	252	273	1051
Level III						
Emission kg/h	.00946	.0291	.000825	.0113	.0243	.000987
Ca pg/m^3	192	391	353	89	111	430
Cw ng/L	20.7	49.4	1.56	56.9	124.8	2.26
Csed ng/g fw	0.0116	0.147	0.283	.106	1.05	1.60
Cb ng/g fw	5.03	96.9	67.6	13.9	245	97.8
Observed						
Cw	38.0	65.5		75.5	104	
Csed fw*	5.6	16.3	27.4			
Cb roach	13.0	1.03		1.99	1.02	
Cb pike	6.98	2.02	19.7	4.70	3.18	
Cb mussel			5.4			

* From Paasivirta *et al.* (1990): Csed fw = 0.37 x Csed dw.

5.2 Regional fugacity model

For more accurate studies of the Bay of Bothnia, a regional model was tried. A simulation of chlorinated hydrocarbon distribution in the Bay of Bothnia was done using a regional four-compartment multi-box; steady-state fugacity model developed by D. Mackay and modified by A. Di Guardo. The program for this model has a user-friendly Windows™-interface. It has different input windows for chemical, environment and other data required. It is possible to test the importance of a parameter (*e.g.* by a sensitivity analysis) in the modelling by changing its value and redoing the calculation.

The results are either tabular or in the form of charts, showing the distribution and concentrations in the different compartments. The simulation is based on the most complete environmental observations of concentrations in salmon and trout (Vuorinen *et al.*, 1985, 1993; Paasivirta & Rantio, 1991; Paasivirta *et al.*, 1993). Also, various reports on concentrations in Baltic seals, atmosphere and rainwater are used as aids in the modelling. The observations indicate that practically all chlorohydrocarbons in the area are transported by air from a long distance and deposited on the Bay of Bothnia. The simulation procedures under development take into consideration the special Nordic climate conditions in the region, including ice-coverage for three to five months per year. Details of the concentrations of the Set I compounds, properties of the Bay of Bothnia environment, and simulation procedures and results will be published later.

5.3 *EXWAT and PPEFF models*

The first pulp mill recipient modelled in EDA is the Äänekoski watercourse in Central Finland (Fig. 3). In addition to the simple fugacity model (see above), a German EXWAT model (Brüggemann & Münzer, 1987; Matthies & Trapp, 1988) and a Canadian PPEFF model (Mackay & Southwood, 1992) were tried. The model EXWAT was developed for the characterization of the transport and fate of a chemical in surface water bodies at steady-state. It is a box model with two compartments: fluid and sediment. The following processes were considered: deposition and resuspension of suspended matter; partitioning of chemicals between water and suspended matter in the fluid and between pore water and benthic sediment solids; ionization equilibrium; exchange between pore and fluid water as driven by dispersion; sediment burial; volatilization; degradation; and bioconcentration. PPEFF is a three-segment version of the Quantitative Water-Air-Soil-Interaction (QWASI) fugacity model. Its theory and operations are described by Mackay (1991).

Both EXWAT and PPEFF models could be readily applied to the Äänekoski watercourse. For EXWAT, the 18 km long region downstream from the discharge point can be divided in five segments each containing 1 km long boxes. The model gives concentrations of the chemical in each of the 18 boxes and in each compartment (water, suspended solids, sediments and biota). In the PPEFF run, a three-box version (Lake Kuhnamo, River Kapeenkoski and Lake Vatia) could be applied. Examples of the modelled and observed data for three chlorophenolics in the pulp-mill discharge are presented in Fig. 4 and in Table 6. Assuming that other necessary environment and compound parameters for the model were reasonably true independent data, degradation rates were to be fitted by the model to give the best agreement of measured and modelled concentrations in the water of all 18 boxes.

Degradation rates were derived by model simulation of the obtained rates and half-lives (Table 6). They consisted mainly of biodegradation rates, and were faster than expected (Table 1). This can be explained by the adaptation of the recipient biota to

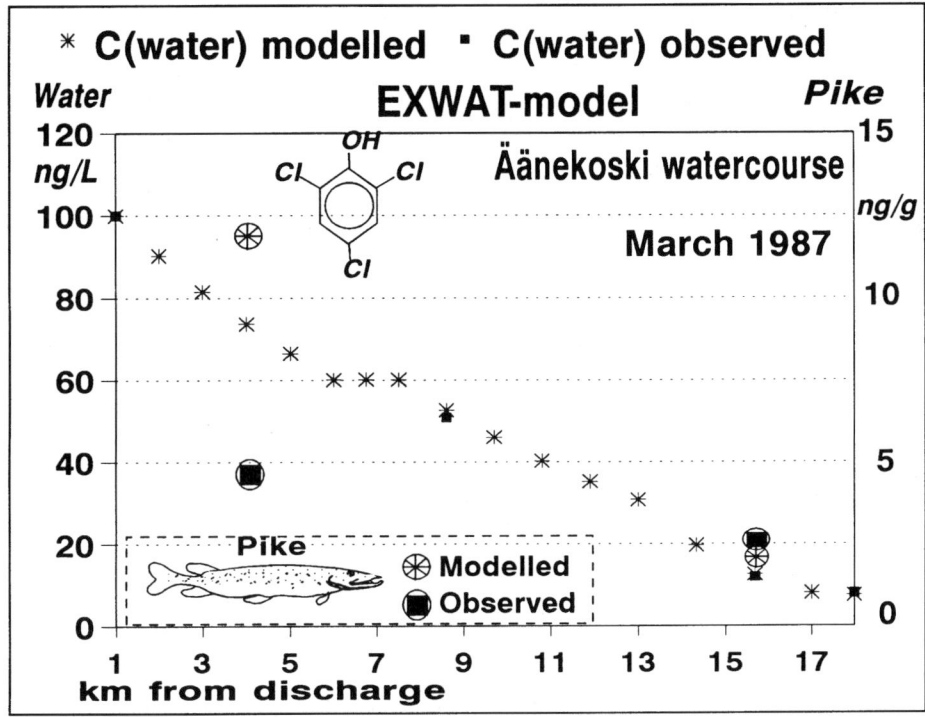

Fig. 4. Example of the modelling results by EXWAT.

the high chlorophenol contents. Also, seasonal variation is very large: For degradation rates, etc., Mikkelson *et al.* (Mikkelson *et al.*, 1993) made laboratory studies in small microcosms on the fate of TeCG using natural sediment and water taken weekly from the Äänekoski watercourse above the discharge point. Each series of measurements lasted four weeks. At 15°C in summer water (June), t1/2 was 1.9, but in winter water (January), 13.6 days. At 10°C in autumn water, during the first three weeks (October), t1/2 was 3.0, but during the fourth week (November, snowfall when water was sampled), it was 19.1 days. Experiments using water and sediment taken in March, was not made. However, the modelling (Table 6) indicated only slightly slower degradation in March than in August. This could be explained by readaptation of the microbial community to degrade OCl's at cold conditions during the season.

Model predictions of the concentrations in fish were in fair agreement with observed levels in pike (Table 6). In the case of EXWAT, this might be a coincidence, because lipid partition should cause lower levels (fat percent in pike muscles is only about 0.5), but this was perhaps compensated by food uptake by this predatory species. In the case of PPEFF, various uptake, growth and metabolism mechanisms are included in the model, and the fish concentration result was selected for 'large piscivores' class.

All the present models tried, assumed steady-state conditions. Trial uses of level IV+ models (Mackay, 1991) for a non-steady state, *e.g.* to obtain predictions about the future fate of chemicals with degreasing emissions, are in progress.

Table 6. Modelled (EXWAT and PPEFF) and observed concentrations in water and in fish (Pike, *Esox lucius*) at the Äänekoski watercourse. Sample places KUH, KAP, VAT and KUU are marked in Fig. 3. Value of pH of water at all sampling places was 6.5.

Compound	246TCP		345TCG		TeCG	
Time	AUG86	MAR87	AUG86	MAR87	AUG86	MAR87
Temp. °C	16	1	16	1	16	1
k *	.063	.058	.087	.08	.077	.070
t1/2 d	11.0	12.0	7.97	8.66	9.00	9.90
Discharge g/d	227	271	600	718	584	699
Waterflow m³/s	60	38	60	38	60	38
Conc. in water μg/L						
KUH EXWAT	.035	.062	.093	.162	.057	.090
KUH obs.	*.038*	*.100*	*.085*	*.362*	*.091*	*.104*
KUH PPEFF	.044		.084			
KAP EXWAT	.024	.036	.062	.093	.019	.021
KAP obs.	*.014*	*.051*	*.026*	*.050*	*.010*	*.043*
VAT EXWAT	.016	.021	.039	.048	.007	.006
VAT obs.	*.019*	*.012*	*.036*	*.057*	*.010*	*.013*
KUU EXWAT	.014	.019	.034	.043	.007	.006
KUU obs.	*.017*	*.008*	*.036*	*.026*	*.010*	*.008*
KUU PPEFF	.035		.051			
Conc. in fish ng/g						
KUH EXWAT	3.75	5.61	6.9	10.2	9.2	9.7
KUH obs.	*6.98*	*4.70*	*8.0*	*9.9*	*2.0*	*3.2*
VAT EXWAT	2.51	3.25	4.3	5.3	3.6	3.0
VAT obs.	6.99	2.52	8.0	4.0	3.0	1.5
KUU EXWAT	2.22	2.94	3.8	4.8	3.3	2.8
KUU obs.	*2.52*	*3.42*	*11.9*	*7.1*	*3.2*	*1.9*
KUU PPEFF	1.48		11.0			

* Degradation rate constant (k 1/d) in water and solids fitted by the model

6. Conclusions

The models seem to give realistic results, provided that the true properties of the environment such as temperature, water amounts and flow rates, amounts of suspended solids, sedimentation rates and organic matter contents of the solids are used. The uncertainty in predictions is mainly due to the poor data on the degradation rates. Their literature values are unaccessible or not valid for Nordic conditions. The watercourse models EXWAT and PPEFF can be used to derive overall degradation rate of the chemical from its measured emissions and concentrations.

The biodegradation rate in Finnish waters depends not only on the temperature but also on the time of year. Also, ice coverage and other large seasonal variations have significant effects on the fate of chemicals. Therefore, these special variables have to be included in models for predictions in Nordic environments. For most compounds studied, the modelling results as well as the analytical observations suggest longer persistency and slightly higher bioaccumulation rates of organochlorines in boreal conditions compared to those at more southern environments.

7. Summary

EDA, a ESF research project is briefly described. As an example of the project activities, environmental fate models for water ecosystems are applied in the Bay of Bothnia and in Finnish freshwater watercourse environments for prediction of the fate of air-transported persistent chlorohydrocarbons (lindane, chlordanes and toxaphene), pulp mill originated chlorophenol compounds 2,4,6-trichlorophenol and tetrachloroguaiacol and pulp bleaching product 2,3,6-trichlorocymene. The predictions are compared with analytical observations including concentrations in air, water, sediment and fish. The results indicate that modelling can give improved estimates of environmental fates and of biodegradation rates in Nordic conditions.

Acknowledgements

European Science Foundation (ESF), Environmental Science Council of the Academy of Finland and Ministry of the Environment, Finland, are acknowledged for the financing of this work.

References

Bidleman, T. F., U. Wideqvist, B. Jansson & R. Söderlund, 1987. Organochlorine pesticides and polychlorinated biphenyls in the atmosphere of Southern Sweden. Atmos. Environ. 21(3): 641–654.

Brüggemann, R. & B. Münzer, 1987. *EXWAT Multicompartiment Modell für den Transport von Stoffen in Oberflächenwässern.* GSF-Bericht 33/87, Neuherberg.

Calamari, D. (ed.), 1993. *Chemical Exposure Predictions,* Lewis L852CJ.

Gaul, H, 1992. Temporal and spatial trends of organic micropollutants in the sea water of Baltic Sea, the North Sea and the Northeast Atlantic. ICES Mar. Sci. Symp. 195: 110–126.

Grain, C. F., 1990. Chapter 14: Vapor pressure. In: W. J. Lyman, W. F. Reehl & D. H. Rosenblatt (eds.), *Handbook of Chemical Property Estimation Methods,* American Chemical Society, Washington, DC, 20 pp.

Jury, W. A., W. F. Spencer & W. J. Farmer, 1983. Behavior assessment model for trace organics in soil: 1. Model description. J. Environ. Qual. 12: 558–564.

Lyman, W. J., 1990. Chapter 2: Solubility in Water. In: W. J. Lyman, W. F. Reehl & D. H. Rosenblatt (eds.), *Handbook of Chemical Property Estimation Methods,* American Chemical Society, Washington, DC, 52 pp.

Mackay, D., 1991. *Multimedia Environmental Models.* Lewis, Chelsea, MI, L-242.

Mackay, D. & J. M. Southwod, 1992. *Modelling the Fate of Organochlorine Chemicals in Pulp Mill Effluents.* Report of the Pulp and Paper Centre of the University of Toronto, October 1992.

Matthies, M. & S. Trapp, 1988. Environmental models for exposure and hazard assessment of chemicals. In: H. Lokke, H. Tyle & F. Bro-Rasmussen (eds.), *1st European Conference on Ecotoxicology,* October 17–19, 1988 Copenhagen, Denmark. pp. 420–438.

Mikkelson, P., S. Herve, P. Heinonen & J. Paasivirta, 1993. Kemikaalin Ympäristökäyttäytymisen Testaus Laboratoriossa (report of ECOTEST project, in Finnish). Mimeograph of Board of Waters and the Environment, Finland, Nro 507, ISBN951-47-7373-X: 53 pp.

Paasivirta, J., J. Knuutinen, P. Maatela, R. Paukku, J. Soikkeli & J. Särkkä, 1988a. Organic chlorine compounds in lake sediments and the role of the chlorobleaching effluents. Chemosphere 17: 137–146.

Paasivirta, J., J. Knuutinen, M. Knuutila, P. Maatela, O.Pastinen, L. Virkki, R. Paukku, & S. Herve, 1988b. Lignin and organic chlorine compounds in lake water and the role of the chlorobleaching effluents. Chemosphere 17: 147–157.

Paasivirta, J. 1989. Organochlorines in Finnish and Baltic environment. The role of the forest industry. In: S. Rekolainen & J. Zeyer (eds.), *Biotransformation of Organic Pollutants in the Aquatic Environment.* CEC Water Pollution Report 14: 1–10.

Paasivirta, J. 1990. Predicted and observed fate of selected persistent chemicals in the environment. In: O. Hutzinger & H. Fiedler (eds.), *Dioxin 90. Organohalogen Compounds,* Vol. 1: 367–376. Eco Informa Press, Bayreuth, F.R.G.

Paasivirta, J., H. Hakala, H. Knuutinen, T. Otollinen, J. Särkkä,, L. Welling, R. Paukku and R. Lammi, 1990.

Organic chlorine compounds in lake sediments. III. Chlorohydrocarbons, free and chemically bound chlorophenols. Chemosphere 21: 1355–1370.

Paasivirta, J. & T. Rantio, 1991. Chloroterpenes and other organochlorines in Baltic, Finnish and Arctic wildlife. Chemosphere 22: 47–55.

Paasivirta, J., T. Rantio, J. Koistinen & P. J. Vuorinen, 1993. Studies on toxaphene in the environment. II. PCCs in Baltic and Arctic sea and lake fish. Chemosphere 27: 2011–2015.

Priha, M & O. Välttilä, 1987. Valkaisussa Syntyneiden Orgaanisten Klooriyhdisteiden Tehdas ja Vesistötaseet (Factory- and Watercourse Balances of Organic Chlorocompounds Formed in Bleaching, in Finnish). Report Z 87189, Centrallaboratorium, Espoo.

Suntio, L. R., W. Y. Shiu, D. Mackay, J. N. Seiber & D. Glotfelty, 1988a. Critical review of Henry's law constants for pesticides. Reviews of Environm. Contamin. Toxicol. 103: 1–59.

Suntio, L. R., W. Y. Shiu & D. Mackay, 1988b. A review of the nature and properties of chemicals present in pulp mill effluents. Chemosphere 17: 1249–1290.

Svenson A. & H. Björndahl, 1988. A convenient test method for photochemical transformation of pollutants in the aquatic environment. Chemosphere 17: 2397–2405.

Vuorinen, P. J., J. Paasivirta, T. Piilola, K. Surma-Aho & J. Tarhanen, 1985. Organochlorine compounds in Baltic salmon and trout. I. Chlorinated hydrocarbons and chlorophenols 1982. Chemosphere 14: 1729–1740.

Vuorinen, P. J., J. Paasivirta, J. Koistinen, & T. Rantio, 1992. Organochlorines in salmon (*Salmo salar L.*) from the Teno river. In: E. Tikkanen, M. Varmola & T. Katermaa (eds.), *Symposium on the State of the Environment and Environmental Monitoring in Northern Fennoscandia and in Kola Peninsula, Extended Abstracts*, Arctic Centre Publications, Rovaniemi, Finland 4:186–188.

Vuorinen, P. J., J. Paasivirta, M. Vuorinen, S. Peuranen & J. Hoikka, 1993. *Organochlorines in Salmon and Sea Trout and the Mortality of the eggs and Yolk Sac Fry of Salmon* (in Finnish), Finnish Game and Fisheries Research Institute, Fish Studies, Vol 65: 71 pp.

Xie, T. M., B. Hulthe & S. Folestad, 1984. Determination of partition coefficients of chlorinated phenols, guaiacols and catechols by shake-flask GC and HPLC. Chemosphere 13: 445–459.

Yoshida, K, T. Shigeoka & F. Yamauchi, 1987. Evaluation of aquatic environmental fate of 2,4,6-trichlorophenol with a mathematical model. Chemosphere 16: 2531–2544.

Cadmium and mercury in north European forest ecosystems

Martin Lodenius

Department of Limnology and Environmental Protection
P.O.Box 27, FIN-00014 University of Helsinki, Finland

Keywords: plant, soil, acidification

1. Introduction

Cadmium and mercury are among the most toxic of the heavy metals. Both metals may bioaccumulate readily in the biota and cause adverse effects in terrestrial and aquatic ecosystems. Cadmium is known to be mobilized from the soil and taken up by plants, especially under acidic conditions. The bioaccumulation of organic mercury in aquatic ecosystems is well known and fish is, in most cases, the main dietary source of this metal for humans. Our knowledge of the occurrence and accumulation mechanisms of mercury in terrestrial ecosystems is much more sparse than for aquatic systems.

While the emissions of cadmium and mercury are now strictly controlled in the north European countries, the occurrence and mobilization of these metals in forest ecosystems are also influenced by many other factors. Changes in chemical or biological conditions may increase the concentrations of heavy metals in some part of forest biota, even if the total anthropogenic input is decreasing.

The forest damages observed in central and northern Europe have been explained on the basis of many different theories. According to one hypothesis all stress factors are summed together so that the limits of biological stress tolerance of different organisms can be exceeded (Krause *et al.* 1986, Schulze 1989). One potential stress factor is soil acidification, which is known to increase the solubility of many metals. This raises the question of how important metals are as stress factors for forest organisms.

2. Use and emissions

Both cadmium and mercury have been used for a wide range of industrial purposes. Nowadays, the use of these metals is restricted and the emissions have decreased significantly (Table 1). In addition to the industrial use, considerable amounts of metals have been added to agricultural soils; cadmium through fertilizers and mercury through seed dressing.

3. Soil and soil invertebrates

In soil, cadmium and mercury are strongly bound to organic matter and particles. The mobilization of metals from the soil is affected by a great number of factors, some of which are interacting. The mobility of cadmium is strongly dependent on soil quality and pH. The sorption decreases at decreasing pH and increasing particle size.

Table 1. Industrial emissions of Cd and Hg (t yr^{-1}) into the Finnish environment in 1987 (Mukherjee 1994).

	Cd	Hg
To air	3.2	3.5
To water	1.0	0.15

Even if acidification increases the mobility of cadmium, it also increases the adsorption of mercury to humic substances. The importance of humic matter for the sorption of mercury is well documented. This binding is almost unaffected by fluctuations in pH. The humic matter is obviously the main carrier in the transport of mercury from terrestrial to aquatic ecosystems and it is of principal importance for the bioaccumulation in aquatic food chains.

Data concerning the effects of pH on the sorption of mercury in soils are partly contradictory. The maximum sorption of mercury for several soil types is in the pH range 4.75 – 6.50. In many soils, the leaching of mercury decreases with decreasing pH (Lodenius, 1990b). The chloride ion reduces the sorption of mercury significantly and the influence of chloride ions on the adsorption of mercury may be more important than that of hydrogen ions. Addition of chloride also shifts the region of maximum sorption to higher pH values.

A strong uptake of mercury has been found in earthworms and some other soil organisms in agricultural soils but there also seems to be an effective excretion mechanism which prevents any further biological accumulation. In forest earthworms, Cd is clearly accumulated, while no difference between topsoil humus and earthworms is detected for Hg (Table 2).

Table 2. Cadmium and mercury concentrations (µg g^{-1} fresh wt) of earthworms (Dendrobaena octaëdra) in Finnish forest soils (Braunschweiler 1992).

	Cd	Hg
topsoil humus	0.16	0.32
earthworm	6.0	0.31

4. Plants

The uptake of metals from the soil into plants is determined by several soil quality properties including physico-chemical factors like soil type, pH, organic matter and ion exchange capacity, and biological factors like microbial activity and plant species. Physical factors like ploughing or tilling may mobilize several elements, including Cd and Hg, and make them available to plants. In soil solution, both metals are phytotoxic, even in small concentrations, affecting root growth and uptake of nutrients and water (Godbold & Hüttermann, 1985, 1986).

Acidification may influence the mobilization of cadmium from the soil and cadmium uptake by plants (Lodenius, 1990a). This is a potential risk for example in Finnish forest soils which are typically coarse soils poor in nutrients and organic matter.

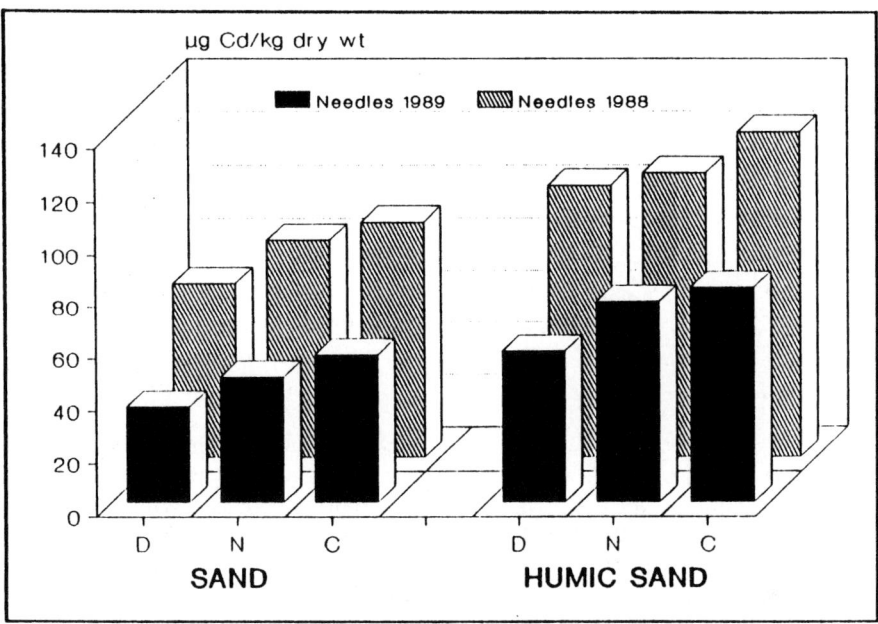

Fig. 1. Uptake of cadmium in needles of current (1989) and previous (1988) year of pine seedlings after experimental acidification in two different soils (sand with and without addition of peat). The treatments were: distilled water "D" (pH 6), artificial acid rain "N" (pH 5.4) and artificial acid rain "C" (pH 3.6) (Lodenius, 1990a).

The mobilized cadmium may be taken up by trees and other plants. The uptake is greater at low pH and in the presence of humic matter (Fig.1). The mechanisms of metal uptake by different plant species, especially trees, is insufficiently known. For example, the synergistic and antagonistic effects of different metals in plant tissues and the availability of organic metal complexes are still unknown. Terrestrial plants normally contain only small amounts of mercury. The uptake of this metal from the soils is poor and only small amounts of organic mercury compounds have been detected.

The uptake is most efficient in young specimens and mercury forms which may be absorbed include Hg^0, Hg^{2+}, CH_3Hg^+ and $C_2H_5Hg^+$. Plant roots also play an important active role in mobilizing metals from soil particles by producing organic compounds that are effective in releasing substances bound to soil particles (Kabata-Pendias 1992).

Vascular plants may absorb Hg^0 directly from the air and therefore can be used in monitoring airborne mercury pollution. Different plant species have different capabilities of absorbing mercury from the air. Most of this mercury is absorbed through the leaves but a small part is translocated to the roots. Vascular plants may also release mercury and this ability is highly dependent on the plant species. This re-emission reduces the concentrations in or on the leaves. The uptake of inorganic mercury seems to be more effective than that of organic mercurials. Catalase appears to play an important role in mercury vapour uptake and this is true especially for gramineous C_3 species, while this mechanism is of lesser importance for C_4 species.

Mercury concentrations are usually lower in fruits, grain and leaves than in other parts of plants. The background concentrations of mercury in plant foodstuffs usually vary within the range 0.003 – 0.1 µg g^{-1} (Barudi & Bielig, 1980).

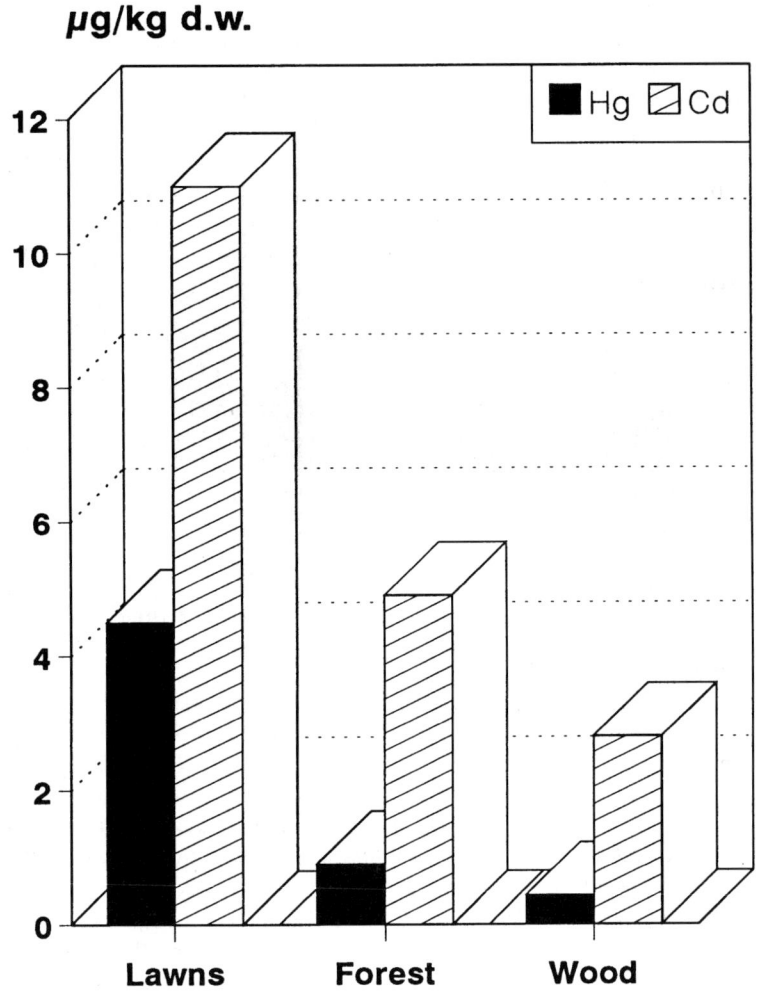

Fig. 2. Cadmium and mercury concentrations (µg g^{-1} dry wt) in fruiting bodies of macrofungi according to the growth place (Kojo & Lodenius, 1989).

5. Fungi

The uptake of metals in fungi is, in many respects, different from that of plants and the strong accumulation of mercury and cadmium in certain species of macrofungi is one example of these differences. There are pronounced differences between various ecological (Fig. 2) and systematic groups of fungi. The highest concentrations of heavy metals have been recorded in species growing on lawns. Very high concentrations may be found in certain species even in uncontaminated environments. Mushrooms growing in forests, however, usually have low contents of these metals.

Low concentrations of mercury are found in fruiting bodies of most macrofungi: usually less than 5 µg g^{-1} (dry wt). The mean concentration of lawn decomposer species is clearly higher than that of mycorrhizal fungi or wood decomposer fungi. High concentrations have been recorded in species such as *Agaricus* and in *Boletaceae*. Moreover, there is a very broad variation of mercury concentrations within the genus *Agaricus*: the mean mercury concentration of *A. edulis* is less than 5 µg g^{-1} dry wt while that of *A. comtulus* is almost 100 µg g^{-1} (Kojo & Lodenius, 1989).

Cadmium and mercury are quite unevenly distributed within the fruiting bodies. Often the lamellae contain significantly more of these metals than the rest of the cap and the stalk in both young and fully developed specimens. The uptake mechanisms are poorly known, but seem to be different for these two metals. It has been suggested that the accumulation of these metals could depend on sulphur containing protein groups (Kojo & Lodenius, 1989).

6. Mammals and birds

The heavy metal concentrations of terrestrial mammals are typically low. However, the accumulation of Cd in kidney and liver of elk (*Alces alces*) and certain other species is a well known phenomenon. The Cd concentration in the kidney of old specimens of elk may be as high as over 20 µg g^{-1} (fresh wt) thereby exceeding the safety limits as a food stuff. Mercury is seldom found in high concentrations in terrestrial mammals. Background mercury values are usually below 0.2 µg g^{-1} (fresh wt) in soft tissues (Wren, 1986). For example, the mercury concentrations of Finnish reindeer are very low: approximately 0.01 µg g^{-1} in muscle and fat, 0.09 µg g^{-1} in the liver and 0.16 µg g^{-1} in the kidney (Lodenius, 1984).

Factors like species, sex, age, diet and season affect the uptake and excretion of mercury. Normally the diet seems to be more important with respect to the mercury concentration of mammals than the species. Elevated concentrations in terrestrial mammals due to the use of mercurial fungicides have been reported in Poland and Austria (Krynski *et al.*, 1982). Elevated mercury concentrations in the terrestrial environment near chlor-alkali plants has been found to affect the developmental stability (measured as genetic aberrations and asymmetry) of shrews (Pankakoski *et al.*, 1992).

Although biomagnification in terrestrial ecosystems is much less pronounced than in aquatic food chains, the mercury concentrations are normally higher in carnivores than in herbivores. Few investigations have dealt with methylation processes in terrestrial ecosystems. In piscivorous mammals (otter and mink), methyl mercury represents 30 to 50 per cent of the total mercury concentration while only 0 to 8.5 per cent was found in small terrestrial mammals (Bull *et al.*, 1977). Typically, the total concentrations (calculated on fresh wt basis) follow the order: hair > liver > kidney > muscle = brain.

The use of methyl mercury in agriculture was responsible for the severe contamination of terrestrial food chains in several countries in the 1950s and 1960s until the use of this compound was banned. Until recently, in many countries, substantial amounts of methoxyethyl mercury have been used as a seed dressing. While the agricultural use of mercury can not be seen in the mercury concentrations of feathers of the Kestrel (*Falco tinnunculus;* Lodenius & Kuusela, 1985), there is a clear peak in feathers of the Sparrowhawk *(Accipiter nisus)*, with the highest concentrations being found in feathers collected in the 1960s. These mercury concentrations, however, seem to follow the industrial use of mercury much more closely than the agricultural use (Solonen & Lodenius, 1984; Fig. 3).

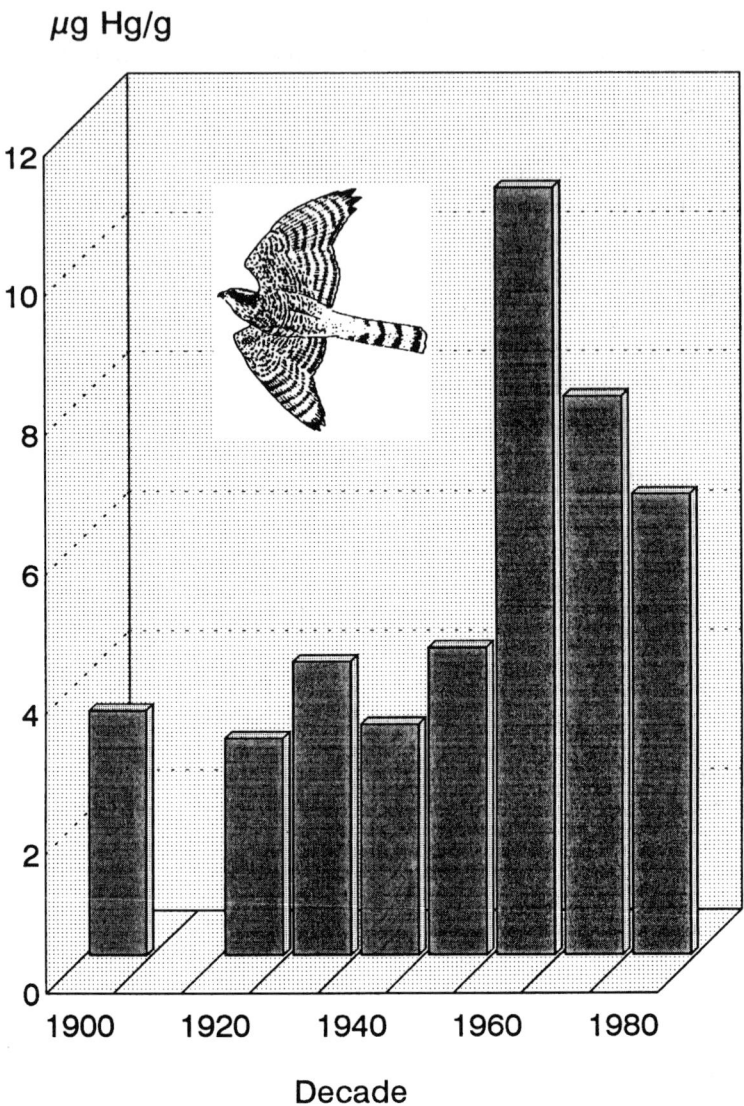

Fig. 3. Mercury concentrations (µg g^{-1} dry wt) in feathers of Finnish Sparrow Hawks *(Accipiter nisus)* (Solonen & Lodenius, 1984).

7. Insects

Many insect species have a very effective mechanism for the excretion of heavy metals, so the concentrations of Cd and Hg are usually low. However, in some species of ants *(Formica spp.)*, the Cd concentrations are surprisingly high: 6–10 µg/g (dry wt). Elevated Hg concentrations are found from blowflies eating Hg-contaminated fish but this metal is excreted rapidly. It is this effective excretion which seems to inhibit bioaccumulation of heavy metals in many terrestrial invertebrates (Nuorteva *et al.*, 1980).

It has been suggested that heavy metals could have an adverse effect on the natural regulation of insect pests in forests and diminish the natural pest resistance of trees. Thus, the metals would be a key factor in the multi-stress disease. Studies concerning the metal concentrations of forest insects show considerable interspecific differences in heavy metal concentrations. These differences can not always be explained by the trophic levels in the food chains.

Although insects seem to have an effective excretion system for heavy metals, metals may be toxic to insects. For example, exposure of dietary methyl mercury clearly affects the activity of Meal bugs (*Tenebrio molitor*) even if it does not shorten their life span (Schütt & Nuorteva, 1983).

8. Conclusions

It is important to study the quantities of metals taken up by forest plants because plants are one way for soil metals to enter forest food chains. Plant roots and mycorrhizas obviously play an active role in mobilizing metals from soil particles. It is also important to know whether metals can enter plants and animals in concentrations harmful to them.

The importance of a single heavy metal in the function of boreal ecosystems is normally small. However, there may be some 'key points', where the combination of acidification, aluminium and heavy metals could cause adverse effects on conifers and other plants. In Arctic and subarctic conditions, even small changes from the optimum may cause considerable ecological changes. The effects of air pollutants on forest ecosystems are based on complex interactions where probably no single dominating factor can be found. The reasons for forest damages are difficult to determine because of the lack of long-term data and the difficulties in determining the 'normal' state of the forest ecosystems. Although it seems obvious that metals are not the main factor in forest decline, it is important to understand the role of metals as a stress factor on forest ecosystems.

9. Summary

Soil acidification, air pollution and ditching of peatlands or forests may influence the sorption, mobilization and bioaccumulation of Cd and Hg within terrestrial ecosystems. The effects of air pollutants on forest ecosystems are based on complex interactions where no single dominating factor has been found. Even though heavy metals are probably not the main factor in forest decline, it is important to understand the role of these elements as a stress factor on forest ecosystems.

Cadmium may be mobilized from the forest soil as a result of environmental acidification with increasing concentrations in the biota as a result. In forest ecosystems, the environmental significance of mercury is less than in aquatic ecosystems. The strong accumulation of cadmium and mercury into certain species of fungi is a potential risk for human health, but it also reminds us that our knowledge concerning the physiology and role of fungi in terrestrial ecosystem is poor.

References

Barudi, W. & H. J. Bielig, 1980. Gehalt an Schwermetallen (Arsen, Blei Cadmium, Quecksilber) in oberirdish wachsenden Gemüse und Obstarten. Z. Lebensm. Unters. Forsch.,170, 254.

Braunschweiler, H., 1992. The effects of acidification and season on the metal contents of the earthworm (*Dendrobaena octaëdra*) in Finnish forest soils. Univ Jyväskylä, Congress Publ 3, XI Int Coll Soil Zool, Jyväskylä, Finland, 10–14 Aug 1992, p. 215.

Bull, K., R. Roberts, M. Inskip & G. Goodman, 1977. Mercury concentration in soil, grass, earthworm and small mammals near an industrial emission source. Environ. Pollut., 12: 135.

Godbold, D. L., A. Hüttermann, 1985. Effect of zinc, cadmium and mercury on root elongation of *Picea abies* (Karst.) seedlings, and the significance of these metals to forest die-back. Environ. Pollut Ser A 38: 375–381.

Godbold, D. L., A. Hüttermann, 1986. The uptake of mercury and lead to spruce (*Picea abies* Karst.) seedlings. Water Air Soil Pollut. 31: 509–515.

Kabata-Pendias, A., 1992. *Trace elements in soils and plants* (2. ed). CRC Press, Boca Raton, Florida, 365 pp.

Kojo M.-R. & M. Lodenius, 1989. Cadmium and mercury in macrofungi – mechanisms of transport and accumulation. Angew. Bot. 63: 279–292.

Krause, G. H. M., U. Arndt, C. J. Brandt, J. Bucher, G. Kenk, E. Matzner, 1986. Forest decline in Europe: development and possible causes. Water Air Soil Pollut. 31: 647–668.

Krynski, A., J. Kaluzinski, M. Wlazeiko & A. Adamowski, 1982. Contamination of roe deer by mercury compounds. Acta Theriol. 27: 499.

Lodenius, M.,1990a. Environmental mobilization of mercury and cadmium. Final report to the research council for the environmental sciences, Academy of Finland. – Publ. Dept Environ. Conserv., Univ. Helsinki 13: 1–32.

Lodenius, M., 1990b. Sorption of mercury in soils. In Cheremissinoff, P. (ed.), *Encyclopedia of Environmental Control Technology 4, Hazardous waste containment and treatment:* pp. 339–353. Gulf Publ. Co., Houston, London, Paris, Zürich, Tokyo.

Lodenius, M., 1984. Poroilla alhaiset elohopeapitoisuudet (Summary: Low mercury concentrations in Finnish reindeer). Ympäristö ja Terveys 15: 162.

Lodenius, M. & S. Kuusela, 1985. Mercury content in feathers of the Kestrel (*Falco tinnunculus* L.) in Finland. Ornis Fenn. 62: 158.

Mukherjee, A., 1994. Fluxes of lead, cadmium and mercury in the Finnish environment and the use of biomonitors in checking trace metals. Environ. Fenn. 18: 1–59.

Nuorteva, P., S.-L. Nuorteva & S. Suckcharoen, 1980. Bioaccumulation of mercury in blowflies collected near the mercury mine of Idrija, Yugoslavia, Bull. Environ. Contam. Toxicol. 24: 515.

Pankakoski, E., H. Hyvärinen & I. Koivisto, 1992. Äetsän klooritehtaan vaikutus ympäristön pikkunisäkkäiden elohopeapitoisuuteen [Influence of a chlor-alkali plant in Äetsä on the mercury concentrations of small mammals]. – Ympäristö ja Terveys 23, 458.

Schulze E. D., 1989. Air pollution and forest decline in a spruce (*Picea abies*) forest. Science 244: 776–783.

Schütt, S. & P. Nuorteva,1983. Metylkvicksilvrets inverkan på aktivitet hos Tenebrio molitor (L. (Col., Tenebrionidae) (Decrease in activity caused by methyl mercury in adult Tenebrio molitor beetles), Acta Entomol. Fenn., 42: 78.

Solonen T. & M. Lodenius, 1984. Mercury in Finnish Sparrow hawks, *Accipiter nisus*. Oris Fenn. 61: 58–63.

Wren, C.,1986. A review of metal accumulation and toxicity in wild mammals. Environ. Res. 40: 210.

Swedish risk assessment of antifouling products: pleasure boats and commercial vessels

C. Lye, L. Törnquist, C. Debourg, A. Johnson & C. Unger

*National Chemicals Inspectorate, Pesticides Approval,
P.O. BOX 1384, S-171 27 Solna, Sweden*

Keywords: analysis of need, salinity, leakage of copper

1. Introduction

The term 'antifouling product' refers to paints and other chemical products which contain growth-inhibiting substances to control aquatic growths on underwater surfaces. If the hull is not protected by antifouling paint or treated in some other way, barnacles and other fouling organisms will take a firm hold, resulting in loss of speed and increased fuel consumption by the boat.

Sweden has many vulnerable waters such as big lakes and the brackish water of the Baltic. It is therefore especially important to protect sealife in our costal waters. Antifouling products have been added to the Pesticides Ordinance (1985:836) in order to reduce health and environmental risks caused by the products in Sweden. This means that, since 1992, they may not be imported or handled without approval from the National Chemicals Inspectorate (KEMI). The registration of a large number of applications was finalized in autumn of 1992.

A condition of approval is that the benefit of a product should outweigh its risks. In order to determine the hazards of health and environment the various products might pose, KEMI commissioned both toxicological and ecotoxicological evaluations of the active ingredients, as well as analysis of the need for antifouling products.

Antifouling products considered, contain one or two of the following active ingredients: Copper, copper oxide, copper thiocyanate, 2-tert-Butylamino-4-cyclopropylamino-6-methylthio-1,3,5-triazine (irgarol), 4,5-dichloro-2-n-octyl-4-isothiazolin-3-one (isothiazolin), tin methylacrylate, tributyl tin oxide (TBTO) and diuron.

Antifouling products are assessed on the basis of their content of active ingredients, the type of paint, operating principle and leaching rate.

2. Analysis of need

In order to arrive at a conclusion regarding the need to use antifouling products, KEMI sent a consultation document to 25 different agencies:

There is a need to combat encrustations on both pleasure boats as well as commercial vessels. The need is greatest on the West Coast and decreases with diminishing salinity. The need is slight in fresh water and in the northern part of the Gulf of Bothnia. In the Baltic Sea area, less toxic antifouling products may be sufficient.

The long periods for which pleasure boats lie unused pose especially severe problems of growth. The main alternative to chemical antifouling products for pleasure boats is mechanical cleaning, such as brushing and high-pressure cleaning (Wisslêr et al., report no. 9/92). The need for chemical antifouling products is slight for small pleasure boats which can easily be removed from the water and cleaned mechanically.

The need to combat aquatic growths on ship hulls is considered very great in the North Sea and particularly acute for ocean-going vessels sailing in warm tropical waters. For the majority of ships, adequate tin-free antifouling alternatives are available today. For vessels sailing in warm tropical waters, however, TBTO paints will continue to be necessary. It is claimed that the negative consequences of a general ban would sinclude increased fuel consumption, increased exhaust emissions, poorer manoeuvrability and hence reduced safety at sea, increased operating and maintenance costs and diminished recreational value for the boat.

3. Toxicological risk assessment

3.1 Copper, irgarol

The major risk of antifouling products containing copper and/or irgarol is that they are assumed to produce sensitization, which increases the risk of allergy (Melsäter, 1992; Sloof et al., 1989). However, according to several studies, no serious health effects are to be expected from exposure to copper, since the body has its own ability to regulate copper levels (Sloof et al., 1989).

3.2 Isothiazolin, TBTO, diuron

The major risk from using products containing isothiazolin is that the substance can cause eye irritation, corrosion damage and sensitization. Its high toxicity on inhalation should also be taken into account. (Berzins, 1992).

The health risks associated with inorganic compounds are considerable. The LD_{50}-values after oral and dermal exposure of TBTO indicate high acute toxicity. TBTO causes skin and eye irritation. Long-time exposure to TBTO results in damage to the immune defence system, nervous system as well as the reproductive system (Ministry of Agriculture, Fisheries and Food, Ergon House, London. Food and Environment Act 1985, part III).

KEMI has classified diuron as a carcinogen with low carcinogenic capacity, according to the authority's criteria. The malignant tumours induced by diuron are found in the urin-bladder and uterus of rats. At low doses, diuron also causes disturbances to the blood-forming organs in several species (Mohammed, 1992a, 1992b, 1992c). Because of these findings, KEMI has banned the use of diuron in antifouling products.

4. Ecotoxicological risk assessment

4.1 General

In general, the active ingredients contained in antifouling products are deemed to be not readily biodegradable (with decomposition products that may continue to be biologically active and with a mineralization that is generally very slow), strongly adsorbed, highly toxic and sometimes also bioaccumulating to aquatic organisms.

(Andersson, 1977; Attaway *et al.*, 1982; Balsberg *et al.*, 1981; Bard, 1992; Björk *et al.*, 1992; Björklund, report no 6/88; Call *et al.*, 1987; Ellis & Camper, 1982; Elzvik *et al.*, 1992; HydroQual. Inc, 1987; Spear *et al.*, 1979; Peck, 1980).

4.2 Non-target organisms

The intention of an antifouling product is that the active ingredients should be gradually released from hulls and other underwater structures in order to inhibit the growth of aquatic organisms. This means that the aquatic environment in the vicinity of the vessel will be exposed to the substance. Copper is released as Cu^{2+} from antifouling products (Woods hole oceanographic inst. U.S Naval Inst). Besides, copper is mainly bioavailable as the ions of Cu and CuOH (Harrison, 1985).

In addition to being an essential element, copper is also highly toxic to most aquatic organisms, with a narrow range between essential and toxic concentrations (Balsberg *et al.*, 1981). The toxicity depends on the bioavailability of copper, which is related to the constitution of the water (Harrison, 1988; Spear, 1979; Gambrel, 1991). Effects of copper arise over a broad concentration range and can, under some conditions, occur at levels that are close to the background levels (National Environmental Protection Board (NEPB), report no 3696, 1991). The risk that non-target organisms in the aquatic ecosystem will be exposed and damaged as a result of the normal use of all relevant antifouling products is therefore considered to be high. The risk is greatest in heavily exposed areas such as marinas in calm bays with slow rates of water renewal, harbour areas and waters close to boat-and shipyards.

4.3 Levels of copper

Registered products contain, among other biocides, copper compounds. Concentrations of copper causing severe effects to organisms are only two to three times higher than those concentrations found in Swedish waters. Furthermore, concentrations of copper in Swedish waters are already increased compared to the background levels (NEPB, report no 3696, 1991).

The bulk of the copper is adsorbed by sediment (HydroQual, Inc., 1987). Sediment along the Swedish coast is contaminated by copper in many places. High levels are found locally (NEPB, report no 3696, 1991).
Due to high density of small pleasure boats and poor water exchange in inshore water, special concern has been focused on areas where reproduction of many aquatic species is localized (Elzvik *et al.*, 1992).

4.4 Leaching of copper

The leaching behaviour of copper from paint can be divided into an initial leaching rate and a steady state leaching rate. In a typical antifouling paint, leaching begins at once when the freshly painted surface is immersed in natural waters. The initial leaching is positively correlated to increased temperature, salt concentration and the concentration of hydrogen ions in the water as well as to the active ingredient on the surface of the paint (Woods hole oceanographic inst. U.S Naval Inst). Launching of freshly painted pleasure boats takes place in a defined period of time: May through June, which is a very sensitive reproduction period. The sudden increase of copper in water constitutes a considerable risk for sensitive species which can affect/alter the composition of the ecosystem (Elzvik *et al.*, 1992).

5. New regulations on antifoulants

5.1 *Pleasure boats*

For pleasure boats in fresh water and the northern Gulf of Bothnia as well as for small pleasure boats, none of the antifouling products in question is considered to offer advantages that outweigh the risks. None of the products is thus acceptable for these applications. Neither is approval given for use on small pleasure boats which are easy to haul ashore (i.e. with a wt less than 200 kg).

Antifouling products for pleasure boats containing TBTO and isothiazolin are considered unacceptable from the point of environmental and health protection. Therefore the benefit of these products is thought not to outweigh the risks. Moreover, it should be possible to avoid TBTO paints for screws and aluminium hulls since modern copper-based antifouling paints in combination with effective primers are normally not considered to pose any greater risk of corrosion on aluminium.

It is of great importance that the quantity of copper in the aquatic environment be reduced, since the total anthropogenic burden of copper is very high. The products containing high levels of copper are only considered to offer advantages that outweigh the risks on the West Coast, where fouling causes severe problems. Therefore, the Inspectorate has accepted different limits of leakage rate of copper into the water based on the first 14 days after launching when the leakage is large, in terms of 'less toxic' products on the East Coast compared to the West Coast.

5.2 *Commercial vessels*

For ships sailing mainly in fresh waters and in the northern Gulf of Bothnia the benefit cannot be deemed to outweigh the risks for any of the relevant antifouling products. Thus, none of the products are acceptable for these applications.

Antifouling products for commercial vessels containing TBTO and isothiazolin are considered unacceptable from the point of environmental protection. In addition, the use of products containing TBTO poses major health risks. The benefit of these products is only thought to outweigh the risks when used on ocean-going vessels. Furthermore, a ban on the use of tin-based products on ocean-going vessels would pose the risk of Swedish yards losing out to competition. As well, it would still be impossible to prevent vessels painted abroad from entering Swedish waters and visiting Swedish ports. It is therefore necessary to pursue coordination with other countries in getting further restrictions on TBTO. However, Sweden should follow the international limitation of TBTO leaching to a maximum of 4 $\mu g/cm^2$ per day.

Diuron meets the criteria of the National Chemicals Inspectorate for classification as carcinogenic. The benefit of products containing diuron is therefore not considered to outweigh the health risks posed by the substance. With the exception of TBTO paints, the alternatives are judged to have better health-related properties.

For ships in the Baltic and the North Sea, only in 'less toxic' antifouling products is the benefit deemed to outweigh the risks. This means that, for commercial vessels, antifouling products containing irgarol and copper as active ingredients are judged acceptable for use in the Baltic (apart from the northern Gulf of Bothnia) and the North Sea.

6. Summary

Since 1992, marketing as well as use of antifouling products has had to be authorised and approved by the Swedish National Chemicals Inspectorate. This paper deals with the health and environmental aspects of the assessment as well as regulations applicable to antifouling products.

In order to make adequate risk-benefit evaluations, the Inspectorate has analyzed the need for antifouling products in Swedish waters as well as performed both toxicological and ecotoxicological evaluations. For an approval to be made, the need for the product has to compensate the risk of using it.

The products are classified depending on the operating principle. Antifouling products are assessed on the basis of their content of active ingredients, the type of paint, operating principle and leaching rate. A great deal of attention has been paid to the salinity of the water where the products are intended to be used since the organisms causing severe fouling problems only live in marine ecosystems, whereas fouling is usually minimal in freshwater.

Registered products contain, among other biocides, copper compounds. Concentrations of copper causing severe effects to organisms are only two to three times higher than those found in Swedish waters. Furthermore, the concentrations of copper in Swedish waters are already increased compared to the background levels. The National Chemicals Inspectorate has accepted different limits on the leakage rate of copper into water for products intended for pleasure boats used in the Baltic Sea and the North Sea. In general, no antifouling products are accepted in lakes or in the Gulf of Bothina, neither for pleasure boats or commercial vessels.

References

Andersson, A., 1977. Heavy metals in Swedish Soils: On their Retention, Distribution and Amounts. Swedish J. agric. Res. 7:7–20.

Anon., 1985. Evaluation of tributyltin oxide, Ministry of Agriculture, Fisheries and Food, Ergon House, London. Food and Environment Act, part III.

Attaway, H. H., N. D. Camper & M. J. B. Paynter, 1982. Anaerobic microbial degradation of diuron by pond sediment, Pestic. Biochem. Physiol. 17:96–101.

Balsberg, A. M., G. Lithner & G. Tyler, 1981. Copper in the environment. National Environmental Protection Board, message 1424.

Bard, J. & A. Pedersen, 1992. Supplement 1 to the ecotoxicological evaluation of the antifouling compound 2-(tert-Butylamino)-4-(cyklopropylamino)-6-(methylthio)-1,3,5-triazine, National Chemicals Inspectorate.

Bard, J., 1992. Supplement 1 to the ecotoxicological evaluation of the antifouling compound 2-(tert-Butylamino)-4-(cyklopropylamino)-6-(methylthio)-1,3,5-triazine, National Chemicals Inspectorate, 1992.

Berzins, T., 1992. Toxicological evaluation of 4,5-dichloro-2-n-octyl-4-isothiazoline-3-one, National Chemicals Inspectorate.

Björk, M. & S. Karlsson, 1992. Ecotoxicological evaluation of 4,5-dichloro-2-n-octyl-4-isothiazoline-3-one, National Chemicals Inspectorate.

Björklund, I., 1988. Environmental Effects of Organotin in Antifouling Paints (*English summary*), Report no. 6/88, National Chemicals Inspectorate.

Call, D. J., L. T. Brooke, R. J. Kent, M. L. Knuth, S. H. Poirier, J. M. Huot & A. R. Lima, 1987. Bromacil and diuron herbicides: toxicity, uptake and elimination in freshwater, Arch. Environ. Contam. Toxicol. 16:607–613.

Ellis, P. A. & N. D. Camper, 1982. Aerobic degradation of diuron by aquatic microorganisms, J Environ. Sci. Health B 17:277–289.

Elzvik, A., & K. Hanze, 1992. Ecotoxicological evaluation of copper in antifouling paints (copper, cuprous oxid, cuprous thiocyanate), National Chemicals Inspectorate.

Gambrel, R. P., W. H. Wiesepape, Patrick, Jr. & M. C. Duff, 1991. The Effect of pH, Redox, and Salinity on Metal Release from Contaminated Sediment. Water, Air and Soil Pollution 57–58: 359–367.

Harrison, F.L., 1985. Effect of physicolochemical form of copper availability to aquatic organisms. Aquatic toxicology and hazard assessment: seventh symposium, ASTM STP 854, pp 469–484.

HydroQual, Inc., 1987. 1 Lethbridge Plaza, Mahwah, New Jersey, USA, for Copper Sulphate Task Force, c/o Tennessee Chemical Company, 3400 Peachtree Road N.E., Atlanta, Georgia, USA; Literature Review to Characterize the Adsorption potential of Copper in the Environment. *Marine fouling and its prevention.* Chapter 16: Mechanism of release of toxics from the paints. Corp. author: Woods hole oceanographic inst. U.S. Naval Inst.

Melsäter, B., 1992. Toxicological evaluation of the antifouling compound 2-(tert-Butylamino)-4-(cyklopropylamino)-6-(methylthio)-1,3,5-triazine, National Chemicals Inspectorate.

Mohammed, A., 1992a. Toxicological evaluation of supplementary studies on the carcinogenicity of Diuron, National Chemicals Inspectorate, August.

Mohammed, A., 1992b. Diuron-induced urothelial hyperplasia and neoplasia in the rat, National Chemicals Inspectorate, September.

Mohammed, A., 1992c. Diuron – the assessment for classification as carcinogenic (*in Swedish*), National Chemicals Inspectorate, September.

National Environmental Protection Board, 1991. Metals in coastal areas of Sweden (*English summary*), report 3696.

Peck, D.E., 1980. Adsorption-desorption of diuron by freshwater sediments, J Environ Qual 9:101–106.

Sloof, W., R. F. M. J. Cleven, J. A. Janus & J. P. M. Ros, (eds.), Integrated Criteria Document Copper, National institute of public health and environmental protection. Bilthoven, The Netherlands. Report no. 758474009.

Wissler, A., & B. Bengtsson, 1992. Cleaning stations to protect pleasure boats from the effects of fouling, National Chemicals Inspectorate. Report no. 9/92.

Management of environmentally hazardous chemicals in Finland

L. Ylä-Mononen

*Finnish Environment Agency, P.O. Box 140,
FIN-00251 Helsinki, Finland*

Keywords: chemicals control, risk assessment, international cooperation

1. Background

Thousands of different products of the chemical industry are in use in Finland today. During recent decades, the production volumes and amounts of chemicals in use have increased continuously. Simultaneously, the need for systematic risk management of hazardous chemicals has increased.

The prevention of health hazards caused by chemicals has been the subject of attention for decades in Finland, as in most industrialized countries. Less attention has been paid to the prevention of environmental risks, although chemicals have been released to the environment directly and with effluents and wastes.

There are numerous examples of environmental problems caused by chemicals, even in Finland. Persistent organochlorine compounds have caused damage to seals, sea eagles and other animals at the top of the food webs. Hundreds of contaminated soil areas have been found, most of them resulting from careless handling of chemicals, e.g. chlorinated phenols. Global pollution problems, like the depletion of the ozone layer, are also affecting Finland. (Wahlström *et al.*, 1993)

Pesticides were the first group of chemicals to be controlled also from the environmental point of view. The cause for this is clear: pesticides are spread directly to the environment and they are developed to be toxic for certain organisms. The inspection and authorization procedure of pesticides was supplemented with the evaluation of environmental effects in 1984.

The new Chemicals Act was under preparation for a long time. It came into force in 1990 and replaced the old Poisons Act from 1969. The Chemicals Act gave measures for the prevention of harm caused by chemicals to human health, but also to the environment.

2. Aims and measures of chemicals control

Through the control of chemicals, the Chemicals Act aims to prevent environmental problems beforehand. It complements the control of discharges and emissions regulated by other environmental legislation. The Chemicals Act enables the control of chemicals hazardous to health and the environment, throughout their production, importation, marketing and use.

This legislation provides effective measures to prevent environmental risks. The use of the most dangerous chemicals can be banned or restricted. New chemical substances are notified and evaluated before their introduction to the market. The advance approval systems of pesticides, wood preservatives and slimicides aims to prevent the unacceptable risks related to the use of biocides. The classification and labelling of dangerous chemicals enhances the safe and careful use of chemicals.

Important measures other than legislative ones are also in use in Finland: Publicity of risks related to the use of chemicals is often effective enough to reduce the risks. Voluntary agreements with industry and trade have also been used in some cases in order to diminish the use of certain harmful chemicals. The joint Nordic positive environmental labelling of products helps consumers choose products which are least harmful to the environment. Also, the use of economical measures like environmental levies and taxes has been considered in recent years. (NBWE, 1994).

3. Environmental risk assessment of chemicals

The National Board of Waters and the Environment (NBWE) is the supreme supervising authority for preventing damage to the environment by chemicals. It is also responsible for the development of methods by which to evaluate ecotoxicological effects and risks of chemicals. Naturally, this development work is closely related to the work done by other countries and in international organizations.

The basis for the risk assessment is formed by hazard identification and assessment (e.g. ECETOC, 1993). The hazard assessment is based on the intrinsic properties of chemicals, i.e., information on physical and chemical properties, degradability, potential for bioaccumulation and toxicity to different organisms. On the ground of effect related data and using a suitable safety factor, a predicted no-effect concentration (PNEC) can be assessed. The next step is to identify the actual use and release of the chemical. Combined with the data on the chemical's fate, theoretical predicted environmental concentrations (PEC) in different environmental compartments can be calculated or estimated. The final task is to assess the possibility of the concentration in the environment exceeding the acceptable level. Usually neither data on measured concentrations, nor actual effects in the environment, is available.

In practise, in very few cases can the ideal risk assessment scheme described above be followed. The most fatal problem is usually the lack of data. Also, the quality of the data can often be questioned. Much international work is in progress on this field but presumably it will take years before the basic data on intrinsic properties, even on high production volume chemicals, is available.

The most central concern related to risk assessment and management is the question of the special characteristics of our northern environment. Too little is known of the effects of, for example, a cold climate on the fate and effects of organic chemicals. The need for research work in this field is becoming more and more evident due to European integration. In order to protect our vulnerable northern environment, a more stringent chemicals control policy would probably be needed in Finland, relative to Central European countries. Restriction measures that can cause technical barriers to trade must, however, be justified scientifically.

4. Conclusions: Challenges for Finnish chemicals management

Increasing international cooperation, and especially European integration, sets new challenges for Finnish chemicals management. Closer international cooperation on the chemicals control field is essential, but at the same time, regional environmental characteristics should be taken into account. Nordic cooperation is going to have a central role in the future. More information on the effects of northern environmental conditions on the fate and effects of chemicals is needed in order to properly assess the risks caused by chemicals.

Improvement of the databases and statistics on chemicals and development and effective use of different risk reduction measures (legislative, economical, informative, etc.), are also central items in order to reduce risks caused by chemicals in a systematic and integrated way.

5. Summary

The prevention of health hazards caused by chemicals has been the subject of attention for decades in Finland. Less attention has been paid to the prevention of environmental risks. The 1990 Chemicals Act made it possible to effectively avert harm to the environment caused by chemicals.

Through the control of chemicals, the Act aims to prevent environmental problems beforehand. Thus it complements the control of discharges and emissions. The Chemicals Act enables the control of chemicals hazardous to health and the environment, throughout their production, importation, marketing and use.

The main challenges for the Finnish management of chemicals are closer international cooperation, especially European integration, and increased understanding of the significance of northern environmental characteristics on the fate and effects of chemicals. Improvement of the databases and statistics on chemicals and the development, as well as the effective use of, different risk reduction measures (legislative, economical, informative, etc.), are also central items in order to reduce risks caused by chemicals hazardous to the environment.

Acknowledgements

I wish to thank Mr. Esa Nikunen for valuable comments on the manuscript.

Reference list

ECETOC (European Centre for Ecotoxicology and Toxicology of Chemicals), 1993. *Environmental Hazard Assessment of Substances.* ECETOC. Technical Report No. 51. Brussels. 92 pp.
NBWE (The National Board of Waters and the Environment), 1994. *Management of Environmental Risks Caused by Chemicals. Action Programme of NBWE.* (In Finnish). NBWE, Publications of the Waters and Environment Administration – series A 187. Helsinki. 83 pp.
Wahlström, E., E-L. Hallanaro & T. Reinikainen, 1993. *The State of the Finnish Environment.* Environment Data Centre & Ministry of the Environment. Forssa. 163 pp.

Use and emission scenarios for methyl chloroform and carbon tetrachloride in Finland

Arun B. Mukherjee

Department of Limnology and Environmental Protection, P.O. Box 27, (A-Bldg), FIN-00014 University of Helsinki, Finland

Keywords: halocarbons, atmosphere, environmental effects

1. Introduction

In recent years, an increase in the load of volatile organic compounds (VOC) in the atmosphere has been documented. Production volume, their volatility and long persistence in the atmosphere of many VOC compounds, have resulted in global pollution problems (Wania & Mackay, 1993). These man-made chemicals are environmentally quite significant, due to their toxicity and their ability to react with the stratospheric ozone (Monola & Rowland, 1974; Surgeon General, 1980; WMO, 1982). Potential ozone depleters such as CFCs, halons, methyl chloroform, carbon tetrachloride, chloroform and related chemicals are responsible for depletion of the stratospheric ozone layer (NAS, 1982; Seidel & Keyes, 1983).

Recently, it has been confirmed that in the troposphere, UV-rays trigger reactions in these air-borne halocarbons, forming secondary pollutants, i.e. the herbicide trichloroacetic acid (TCA), which is phytotoxic, hygroscopic and water soluble. In addition, the occurrence of this compound in the forest ecosystems in Germany and Northern Finland has been documented (Frank *et al.*, 1992).

In this paper, the industrial use of, and the atmospheric and aquatic emissions of methyl chloroform and carbon tetrachloride from industrial sources in Finland have been considered for the year 1992. In addition, the formation of secondary pollutants in the atmosphere and their concentration in coniferous forests in Finland are touched upon.

2. Production, import and uses of methyl chloroform and carbon tetrachloride

2.1 Methyl chloroform (CH_3CCl_3)

Methyl chloroform, also known as 1,1,1-Trichloroethane, is widely used for metal degreasing. It is manufactured by a number of different processes, including chlorination of ethane, hydrochlorination of 1,1-dichloroethylene and vinyl chloride (Kirk-Othmer, 1978; Jordan, 1979). World production of CH_3CCl_3 is shown in Fig. 1. The production capacity of CH_3CCl_3 is largest in the United States, followed by Western Europe and Japan.

In Finland this chemical is imported but not produced. The imported amount of CH_3CCl_3 has declined from 1100 t in 1985, to 640 t in 1992. In addition, several mil-

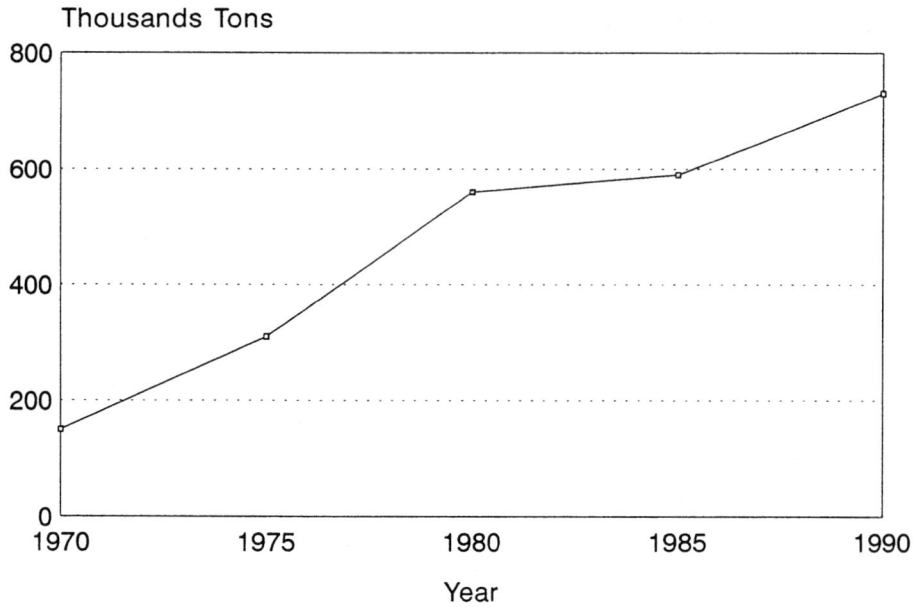

Fig. 1. World production of Methyl chloroform, 1970–90 (H. Frank, personal communication).

lion aerosol units containing CH_3CCl_3 are imported annually into Finland, but the exact amount imported through these units is unknown (Mukherjee, 1991).

CH_3CCl_3 is still widely used as a cleaning solvent in industry today. Due to its low toxicity and other favourable properties, metal industries use this chemical to remove grease and oil from the surface of metal parts. In addition, it is used to clean electrical equipment, semiconductors, printed circuit boards and also used as a solvent in textile processing, in cutting fluids, aerosols, lubricants, shoe polishes, printing inks and many other materials (Syracuse Research Corp., 1990). Current uses of CH_3CCl_3 in Finland include: cold cleaning and vapour degreasing 72 percent (460 t), adhesives three percent (20 t), aerosols nine percent (60 t), electronics eight percent (50 t) and others eight percent (50 t).

2.2 Carbon tetrachloride (CCl_4)

The maximum import of CCl_4 was noted to be 110 t in 1990. But its import has ceased completely in Finland since the middle of 1992.

In the early part of this century, CCl_4 was used for the treatment of intestinal worms (Hall, 1921). But the current major use of CCl_4 is in the production of chemicals (F-11 and F-12), which are used as refrigerants (CEH, 1985). Due to restrictions on the use of CFC-113 under the Montreal Protocol, it seems probable that, in the future, use of CCl_4 will be reduced. In Finland, its use in industry will soon cease completely.

3. Release to the Finnish environment

3.1 *Methyl chloroform*

CH_3CCl_3 is released to the environment as a result of anthropogenic activity. Due to its volatility, a large proportion of the CH_3CCl_3 used finds its way into the atmosphere. Literature studied indicates that about 95 to 99 percent of the CH_3CCl_3 used is released to the atmosphere (Torslov, 1988; Midgley, 1989). Here, in this paper, it is estimated that about 590 t (92 per cent) of CH_3CCl_3 enter into the Finnish atmosphere and the rest is distributed as follow: wastes distilled and resold – 32 t (5 per cent), wastes – treated in the hazardous waste treatment plant – 12 t (2.0 per cent) and the rest 6 t (1.0 per cent) enter into the aquatic environment and landfills (based on the study – Mukherjee, 1991).

In Finland, there is very little information on the concentration of CH_3CCl_3 in the aquatic environment. Paasivirta *et al.* (1985) observed 0.19–1.23 µg CH_3CCl_3 l^{-1} in sea water near the biggest oil refinery in Finland. In Europe, the concentration of CH_3CCl_3 of halogenated hydrocarbons in water varies between a fraction of a µg l^{-1} to 100 µg l^{-1} (CEFIC, 1986).

Henry's Law constants for this compound, $6.3 \times 10^{-3} – 17.2 \times 10^{-3}$ atm m^3 mol^{-1} at 25 °C (Gossett, 1987), suggest that CH_3CCl_3 volatilizes quickly from the aquatic environment, which has also been confirmed by laboratory and field tests (Kincannon *et al.*, 1983; Piwoni *et al.*, 1986).

3.2 *Carbon tetrachloride*

There is no recent information about the release of CCl_4 into the Finnish environment. But its emission may occur from oil refineries, chemical and pharmaceutical manufacturing industries, landfills and water surfaces; this last being still unknown (Mukherjee, 1991). In addition, its long range transport history, precipitation into the forest ecosystem and volatilization should also be taken into consideration.

In 1989, about 2 t of CCl_4 was discharged from the chemical industry to the aquatic environment. Paasivirta *et al.*, (1985) observed 0.8 to 26 µg CCl_4 l^{-1} in sea water, whereas in drinking water in Finland, the mean value was reported to be below 1 µg l^{-1}.

4. Environmental Fate

Due to high volatility and small Henry's constant, about 99 percent of the CH_3CCl_3 will find its way into the atmosphere. The lifetime of this compound in the troposphere is quite uncertain. WMO (1982) estimated a lifetime of CH_3CCl_3 ranging from five to ten years whereas Prinn *et al.*, (1987) estimated it to be five to seven years against the average OH-mixing ratio of $8 \pm 1 \times 10^5$ molecule cm^{-3}. The later value would seem to be more realistic. On the other hand, CCl_4 released to the environment exists in the atmosphere and as it does not degrade readily, cross country transport is expected (Life Systems Inc., 1992). Its lifetime has been estimated by many authors, based on model calculations and field work (Singh *et al.*, 1976; Borchers & Fabian, 1983) and the most realistic value is reported to be 67 years (Hammitt *et al.*, 1987).

Table 1. Trichloroacetic acid concentration in birch leaves and defoliation degree of the birch trees in Rovaniemi, Finland (Norokorpi & Frank, 1992).

Tree No.	Tree ht. m	TCA (ng/g*)		Defoliation Aug 92 dgree(%)
		Aug 91	Aug 92	
Birch 1	14	44.0	8.0	35
Birch 2	15	2.6	6.0	20
Birch 3	16	5.9	5.0	25
Birch 4	14	3.4	–	20
Birch 5	13	<1.0	–	0
Birch 6	13	<1.0	–	0

*Fresh weight
– Not analysed

5. Secondary airborne halocarbons and their effects on deciduous forests

The spread of organic pollutants and the emission of semi-persistent VOC into the atmosphere are both increasing. The atmospheric oxidation products of C_2-chlorocarbon solvents generally include: chloral, trichloracetyl chloride or trichloroacetic acid (TCA) (Frank, 1984; 1991; Frank et al., 1992). In the Finnish troposphere, conversion of CH_3CCl_3 by OH-radicals to TCA-herbicide is to be expected.

TCA is hygroscopic, water soluble, mobile in soil and easily leached out by precipitation (Kearney et al., 1965). Recently, the presence of TCA was observed in spruce needles in the northern part of the Black Forest in Germany (Frank, 1991) and also in birch trees (*Betula pubescens* Ehrh.) and conifer needles (*Picea abies*, *Pinus sylvestris*) in Northern Finland (66°23′N, 25°00′E) (Frank et al., 1992; Norokorpi & Frank, 1992). The latter authors observed the presence of TCA in birch leaves which had turned brown along their edges, causing 20 to 35 percent defoliation (Table 1). They believe that atmospheric herbicide damage to forests in the Arctic region is increasing, although study of the ecotoxicology of northern areas is still quite new.

6. Conclusions

This study assesses the scenarios of use and emissions of CH_3CCl_3 and CCl_4 in Finland. There is no doubt that these chemicals are useful to us as well as being harmful to our ecosystems. The import of CH_3CCl_3 has decreased from 1100 t in 1985 to 640 t in 1992 and the import of CCl_4 ceased completely from the middle of 1992. The largest use of CH_3CCl_3 was observed in the field of cold cleaning and vapour degreasing. At present, there is no industrial use of CCl_4 in Finland.

It is estimated that atmospheric and aquatic emissions of CH_3CCl_3 are about 92 percent and one percent, respectively of the total used. This chemical will also further evaporate into the atmosphere from the aquatic environment.

It is observed that the burden of organic compounds and formation of secondary pollutant such as TCA by the oxidation of C_2-chlorocarbon solvents (e.g. CH_3CCl_3) are both increasing in the atmosphere, causing high concentrations to be found in the forest ecosystems in northern Finland.

7. Summary

In recent years, an increase in the load of volatile organic compounds in the environment has been documented. These compounds are widely used as raw materials in the production of chemicals and solvents. Methyl chloroform (CH_3CCl_3) is generally used as a degreasing agent whereas the majority of the world production of carbon tetrachloride (CCl_4) is used by the chlorofluorocarbon industry.

In Finland, the import of CH_3CCl_3 has declined from 1100 t in 1985 to 640 t in 1992 and the import of CCl_4 ceased completely since the middle of 1992. In 1992, total estimated emission of CH_3CCl_3 to air and water was about 93 percent (i.e. 600 t).There is no information on the release of CCl_4 to the atmosphere for 1992. CCl_4 emission to the atmosphere was, however, reported to be only 0.005 t for 1990.

The lifetime of CH_3CCl_3 in the troposphere is reported to be 6 ± 1 years, whereas for CCl_4, it is about 52 ± 15 years. Problems of stratospheric ozone depletion and greenhouse effect due to release of these halocarbons have been confirmed by climate modellers.

Recently, it has also been confirmed that, in the atmosphere, air-borne halocarbons are oxidized to secondary toxic pollutants such as herbicide trichloroacetic acid (TCA) which is phytotoxic, hygroscopic and water soluble. Concentrations of this chemical have been found in the forest ecosystems in Germany and Northern Finland and it is believed that the damaging impact of airborne herbicides on forest trees is increasing.

Acknowledgements

The linguistic assistance of Mr. Ken Himsworth is gratefully acknowledged.

References

Borchers, R. & P. Fabian, 1983. First measurements of the vertical distribution of CCl_4 and CH_3CCl_4 in the stratosphere. Naturwissen 70: 514–516.
CEFIC-BIT Solvent Chlores, 1986. The occurrence of chlorinated solvents in the environment. Chemistry & Industry: 861–869.
CEH, 1985. CEH products review: Chlorinated methanes. Chemical Economic Hnadbook-SRI International: 635.2020A-635.2022B.
Frank, H., 1984. Waldschaden durch photooxidants ? Nachr. Chem. Tech. Lab., 34: 15–20.
Frank, H., 1991. Airborne chlorocarbons, photooxidants and forest decline. Ambio 20(1): 13–18.
Frank, H., H. Scholl, S. Sutinen & Y. Norkorpi, 1992. Trichloracetic acid in conifer needles in Finland. Ann. Bot. Fennici 29: 263–267.
Gossett, J. M., 1987. Measurement of Henry's Law Constant for C1 and C2 chlorinated hydrocarbons. Environ. Sci. Tech., 21: 202–206.
Hall, M. C., 1921. The use of carbon tetrachloride for the removal of hookworms. J. Am. Med. Assoc., 77: 1641–43.
Hammitt, J. K., F. Camm, P. S. Connell, W. E. Mooz, K. A. Wolf, D. J. Wuebbles & A. Bamezai, 1987. Future emission scenarios for chemicals that may deplete stratospheric ozone. Nature 330: 711–716.
Jordan, J. I., 1979. *Encyclopedia of Chemical Processing and Design*. Vol. 10, Chapter 8, J. J. McKetta & W. A. Cunningham (eds).
Kearney, P. C., C. I. Harris, D. D. Kaufmann & T. J. Sheets, 1965. Behaviour and fate of chlorinated aliphatic acids in soils. In: R. L. Metcalf (ed), *Advances in Pest Control Research*, Vol. VI, 1–30. Wiley-Interscience, New York.
Kincannon, D. E., A. Weinert, R. Padorr & E. L. Stover, 1983. Predicting treatability of multiple organic priority pollutant waste water from single-pollutant treatability studies. In: J. M. Bell (ed), *Proceeding of 37th Conference: Industrial waste*, 641–650, Ann Arbor Science Pub.
Kirk-Othmer, 1978. *Encyclopedia of Chemical Technology*, 5: 728–730, John & Wiley & Sons.

Life Systems Inc., 1992. Toxicological profile for carbon tetrachloride. Prepared for U.S. Dept. of Health & Human Services, Public Health Service, Atlanta, GA 30333, USA: pp. 133.
Midgley, P. M., 1989. The production and release to the atmosphere of 1,1,1-Trichloroethane (Methyl chloroform). Atms. Environ., 23(12): 2663–65.
Monila, M. J. & F. S. Rowland, 1974. Stratospheric sink of chlorofluromethanes: Chlorine atom catalyzed destruction of ozone, Nature 249: 810–812.
Mukherjee, A. B., 1991. The use of 1,1,1-trichloroethane and carbon tetrachloride in Finland and their impact on the environment. Report No. 294, National Board of Waters and the Environment, Helsinki, Finland: pp 94.
NAS (National Academy of Science), 1982. Causes and effects of changes in stratospheric ozone reduction: An update. National Academy Press, Washington, D.C.
Norokorpi, Y. & H. Frank, 1992. Effect of stand density on damage to birch (*Betula pubescens* E.) caused by airborne herbicides. Symposium on the State of the Environment & Environmental Monitoring in Northern Fennoscandia and the Kola Peninsula, Oct. 6–8, 1992, Rovaniemi, Finland: 1–12.
Paasivirta, J., R. Paukku & K. Mäntykoski, 1985. Haitalliset aineet Sköldvikin meriympäristössä. Helsingin vesipiirin vesitoimisto: 1–29 (in Finnish).
Piwoni, M. D., J. T. Wilson & D. M. Walters, 1986. Behavior of organic pollutants during rapid-infiltration of wastewater into soil. I. Processes, definition, and characterization using a microcosm. Hazard Waste. Hazard Material 3: 43–55.
Prinn R., D. Cunnold, R. Rasmussen, P. Simmonds, F. Alyea, A. Crawford, P. Fraser & R. Rosen, 1987. Atmospheric trend in methyl chloroform and the global average for the hydroxyl radical. Science 90: 5547–67.
Seidel, S. & D. Keyes, 1983. Can we delay a greenhouse warming. USEPA, Washington, D.C.
Singh, H. B., D. P. Fowler & T. O. Peyton, 1976. Atmospheric carbon tetrachloride: Another man-made pollutant. Science, Vol. 192: pp. 1231–1234.
Singh, H. B., L. J. Salas & R. E. Stiles, 1983. Selected man-made halogenated chemicals in the air and oceanic environment. J. Geophys. Res., 88(C6): 3675–3683.
Surgeon General, 1980. Health effects of toxic pollutants: A report from the Surgeon General, Rep., 96–15, Dept. Health Human Serv., Washington, D.C.
Syracuse Research Corporation, 1990. Toxicological profile for 1,1,1-Trichloroethane. Prepared for Agency for Toxic Substances and Disease Registry, U.S. Public Health Service, Atlanta, GA 30333, USA: 105–133.
Torslov, J., 1988. Hazard assessment of 1,1,1-Trichloroethane. Miljoprojekt Nr. 100,Danish National Agency of Environmental Protection, Copenhagen, Denmark.
Wania, F. & D. Mackay, 1993. An approach to modelling the global distribution of Toxaphene: A discussion of feasibility and desirability. Presentation given at the Workshop on the Analytical Chemistry of Toxaphene held in Burlington, Ont., Canada, Feb. 1993: 1–15.
WMO (World Meterological Organization), 1982. The stratosphere 1981: Theory and measurements, Rep. 11, Geneva.

Ecological situation in the Leningrad Region: Emissions, transboundary air pollution, decline of the conifer forest (results of a three–year assessment and monitoring survey of the effects of air pollution on forests)

N. I. Goltsova,[1] T. V. Vasina[2] & B. G. Popovichev[3]

[1]*Biological Research Institute of Saint Petersburg State University Oranienbaumskoe Schosse 2, 198904 Petrodvoretz, St. Petersburg, Russia*
[2]*Leningrad Region Committee of Environment, St. Petersburg, Russia*
[3]*Forestry Academy, St. Petersburg, Russia*

Keywords: Forest, pine, lichens, bioindicators

1. Introduction

This work is based on international methodologies prepared by the Centre of Forest Monitoring of the United Nations' Economic Commission for Europe (UN/ECE), together with the UNEP and the Secretariat of the UN/ECE (Manual, 1992), Helsinki University (Manual, 1993),and Finnish Standards SFS-5670 (1990), SFS-5671 (1990), SFS-E221(1992). Part of it was published in international reports of UN/ECE (Forest Condition in Europe, 1992, 1994) and Nordic Council (Ruhling et al., 1993, 1994).

This study involved regular collection of representative and comparable (neighbouring states) data on changes in forests on permanent sample plots under the impact of air pollution and other anthropogenic factors. The purpose of this study was to determine the condition of forests in the Leningrad region, so it would then be possible to find out the history of the forest condition, and to determine forest pollution from air fallout.

2. Materials and methods

Permanent sample plots were established in pine stands far from emission sources (at a distance of more than 2 km) and roads (at a distance of more than 500 m) as prescribed by the regulations of international monitoring on the model network of 32 by 32 km. The age of trees is over 45 years (45 to 80 year-old trees, developmental class I-III). Material was collected in the period July through September, in 1990-1993. The tree condition indices were assessed by criterion according to the UN/ECE methodology: defoliation of the whole crown, loss of needle colour in a percent of the whole crown, needle longevity, diseases of trees and their injury by insects. The condition of epiphyte lichens, their species composition and projective cover on pine trunks were counted at four trunk heights (0.6, 0.9, 1.2, 1.5 m) and at the trunk base. Ten pine individuals were examined at each point. Moss samples were taken for a chemical analysis of eight heavy metals: Cd, Cu, Ni, Zn, V, Fe, Pb, Cr – as prescribed by the international regulations (Ruhling *et al.*, 1992). Data intercalibration was performed. The analysis of sulphur in needles was made in the Universities of Helsinki and Oulu. Samples were collected and prepared in accordance with the Finnish standard SFS-5669.

3. Results of coniferous-forest condition analysis

3.1 *Defoliation*

About 44 percent of the trees examined have needle loss of over 11 percent. A more significant needle loss in the crown was noted in 5.4 percent of the trees examined. In ripening and ripe stands, average regional indices of defoliation are higher than in young ones. The average index of coniferous stand defoliation in the region is 14.2 percent. Sample plots located in forests tracts of the west-southern and north-southern parts of the Leningrad region and near St. Petersburg, reveal features of a high degree of crown defoliation (Fig. 1).

3.2 *Discoloration*

In 1992, over half the territory of the region was covered by forests with changed needle colour (Fig. 2). Since, in 1992, 19527 ha out of 5,800,000 ha of forest land were covered by fires (official data, GosKomStat, 1992). They were excluded from the analysis. The highest needle discolouration was characteristic for the forests of the Karelian Isthmus, and of the extreme south-western part of the region. On average, on all plots of the region, tree discolouration accounts for 7.3 percent.

3.3 *Needle longevity*

According to the data on this years' observations, the pine needle longevity for all the sample areas of the Leningrad region is, on average, 3.0 plus 0.4 years. Short needle longevity, as compared to normal longevity, has been noted in two–fifths of the region's territory (Fig. 3): shoots retained needles for two years or less.

Zoning of the territory of the Leningrad region by wood vegetation condition (Figs. 1-3) has demonstrated that appreciable weakening of the stands is observed in vast territories, as a result of high anthropogenic loads – including air pollution. The territories are primarily: the Karelian Isthmus (Svetogorsk-Kamennogorsk zone, the central part of the Karelian Isthmus, a long and wide belt in the direction of St. Petersburg-Vyborg, Vsevolozhsk district), the south-western part of the region (Kingisepp and Slantsy districts, Lomonosov district) as well as the Luga, Kirishi and Volkhov districts of the region, and the Kolpino district adjacent to St. Petersburg. In 1992, according to our data of observations on the same sample plots, the total area of forests tracts showing needle loss increased as compared with 1991 (Fig. 1). Eastern districts (Podporozhye, Tikhvin, Bauxitogorsk) have symptoms of stand injury mainly in western parts (injuries of forest territories as a consequence of felling and inundation were excluded from the analysis). Favourable conditions remained locally in the Karelian Isthmus territory, only in the vicinity of Primorsk and in the southern part of the Priozersk district on the Ladoga side.

3.4 Conditions of pine epiphyte lichens

A lichenoindication survey was conducted on sample plots. A substantial disturbance of pine-trunk epiphyte cover (projective pine trunk cover by lichens; their floristic composition) as a result of air pollution has been noted in Slantsy, Kingisepp, Lomonosov and Sestroretsk districts of the region; in southern and north- western parts of the Karelian Isthmus and in the suburban zone (Kolpino), (Fig. 4 and 5). In the polluted areas, as a rule, there has been a noted decrease in the projective lichen cover of pine trunks and in the number of species, pine-bark inhabitants, as a result of the action of SO2 and NOX. The pathological lichen flora of conifer trees has been observed in the SW districts of the Leningrad region as resulting from transboundary air pollution (from Baltic and Estonian Power Plants, Estonia) and local emissions (from Slantzy, Fig. 8)

4. The concentration of sulphur in needles

This year, amounts of sulphur have come to light, accumulated by needles in unpolluted habitats (860 mg/kg dry needle substances, that is, ppm) and on individual polluted model forest plots (up to 2000 ppm)(Fig. 6). Sulphur accumulation in needles is directly connected to air pollution (Manninen *et al.*, 1991; Haapala *et al.*, 1993). The maximum pollution of SOX in the south-western part of the Leningrad region (Fig. 7) is a result of the influence of the Baltic Power Plant of Estonia (source of oil-shale emission). The biggest doses are in the Northern part of Karelian Isthmus-in the Russian-Finnish Industry zone (Svetogorsk-Kamennogorsk-Imatra); near the Kirishi oil refining plant and around St. Petersburg and its satellites.

5. Results of chemical analysis of mosses

The first attempt to map pollution levels of the region's wooded areas by analyzing the indicator plants, was made by N. Goltsova, published in the Report of Nordic countries (Ruhling *et al.*, 1992, 1994; Goltsova & Vasina, 1993). Similar papers were published from other countries (Makinen, 1993; Zechmeister, 1994). In addition, it is quite possible to assess the amount of deposition of heavy metals (singly or in combination) per unit area (Ruhling *et al.*, 1992). Data on the concentration of heavy metals in mosses can be seen in Table 1.

6. Sources of emission in St. Petersburg region

The major emission sources are situated in the town of St. Petersburg, in it's industrial town satellite Kolpino, in the Vyborg region, in Tikhvin and Kirishi, and on the southwest part of the Leningrad region (Table 2, Fig. 7). The large sulphur emissions from Narva (Baltic and Estonian Energy Plants, Estonia) are twice more than the emissions of the town of St. Petersburg (Haapala *et al.*, 1993). The participation from Narva is three times more than all of Finland (Haapala *et al.*, 1993). It is the main source of transboundary air pollution in the Leningrad region. The damaged area is more than 1024 km2 (Fig.8b).

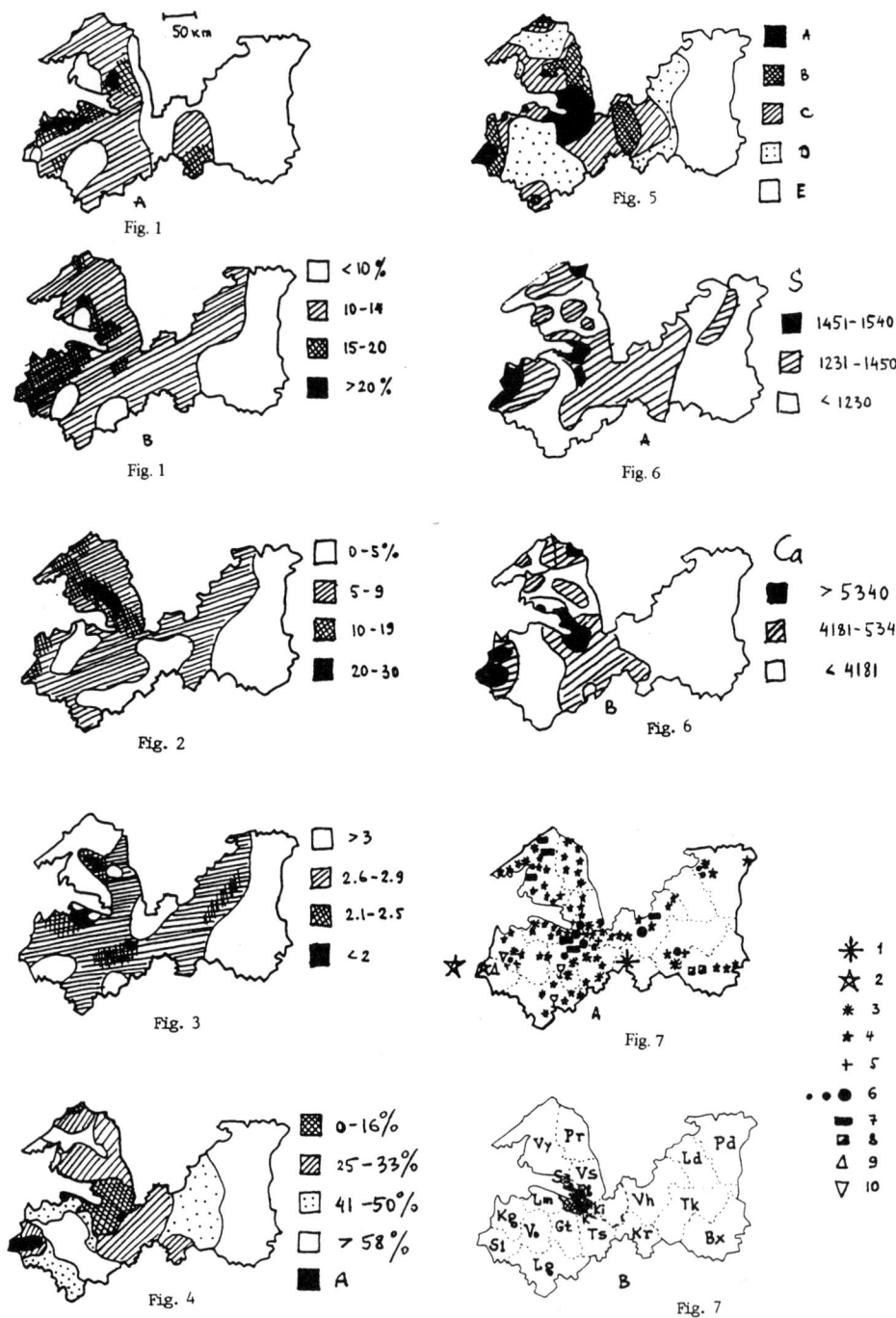

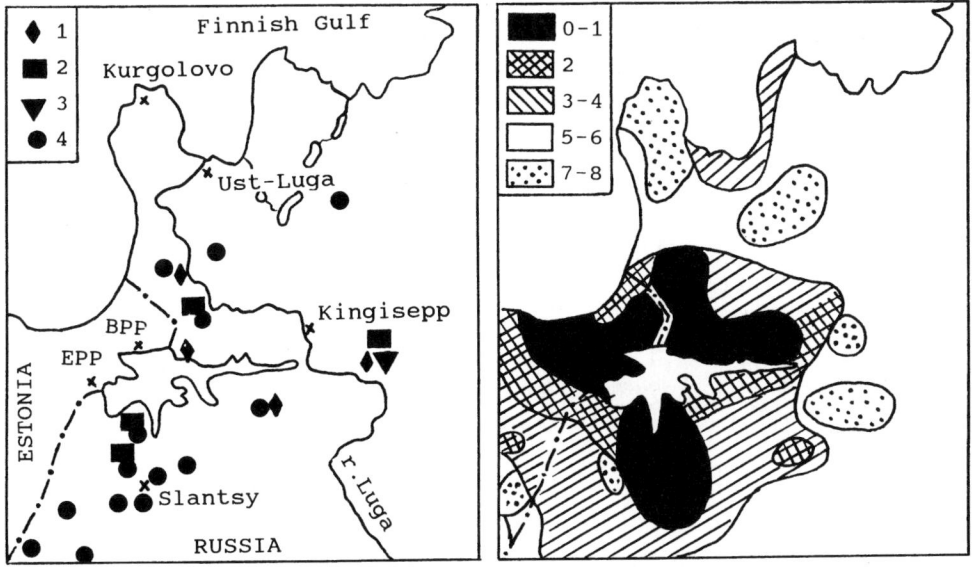

Fig. 8. The lichenoindication of air pollution in the south-western districts of the Leningrad region. A: pathological changes in epiphytic lichen flora of pine bark. Anability of next species: 1: *Physcia spp.*, 2: *Xanthoria parietina*, 3: *Xanthoria polycarpa*, 4: *algae Trentepolia umbrina* BPP: Baltic Power Plant, EPP: Estonian Power Plant. B: mapping of the SW area by number of lichen species. The number of lichen species is written according to Finnish standard methodology, SF 5667, (from 12 species). 0 to 1, 2, 3 to 4, 5 to 6, 7 to 8, the number of species.

←

Fig. 1. Defoliation of Scots pine needles in the Leningrad region in 1991 (A) and 1992 (B). Percentage of needle loss in the crown of forest stands.
Fig. 2. Discolouration of needles in Scots pine forests of the Leningrad region in 1992. The degree of discolouration is shown in percentages.
Fig. 3. Needle longevity of Scots pine in the Leningrad region 1992. White areas (needle age at least 3 years) represent a normal situation in the southern part of taiga forests.
Fig. 4. Estimation of air pollution by flora of pine lichens. The number of species and their percentages from 12 standard species. A: pathological species of lichens on pine bark (they are absent on healthy sample plots)
Fig. 5. Estimation of air pollution by lichen cover (*Hypohymmia physodes*) on pine trunks. An average cover (%) of trunk on high 0.5 m to 1.55 m. From 1991 to 1992. A: 0 to 22 percent (extremely polluted air), B: 22 to 40 percent (polluted air), C: 41 to 49 percent (insignificant amount of pollutants in air), D: 50 to 60 percent (relatively clean air), E: > 60 percent (clean air)
Fig. 6. The concentrations of sulphur (A) and calcium (B) in 2 year-old pine needles, ppm. February, 1993. Leningrad region.
Fig. 7. Major emission locations of the Leningrad region. Power plants: 1: refineries and petrochemical plants; 2: shale-processing plant; 3: electricity power plants using fuel oil; 4: electricity power plants using solid fuels (coal, shale, peat); 5: gas boilers. Other industries: 6: foundries and metallurgical works, 7: paper mills; 8: bauxites mining and processing; 9: cement plants; 10: phosphorites mining and processing; 11: glassworks. Districts: Vy – Vyborg, Pr – Priozersk, Vs – Vsevolozhsk, Ss – Sestroretsk; Lm – Lomonosov, Kg – Kingisepp, Sl – Slantsy, Vo – Volosovo, Gt – Gatchina, Ki – Kirovsk, Ts – Tosno, Lg – Luga, Vh – Volhov, Kr – Kirishi, Tk – Tikhvin, Bx – Bauxitogorsk, Ld – Lodeinoe Pole, Pd –Podporozie. Towns: SPb – St.Petersburg, K – Kolpino

Table 1. The content of heavy metals in moss samples in the Leningrad region and the Karelian Isthmus (min-max; mg/kg d.w., ppm).

	Cu	Fe	Pb	Ni	Cd	V	Zn	Cr	n
Northern districts:									
(The Karelian Isthmus)									
Sestroretsk	5-17	700-1400	10-100	1.1-8	0.1-1.0	4-20	35-70	2.5-8	5
Vsevolozhsk	10-14	1540-6300	2-5	4-19	0.2-0.6	6-20	47-200	8-17	5
Priozersk	3-10	330-3400	0.4-180	1.9-11	0.2-0.7	3-9	32-75	1-21	21
Vyborg	1.2-15	280-3500	0.2-140	4-15	0.1-0.8	1-31	28-92	1-23	34
Western districts:									
Lomonosov	11-12	530-3600	0.6-40	1-18	0.2-1.2	2.5-13	24-100	1-43	77
Kingisepp	1.5-7	660-4200	0.8-20	0.6-10	0.2-0.5	2.4-22	21-170	2.5-32	32
Slantsy	2-10	570-4480	1.6-22	2-11	0.1-0.7	4-9	24-74	8-36	12
Volosovo	4-6	580-640	1.1-3	2-5	0.4-0.5	2.3-4	27-32	9-11	2
Southern districts:									
Gatchina	(3.8)	(700)	(2.3)	(6.3)	(0.5)	(3.4)	(36)	(10)	1
Tosno	3-27	500-4400	1.6-8	3-16	0.2-0.8	4-25	29-180	3-12	7
Luga	1.5-8	540-4130	1.4-100	3-12	0.3-1.6	1-7	22-63	1-18	8
Eastern districts:									
Volhov	5-8	650-1750	1.6-100	6-10	0.2-0.8	4-10	33-53	3-12	7
Kirishi	4-5	700-1050	1.8-2.4	8-11	0.3-0.9	7-15	42-46	6-12	2
Tikhvin	1-13	520-4270	1.4-7	4-19	0.2-1.3	4-12	30-170	5-16	11
Bauxitogorsk	3-8	240-650	1.2-3	2.2-4	0.2-0.4	3-5	37-41	6-10	4
Lodeinoe Pole	4-14	700-3150	1.4-5	5-13	0.2-1.1	4-7	19-77	3-12	9
Podporozhie	(3.1)	(550)	(1.7)	(4.6)	(0.68)	(3.2)	(32)	(6.7)	1
The whole Leningrad region:									
min-max	1.2-40	240-6300	0.2-180	0.6-25	0.1-1.6	1.5-31	19-200	1-43	360
mean	6.4	1500	8.8	6.5	0.51	6.3	49.3	11.1	
variation	3.8	100	19.8	4.4	0.23	3.9	22.8	7.71	

7. Discussion and conclusions

Determination of air pollution by sulphur and nitrogen oxides has been carried out by us on sample plots, through evaluation of the condition of lichen cover. Such monitoring is very effective, and will be necessary later on for validating the statistical data on the emission amount of sulphur and nitrogen oxides in the course of Convention (Helsinki, 1974) implementation. The impact of such a gigantic city as St. Petersburg on the environment is very significant and covers great distances: (60 km on E, NE and SE directions and 40 km on NW and W directions). The effect of local emission sources is great also. It is particularly marked in the vicinity of Slantsy (by the appearance of pathological species of pine lichens and the disappearance of the common ones), in the Karelian Isthmus, in the neighbourhood of Volhov, Kirishi and the Sosnovii Bor industrial zone. The whole southern part of the Karelian Isthmus, up to Otradonoye and Pervomaiskoye, shows degradation of lichen cover and disappearance of a number of indicator species, as a result of high transport and other technological loads.

All the heavy metals studied, could be involved in long-range transboundary transport but atmospheric depositions in the Leningrad region are mostly related to local emission sources. High levels of chromium and nickel are detected in the vicinity of either urban centres (towns and industrial townships) or industrial waste emission

Table 2. The air emission of pollutants in 1990 year (official data) in the districts of Leningrad region and towns, in ton/year. 1-whole district without main town, 2-main town of district.

	Summarased 1	Summarased 2	Solid 1	Solid 2	Gas & Liquid 1	Gas & Liquid 2	SO_2 1	SO_2 2	CO 1	CO 2	NO_2 1	NO_2 2	F 1	F 2
Northern districts: (The Karelian Isthmus)														
Vsevolozhsk	18900	1129	2174	109	16725	1020	9507	407	684	190	5899	220	0.25	-
Priozersk	6179	3404	3995	671	2184	2733	1185	850	781	1414	185	256	-	-
Vyborg	10670	7007	2929	2504	7739	4503	5323	2766	925	952	491	418	-	0.037
Western districts:														
Lomonosov	5516	-	1732	-	3784	-	105	-	68	68	200	20	0.671	0.05
Kingisepp	3869	5450	881	1561	2988	3889	2053	2118	704	410	221	179	0.013	54.3
Slantsy	-	111546	-	62782	-	48764	-	7171	100	38576	6	2364	-	0.023
Volosovo	381	1290	27	389	354	900	152	458	71	255	31.7	68	-	-
Southern districts:														
Gatchina	7696	4502	1212	1284	6484	3218	2541	1563	344	904	761	399	0.022	0.003
Tosno	3497	-	2004	-	1493	-	309	-	-	-	-	-	-	-
Luga	2346	6807	495	2413	1851	4394	802	2249	498	1187	92	587	0.008	-
Eastern districts:														
Volhov	17528	10724	4734	4145	12800	6580	6363	3051	6363	3051	4706	1974	0.04	375.6
Kirishi	356	196438	2	1428	354	195010	-	121807	20	1696	-	13481	0.001	0.008
Tikhvin	179	13733	76	3337	103	10396	23	1888	-	-	-	-	-	-
Bauxitogorsk	40620	42744	21900	22150	18720	20500	16102	16160	1250	2485	1356	1774	9.87	9.90
Lodeinoe Pole	65	3412	31	819	34	2593	-	1732	31.4	534	26	232	-	-
Podporozie	300	3218	51	457	249	2743	164	1779	72	671	12	191	-	-
Kirov	7325	23918	2533	17941	4792	5977	2650	2844	279	1940	458	1138	0.025	0.015
The whole Leningrad region	533255		145961		387294		201762		24640		35926		439	
Kolpino town	28368		6767		21601		6872		8340		2580		12.6	
St. Petersburg town	191491		34582		156909		62700		29670		41761		3.67	

sources. Vanadium distribution follows those of population density and industrial activity. High levels of cadmium, the most toxic of all the metals studied has been detected for practically all industrial and agriculturally developed areas of the Leningrad region. Transboundary transport contribution is significant for the south-western part of the Leningrad region (the Kingisepp region) and the Karelian Isthmus. The gigantic qualities of sulphur, metals and dust precipitation come from North-Eastern Estonia (oil-shale Power Plants and Kohtlajarve industrial plants). Their influence is felt in the terrestrial part of the Kingisepp region, the Kurgal peninsula and the islands of the Baltic Sea (ie. Malii Tjuters). In a 1024 quadrate kilometres area, sensitive species of pine-lichens are dead, some new pathological lichen species have appeared, and the needle-age is reduced. Extensive damage to needle tissues, waxes and stomata are observed.

8. Main conclusions

Forest-territory zoning of the Leningrad region by conifer and lichen condition makes it possible to assess the ecological situation in the region and single out both critical and background zones. The Karelian Isthmus, south-western areas, as well as the Kirishi district, should be considered critical zones. Placed into the background category, there may still only be three eastern districts (the Podporozhye, Tikhvin and Bauksitogorsk ones) – and not over their whole extent — only some parts.

The heaviest influence of transboundary transfer is felt on the south-western part of the Leningrad region. The large, heavy metal deposition and acidification takes place over a wide area. Additional effects come from more than 60 km away. The pathological changes of healthy forest ecosystems take place in this area. The critical condition of forests and heavy pollution takes place in other border areas, ie. the Finnish-Russian paper industry area (Imatra- Svetogorsk-Kamennogorsk). High sulphur emission has a profound effect on pine trees conditions and lichens' flora. The forests of the Karelian Isthmus are in the process of decline. There are many reasons for this decline: high intensity of traffic, increasing density of population and industry, the cutting down of trees and the influence of local and global emissions. The increase of insects that cause tree illnesses, is a result of anthropogenic stress.

To date, the time-history of stand conditions has been followed on the sample plots for three years. Every year, tree decline is noted on the Karelian Isthmus (Svetogorsk-Kamennogorsk and the central part of the Isthmus) and in the south-western districts (Kingisepp, Lomonosov, Slantsy). The changes were particularly drastic during the last year.

The health of the Baltic Sea strongly depends on the health of its terrestrial portion as well. We must not forget about this.

9. Summary

Regular collection for forest monitoring has been carried out with representative and comparable (neighbouring states) data on the network of permanent sample plots with a grid of 32 by 32 km territory of the whole Leningrad region. Defoliation, discoloration, needle longevity and dieback of conifers, as well as species composition and cover of pine epiphyte lichens, have been studied for the purpose of air pollution bioindication. On a denser network of temporary and permanent sample plots, the atmospheric deposition of heavy metals has been surveyed by chemical analysis of mosses for Cd, Cr, Cu, Ni, V, Fe, Zn, and Pb. Intercalibration of collection and analysis methods preceded the work.

Forest territories of the Leningrad region have been zoned by the aforementioned bioindicators. The zoning makes it possible to assess the ecological situation in the region. The Karelian Isthmus, the south-western areas, as well as the Kirishi district, were classified as critical areas.

The history of stand conditions has been followed on the sample plots for three years. Each year, the trees' decline is noted on the Karelian Isthmus (Svetogorsk-Kamennogorsk and central part of the Isthmus) and in the south-eastern districts (Kingisepp, Lomonosov, Slantsy). The changes were particularly drastic during the last year. Emission source data is presented as well.

Acknowledgements

Our special thanks to Dr. H. Haapala, Helsinki University, Kotka Unit, Finland, and to Prof. S. Huttunen, Oulu University, and other colleagues for the sulphur analysis of needles. We would also like to say thanks to Dr. Charles M. Wynn and Sandra Snow, Eastern Connecticut State University, U.S.A., for their help in the preparation of the manuscript.

References

Forest condition in Europe, 1992. International Co-operative programme on assessment and monitoring of air pollution effects on forests. UN/ECE, CEC. 1992 report. Brussels, Geneva. 117pp.
Forest condition in Europe, 1994. Results of the 1993 survey. International Co-operative programme on assessment and monitoring of air pollution effects on forests. UN/ECE, EC. Brussels, Geneva. 93pp.
Goltsova, N. & T. Vasina, 1993. Heavy metals retention by the woodland moss *Pleurozium schreberi*, Leningrad region survey. J. Ecol. Chem., 1 : 51-60.
Haapala H., N. Goltsova, S. Huttunen, R. Seppala, J. Lamppu & J. Kouki, 1993. Mannyn neulasten rikki- ja kalsiumpitoisuudet Suomenlahden itaosan reuna-alueiden ilman laadun indikaattoreina. In: Metsien ekologinen tila Suomenlahden itaosan reuna-alueilla. KOTKA, FINLAND. 51-73p.
Manninen, S., S. Huttunen & H. Torvela, 1991. Needles and lichen sulphur analyses on two industrial gradients. Water, Air and Soil Pollution. 59: 153-163.
Manual for integrated monitoring, 1993. Programme phase 1993-1996. UN ECE Convention on long-range transboundary air pollution. International Co-operative Programme on Integrated Monitoring on Air Pollution Effects. 1993 EDC, Helsinki.
Manual on methodologies and criteria for harmonized sampling, assessment, monitoring and analysis of the effects of air pollution on forests, 1992. Convention on long-range transboundary air pollution. International Co-operative Programme on Assessment and Monitoring of Air Pollution Effects on Forests. Hamburg.
Ruhling, A.,(ed.), 1992. Atmospheric heavy metals deposition in Northern Europe. NORD 1992:12,41pp.
Ruhling, A.,(ed.), 1994. Atmospheric heavy metal deposition in Europe – estimations on moss analysis. NORD 1994:9,59pp.
Zechmeister, H., 1994. Biomonitoring der schwermetalldepositionen mittels moosen in Osterreich. Monographien, band 42. Wien. 168pp.

Pollution-related environmental gradients around the "Severonikel" smelter complex at the Kola Peninsula, Northwestern Russia

M. V. Kozlov & E. Haukioja

Laboratory of Ecological Zoology, Department of Biology,
University of Turku, SF-20500 Turku, Finland

Keywords: aerial pollution, forest decline, ecosystem destruction

1. Introduction

The area surrounding the city of Monchegorsk on the Kola Peninsula is one of the most extreme examples of terrestrial pollution in the boreal zone (Kozlov et al., 1993b). Vast quantities of sulphur and heavy metals emitted during 55 years (Pozniakov, 1993) have caused widespread destruction of soils and vegetation and imposed additional environmental gradients in the region around the smelter.

This paper aims, firstly, to summarize research on how pollution has affected terrestrial ecosystems in the central Kola region. Secondly, some general methodological problems relating to data collection and interpretation along pollution-induced environmental gradients are discussed. In particular, researches must be cautious concerning problems caused by the pre-existing spatial and temporal heterogeneity of the natural environment, other human-induced influences, and the history of the pollution impact.

2. Extent of pollution and characteristics of the impact zone

The "Severonikel" smelter complex was established in 1938 (see Pozniakov, 1993, for the history) near Monchegorsk in the Murmansk district (68° N, 33° E; Fig. 1), about 150 km south of the northern tree line. It is one of the largest sources of aerial emission of SO_2 and heavy metals in Northern Europe: the total amount of pollutants emitted in 1990 was 2.64×10^8 kg, which included 2.33×10^8 kg of SO_2 and 1.58×10^7 kg of dusts containing heavy metals (2.7×10^6 kg of Ni, 1.8×10^6 kg of Cu) (Berlyand, 1991).

The total area influenced by aerial pollution is estimated to exceed 10,000 km^2. Complete destruction of primary ecosystems occurs within 6-10 km from the smelter (Kryuchkov, 1993). Visible effects of the pollution on vegetation have been recorded up to 60-80 km, and measurable traces of pollutants up to 200 km from the smelter (Derome, 1993).

3. Current level of environmental knowledge

The state of the environment around the "Severonikel" smelter has been a subject of investigation from the 1970's, mostly by Russian scientists. To date, over 500 scien-

Table 1. The number and proportion of papers that discuss the effects of pollution on different taxa of plants and animals.

Taxonomical group	Number of papers	Share (%)
Plants	167	82.3
Algae	5	2.5
Fungi	12	5.9
Mosses	13	6.4
Lichens	23	11.3
Conifers	87	42.9
Other	27	13.3
Animals	36	17.7
Arthropoda	15	7.4
Birds	2	1.0
Mammals	19	9.3

tific manuscripts have been published; for reviews, see Kismul et al. (1992), Tikkanen et al. (1992), Alexeyev (1993), Derome (1993) and Kozlov et al. (1993b). The most popular research topic concerned the impact of pollutants on forest soils (44.5 % of publications), while the life-forms most often studied were conifers, especially Scots pine (Table 1).

4. Pollution gradients and terrestrial ecosystems at the Kola Peninsula

4.1. Content of pollutants

4.1.1 Air
Barkan (1993) demonstrated that SO_2 concentrations decreased hyperbolically with increasing distance from the smelter. Furthermore, the aerial sulphur dioxide content was strongly correlated with the content of heavy metals in soils. The decrease in SO_2 gradients south of the smelter showed an irregular pattern, probably because the mountain relief in that region affected the flow and deposition of pollutants. Measurements of the heavy metal content of snow sampled from 22 plots (Tchekhin, 1993) were consistent with the observations of Barkan (1993), including the mosaic distribution of pollutants south of the smelter.

4.1.2 Soils
There was a general trend for the heavy metal content of soils also to decrease hyperbolically from the emission source to unaffected regions. However, a study based on a grid of about 140 plots revealed an underlying mosaic spatial pattern of soil contamination along any gradient: areas of relatively low pollution concentrations were interspersed within heavily polluted zones (Barkan et al., 1993).

4.1.3 Vegetation
The content of sulphur and heavy metals in spruce (Kryuchkov & Eliseev, 1991) and pine (Raitio et al., 1993) needles, birch leaves (Kozlov et al. 1993a) and forest berries (Barkan et al. 1993) followed the same patterns as soils.

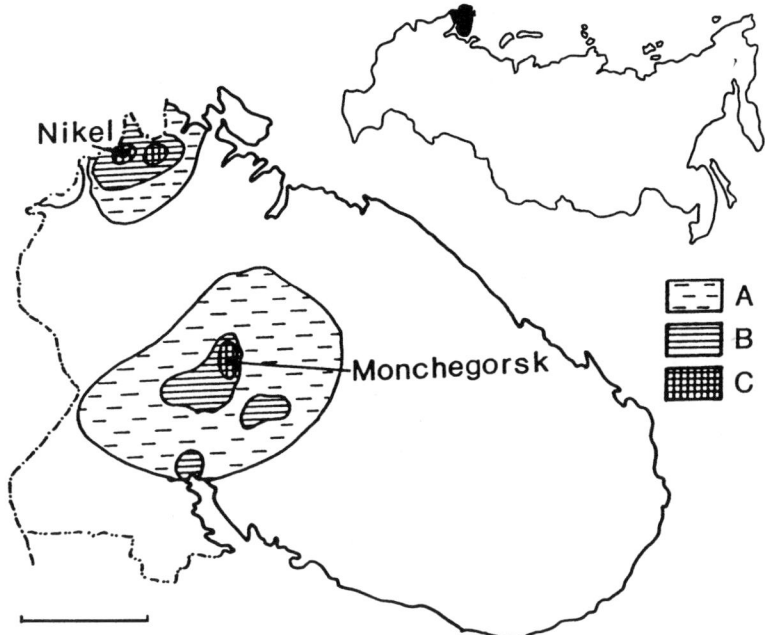

Fig. 1. Accumulation of sulphates in snow cover around the smelters in Kola Peninsula: A – 1–2 g/m², B – 2–5 g/m², C – > 5 g/m² (data collected in March of 1983; regional background < 1 g/m²; modified from Kryuchkov & Makarova, 1989). The position of the Kola Peninsula in Russia is shown in black. Scale: 100 km.

4.1.4 Animals

Data on the heavy metal content in animal tissues from Kola are restricted to some bird species (Gilyasov, 1993) and voles (*Cletrionomys rufocanus*) (Panichev et al., 1993); these samples also showed an increase of metal content with proximity to the smelter.

4.2 Measurement of pollution load

Data on both the annual fallout and the total load of pollutants are needed to monitor the effects of pollution. This is easily achieved for heavy metals present in snow, soil or biological samples. Estimation of SO_2 loads, however, is more difficult because the total sulphur content of living tissue is not indicative of the SO_2 impact. We require methods to determine the extent of SO_2 fallout, which takes into account the variability of aerial SO_2 content and preferably are not laborious, *e.g.* as are the lead-dioxide absorbers (Barkan, 1993). A vigour index of coniferous forests has been calculated, based on the characteristics of stands at different sites, since stunted trees even occur within the wasteland (Alexeyev, 1989). This index (Fig. 2) correlated well with estimated fallouts of sulphur dioxide, although it is difficult to exclude interference from other, co-occurring pollutants. A simpler measure is the maximum age of spruce needles which also corresponded well with the degree of habitat destruction: the lowest values were observed within the wasteland, 5 to 9 km south of the smelter (Fig. 3).

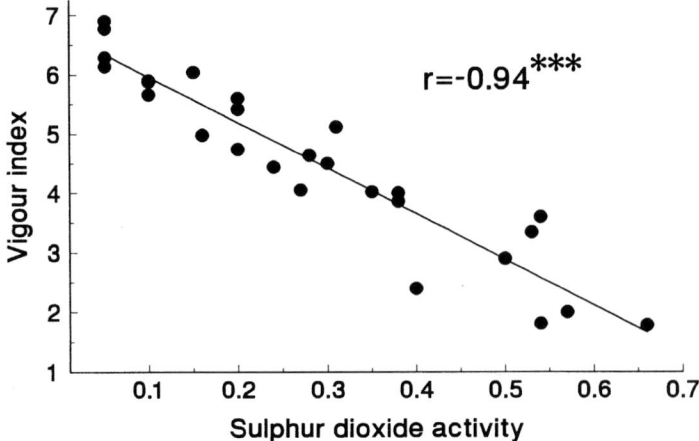

Fig. 2. Correlation between the vigour index of pine stands and the mean activity of sulphur dioxide (as revealed by passive lead-dioxide absorbers; in mg of sulphates per dm^2 per day) around "Severonikel" smelter (based on data from Tsvetkov, 1993).

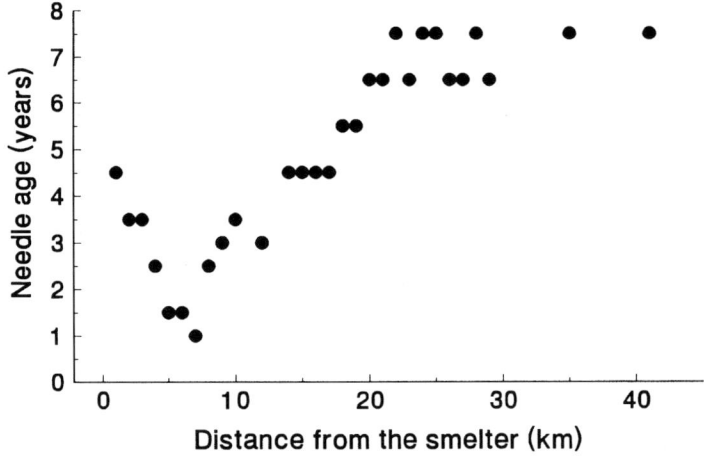

Fig. 3. Maximum age of spruce needles at different distances south of the "Severonikel" smelter complex (original data collected in 1991).

4.3 Successional changes in ecosystems

4.3.1 Soil and soil microorganisms

The depth of the litter layer within the industrial wasteland is 1.5-2 times lower than in virgin spruce forests; from 1983 to 1991, the depth on all study plots decreased by a factor of about 1.5 (Fig. 4). In the industrial wasteland, podzol (A_0) and ferrous-brown illuvial (B) soil horizons were exposed and partly or wholly eroded away, consequently revealing the primary bedrock in many locations. Maximum erosion of soil occurs 1 to 5 km N and 6 to 10 km S of the smelter. Even large stones have eroded surfaces at the plots situated 3 km N and 7 km S of the smelter.

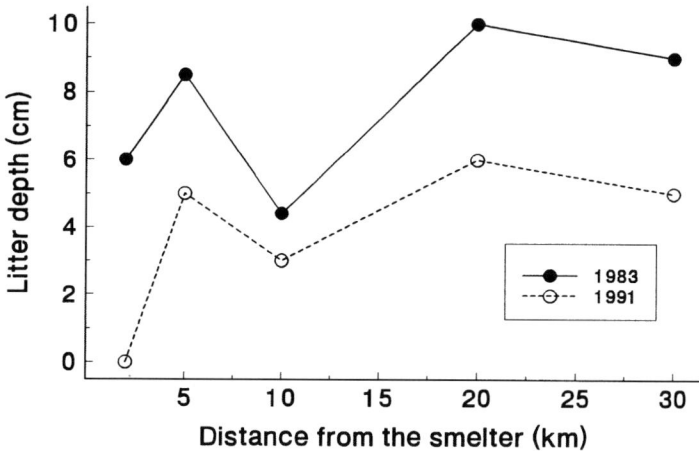

Fig. 4. Depth of the litter layer at different distances south of the smelter in 1983 and 1991 (based on the data of Koneva, 1993).

Structure of microbial communities altered after the content of nickel in podzol soils exceeded 600-700 mg/kg. Fungi were absolute dominants in heavily polluted soils, while the primary communities comprised of a variety of bacteria, actinomycetes and algae (Evdokimova & Mozgova, 1993).

4.3.2 *Vegetation*
The density and indices of health of coniferous trees decreased with proximity to the emission source (Alexeyev, 1990; Norin & Yarmishko, 1990). From 9 to 17 km S of Monchegorsk, coniferous forests survive as sparse patches, mainly on protected slopes, begin to recur about 17 km S of the smelter and show apparently full vigour at about 35 km distance. Understorey is suppressed in the area where the primary forest is destroyed (Deeva & Maznaja, 1993; Gorshkov, 1993), and has completely vanished from the industrial wasteland.

Birch seems to be less sensitive to sulphur dioxide pollution than conifers. At heavily polluted sites, birch trees grow as low bushes, since the upper canopies are dead. Within the industrial wasteland, however, regenerating birches are absent and populations decline steeply or become locally extinct (Kozlov, 1992). Trees growing in the shelter of the high chimneys of the smelter may attain partial protection and look healthy.

Willows are the woody plants most tolerant to pollution: even within the industrial wasteland, where birch trees suffered heavily, the appearance of *Salix caprea* bushes was similar to that in unpolluted areas. Accordingly, willow species dominated the secondary communities.

4.3.3 *Animal communities*
Clear decreases in biomass and population densities of soil mesofauna were observed in relation to pollution (Koneva, 1993; Koneva & Koponen, 1993). Among herbivorous insects, only lepidopterans have been studied in detail. The distribution of Lepidoptera along the pollution gradient varied with the species: some were confined to the surroundings of the smelter, where ruderal vegetation predominated while others

only occurred at the most distal study plots. Most lepidopteran species, however, occurred over the entire gradient, from virgin coniferous forests to industrial wasteland, although density peaks were species- specific (Kozlov, unpublished). Most butterfly species and some noctuids typical of the boreal forests were completely absent from the industrial wasteland and almost absent from moderately polluted areas (Kozlov et al., 1993c).

4.4 *Classification of the polluted areas*

In discussions about pollution-induced changes in vegetation, the concepts of "health of the ecosystem" (*e.g.* Alexeyev, 1993) or "destruction of the ecosystem" (*e.g.* Kryuchkov, 1993) are inconsistently applied by different authors. These terms generally are of little importance without reference to a genuinely healthy ecosystem. In many cases, authors have considered the entire damaged area as being originally covered by coniferous forests (the climax community for the Kola region). This surely was not the case. Pollution-induced development has started from different stages of the natural succession, depending upon the particular site. The "zone of total destruction", for example, is a mixture of dead coniferous forest in which the establishment of birches was prevented by factors other than pollution, and declining secondary birch forests. Standard phytosociological terminology, therefore, is a more appropriate descriptor of community changes along pollution gradients than zonal terminology.

Doncheva's (1978) classification of "landscape degradation" into three zones become very popular among Russian scientists (see Alexeyev, 1993, for a review). However, there are no clear explanations on how to recognize the boundaries between zones and, since classification of a zone often is based on a limited number of plots, the identification of zones and boundaries is somewhat subjective. Furthermore, the boundaries were based on reported concentrations of pollutants, and the environmental effects observed within 1-2 study plots then were extrapolated over whole zone. For example, Kryuchkov (1993), who distinguished five zones, largely based his classification on data from only eight study plots. We feel that an emphasis on classifying polluted territories at Kola has adversely affected ecological research. Overemphasis on the zone as the sole identifying characteristic of study plots implies a greater homogeneity of both the original ecosystem and pollution impact than has been observed.

5. Retrospective analysis of the environmental situation

The data on vegetation state before the smelter was put into operation are quite scanty (Bobrova & Kashurin, 1936; Nekrasova, 1938). The forests within 6 km of the smelter already had perished by 1946. Between 1955 and 1963, the ecosystems within 6 to 7 km were reported to have been destroyed (Kryuchkov, 1993).

Pollutant concentrations and environmental effects first were monitored during the late 1970's - early 1980's (Doncheva, 1978; Alexeyev, 1982), but the publications about the negative impact of human activity on the environment largely were prohibited by the former Soviet Union government. This explains some odd features of the pre-1990 literature: the book by Doncheva (1978), for example, shows a map of the study area without site names. Older Russian papers avoided combining the names of emission sources with concepts like "distance - pollution load - environmental effects". Fortunately, the research groups which collected most of the original environmental data were still active in the 1990's, and could summarize and publish some of the important information collected earlier retrospectively (Alexeyev, 1990; Norin & Yarmishko, 1990; Kryuchkov & Eliseev, 1991; Doncheva et al., 1992).

6. Confounding between gradients caused by pollution and gradients related to other human activities

It is erroneous to attribute all the negative environmental effects around the smelters to pollution because there are co-occurring stress factors caused by other human activities. Logging and forest fires do not operate as gradients, but both are more evident around urban and smelter sites than in undisturbed areas. These factors have increased the mosaic nature of the environment and interfered with attempts to reveal clear pollution-induced succession. They only are rarely taken into account in environmental impact studies about pollution. A model to estimate the relative input of different factors into the destruction of forest ecosystems has been developed (Selikhovkin, 1993), but the data collected at the Kola Peninsula are still too scanty for its proper application.

6.1 *Logging*

Different types of felling have been widely employed around Monchegorsk, even within the Lapland Biosphere reserve. It is probable that the logging of subarctic forests has been most intensive at the Kola Peninsula. Since the regeneration of these forests is slow, and measures to promote regeneration are absent, logging presumably has caused significant changes in vegetation, microclimate and hydrology, thus contributing to the degradation of soils (Selikhovkin, 1993).

6.2 *Fires*

During 1967-1982, about 72 percent of the forests adjacent to Monchegorsk were damaged by fires (Tsvetkov, 1982). Natural regeneration within highly polluted areas following fire is virtually negligible, and the burnt areas are transformed into industrial wastelands (Selikhovkin, 1993).

6.3 *Invaded animals and ruderal vegetation*

Tens of hectares are covered by ruderal and invaded vegetation. The main plant species of these communities were *Epilobium angustifolium, Urtica dioica, Elytrigia repens,* and some other ruderal species mentioned by Ramenskaja (1983), while the most characteristic insect species included the moths *Mompha idaei* and *Anthophila fabriciana*. Some common species in these areas otherwise were rare or absent from natural ecosystems (Kozlov & Jalava, 1994). The densities of synantropic animals (*e.g.* some Calliphoridae, Sepsidae and other flies) are high close (2-5 km) to the city but sparse in natural ecosystems (Zvereva, 1993).

7. Confounding between gradients caused by pollution and other natural environmental gradients

7.1 Geographical gradients

Since the length of the polluted territory around Monchegorsk is over 200 km in the north-south direction, it can be expected that inherent geographical variation contributes to differences among sites. Although this source of error can be accounted for by comparing separate northern and southern transects, much of research is still based on single transects, usually to the south of Monchegorsk. East-west transects are difficult because of the mountain relief and poor road connections.

7.2 Relief and altitudinal gradients

The southern transects usually follow the St. Petersburg - Murmansk road and traverse the lower parts of the Monche and Tshuna mountains, which range in altitude from 130 to 280 m a.s.l. Variations in topography (*e.g.* altitude, slope, aspect) further accentuate environmental heterogeneity by altering the general distribution and amount of deposition of emissions. Furthermore, differences in the initial composition and condition of forests at different altitudes may influence the speed of the pollution-induced succession.

7.3 Types of natural vegetation

North of Monchegorsk, a transect that follows the road to Murmansk for 50 km passes through an area originally covered by pine forests. South of Monchegorsk, the transect first crosses spruce forest, then a belt of pine forest, 7 to 14 km of the smelter. The southernmost parts of the transects, from 40-45 to 100 km S of the smelter, represents pine forests. Thus, the study region around the smelter may represent stages of transition between pine and spruce forests (Koroleva, 1993). This imposes an additional variation that masks the impact of pollution.

8. Discussion

The shortage of data on most ecologically important groups (*e.g.* microbiota including mycorrhizas, worms, ants and vertebrates) strongly restricts discussion on changes of food chains under pollution impact. Reliability of the data is the other general problem, since most of field studies did not comply with the criteria coined by Hurlbert (1984) and developed by Eberhard & Thomas (1991). Furthermore, the spatial pattern is quite poorly known even for the major components of pollution.

The area polluted by the "Severonikel" smelter complex is so large that it encompasses existing natural gradients and heterogeneous regions of both natural and artificial origin; the presence of these additional sources of variation interferes with attempts to identify changes related solely to pollution emanating from the emission source. There is no doubt that pollution has caused the wasteland around Monchegorsk, and that the general trends in pollution and ecosystem structure are clearly visible. However, not all deviations from the reconstructed original vegetation are caused directly by pollution.

Emissions from any smelter consist of numerous substances, most of which are poorly monitored. Government survey data on the "Severonikel" smelter complex show that, besides SO_2 and heavy metals, emissions also include dust, oxides of carbon and nitrogen, chlorine, H_2S, SeO_2, V_2O_5, and formaldehyde (Berlyand, 1991). To our knowledge, the distribution patterns of these compounds have not been studied nor have ozone concentrations or the amounts of organic compounds (*e.g.* chlorinated hydrocarbons) in the impact zone. Thus, exact causal relationships within the polluted areas still are difficult to explain, and further controlled experiments are needed to clarify the situation.

It is of practical importance to determine the future of moderately and slightly polluted ecosystems given further uninterrupted pollution. It is often presumed that temporal impact of continued pollution is analogous to moving closer to the source of emission, *i.e.* the zone of maximum damage will gradually spread more widely. We have been unable to locate any references that raised the question of changes over time rather than in space. Based on subjective assessments from field work around Monchegorsk during the last ten years by the senior author, the outer border of the industrial wasteland has moved not more than 2 to 4 km.

9. Conclusion

In spite of the large pool of data collected during the past decade, the present level of knowledge still does not allow to either clearly outline the environmental situation around "Severonikel" smelter or predict the future ecosystem changes under different options to modify the smelter. The main gaps in knowledge concern spatial pattern of both pollutants distribution and their effects on ecosystems. Among the life forms, microbes, worms, ants and vertebrates were hardly studied at all. However, the impact zone of this smelter represents a unique opportunity to study ecological and evolutionary effects of pollutants on biota. Possible reduction of pollution in the coming years will open opportunities for recovery studies.

10. Summary

The surroundings of the "Severonikel" smelter complex at Monchegorsk on the Kola Peninsula, Russia, are badly damaged by aerial pollutants, predominantly sulphur dioxide and heavy metals, emitted from the smelter. We summarize the results of studies on pollution gradients centring on the smelter complex. The aspects which have been studied most thoroughly are the impact on forest soils and on conifers, especially Scots pine. There is no doubt that pollution is the cause of the huge wasteland surrounding Monchegorsk. However, the wider impact zone covers a much larger ($>10,000$ km^2) and heterogeneous region. The heterogeneity, arising from both natural and anthropogenic factors, complicates and interferes with attempts to demonstrate clear pollution-induced gradients around the emission source.

Acknowledgements

We are grateful to V. Alexeyev, V. Kryuchkov, A. Selikhovkin and E. Zvereva for fruitful discussions. The comments by J. Vranjic greatly improved both the biological content and the language of the manuscript. We thank referees for their constructive criticism. The study was financed by the Academy of Finland.

References

Alexeyev, V. A., 1982. Air pollution and specificity of stand description. In: Martin, J. L. & V. A. Alexeyev (eds.), *Interaction between forest ecosystems and pollutants. Abstr. Rep. 1st Soviet-American Symp.*, pp. 97–115. Tallin.

Alexeyev, V. A., 1989. Diagnostics of trees and stands health. Lesovedenie (4): 51–57. (in Russian)

Alexeyev, V. A. (ed.), 1990. *Forest ecosystems and air pollution.* Nauka, Leningrad, 200 pp. (in Russian)

Alexeyev, V. A., 1993. Air pollution impact on forest ecosystems of Kola Peninsula: history of investigations, progress and shortcomings. In: M. V. Kozlov, E. Haukioja & V. T. Yarmishko (eds.), *Aerial pollution in Kola Peninsula: Proc. Intern. Workshop, April 14–16, 1992, St.Petersburg,* pp. 20–34. Apatity.

Barkan, V. Sh., 1993. Measurement of atmospheric concentrations of sulphur dioxide by passive lead dioxide absorbers. In: M. V. Kozlov, E. Haukioja & V. T. Yarmishko (eds.), *Aerial pollution in Kola Peninsula: Proc. Intern. Workshop, April 14–16, 1992, St.Petersburg,* pp. 90–98. Apatity.

Barkan, V. Sh., R. P. Pankratova & A. V. Silina, 1993. Soil contamination by nickel and copper in area polluted by "Severonikel" smelter complex. In: M. V. Kozlov, E. Haukioja & V. T. Yarmishko (eds.), *Aerial pollution in Kola Peninsula: Proc. Intern. Workshop, April 14–16, 1992, St.Petersburg,* pp. 119–147. Apatity.

Barkan, V. Sh., M. S. Smetannikova, R. P. Pankratova & A. V. Silina, 1993. Nickel and copper accumulation by edible forest berries in surroundings of "Severonikel" smelter complex. In: M. V. Kozlov, E. Haukioja & V. T. Yarmishko (eds.), *Aerial pollution in Kola Peninsula: Proc. Intern. Workshop, April 14–16, 1992, St.Petersburg,* pp. 189–196. Apatity.

Berlyand, M. E. (ed.), 1991. *Annual report on ambient air pollution in cities and industrial centres of Soviet Union. Volume "Emission of pollutants: 1990".* Voeikov Main Geophysical Observatory, St-Petersburg. 576 pp. (in Russian).

Bobrova, L. I. & M. N. Kashurin, 1936. Vegetation of Monchetundra. In: Zinzerling, Y. D. (ed.), *Materials on the vegetation of northern and western parts of Kola peninsula,* pp. 95–121. Acad. Sci. USSR, Moscow & Leningrad.

Deeva, N. M. & E. A. Maznaja, 1993. The state of bilberry in polluted and unpolluted forests of the Kola Peninsula. In: M. V. Kozlov, E. Haukioja & V. T. Yarmishko (eds.), *Aerial pollution in Kola Peninsula: Proc. Intern. Workshop, April 14–16, 1992, St.Petersburg,* pp. 308–311. Apatity.

Derome, J. (ed.), 1993. *The Lapland Forest Damage Project: Russian-Finnish cooperative report.* The Finnish Forest Res. Inst., Rovaniemi. 111 pp.

Doncheva, A. V., 1978. *The landscape in zone of industrial impact.* Lesnaya promyshlennost, Moscow, 96 pp. (in Russian).

Doncheva, A. V., L. K. Kazakova & V. N. Kalutskov, 1992. *Landscape indication of the environmental pollution.* Ecologia, Moscow, 256 pp. (in Russian)9

Eberhard, L. L. & J. M. Thomas, 1991. Designing environmental field studies. Ecol. Monographs, 61: 53–73.

Evdokimova, G. A. & N. P. Mozgova, 1993. Microbiological study of soil contamination by heavy metals. In: M. V. Kozlov, E. Haukioja & V. T. Yarmishko (eds.), *Aerial pollution in Kola Peninsula: Proc. Intern. Workshop, April 14–16, 1992, St.Petersburg,* pp. 184–188. Apatity.

Freedman, B. & T. C. Hutchinson, 1980. Pollutant inputs from the atmosphere and accumulations in soils and vegetation near a nickel-copper smelter at Sundbury, Ontario, Canada. Can. J. Bot., 58: 108–132.

Gilyasov, A. S., 1993. Contents of metals in some birds of the Lapland reserve. In: M. V. Kozlov, E. Haukioja & V. T. Yarmishko (eds.), *Aerial pollution in Kola Peninsula: Proc. Intern. Workshop, April 14–16, 1992, St.Petersburg,* pp. 206–209. Apatity.

Gorshkov, V. V., 1993. The state of moss-lichen cover in polluted and unpolluted pine forests of the Kola Peninsula. In: M. V. Kozlov, E. Haukioja & V. T. Yarmishko (eds.), *Aerial pollution in Kola Peninsula: Proc. Intern. Workshop, April 14–16, 1992, St.Petersburg,* pp. 290–298. Apatity.

Hurlbert, S. H., 1984. Pseudoreplication and the design of ecological field experiments. Ecol. Monographs, 54: 187–211.

Kismul, V., J. Jerre & E. Løbersli (eds.), 1992. *Effects of air pollutants on terrestrial ecosystems in the border area between Russia and Norway. Proc. 1st Symp., Svanvik, Norway, 18–20. March 1992.* Statens forurenssningstilyn, Document 92:04, Oslo, 220 pp.

Koneva, G. G., 1993. Changes in soil macrofauna around "Severonikel" smelter complex. In: M. V. Kozlov, E. Haukioja & V. T. Yarmishko (eds.), *Aerial pollution in Kola Peninsula: Proc. Intern. Workshop, April 14–16, 1992, St.Petersburg,* pp. 362–364. Apatity.

Koneva, G. G. & S. Koponen, 1993. Density of ground-living spiders (Araneae) near smelter in Kola Peninsula. In: M. V. Kozlov, E. Haukioja & V. T. Yarmishko (eds.), *Aerial pollution in Kola Peninsula: Proc. Intern. Workshop, April 14–16, 1992, St.Petersburg,* p. 365. Apatity.

Koroleva, N., 1993. Pollution-induced changes in forest vegetation structure as revealed by ordination test. In: M. V. Kozlov, E. Haukioja & V. T. Yarmishko (eds.), *Aerial pollution in Kola Peninsula: Proc. Intern. Workshop, April 14–16, 1992, St.Petersburg,* pp. 339–345. Apatity.

Kozlov, M. V., 1992. Pollution-induced changes in birch population structure in Kola Peninsula, Northern Russia. In: *Disturbance related dynamics of birch and birch dominated ecosystems: A Nordic Symposium, Illugastathir, Fnjoskadalur, Iceland 18-22 September 1992*, pp. 51-53.

Kozlov, M. V., A. V. Bakhtiarov & D. N. Stroganov, 1993a. Heavy metal contents in birch leaves in surroundings of "Severonikel" smelter complex. In: M. V. Kozlov, E. Haukioja & V. T. Yarmishko (eds.), *Aerial pollution in Kola Peninsula: Proc. Intern. Workshop, April 14–16, 1992, St.Petersburg*, pp. 197–200. Apatity.

Kozlov, M. V., E. Haukioja & V. T. Yarmishko (eds.), 1993b. *Aerial pollution in Kola Peninsula: Proc. Intern. Workshop, April 14–16, 1992, St.Petersburg.* 417 p. Apatity.

Kozlov, M. V., J. Jalava, A. L. Lvovsky, K. Mikkola & L. N. Shvetsova, 1993c. Diversity and abundance of lepidopterans along an air pollution gradient in the Kola Peninsula. In: M. V. Kozlov, E. Haukioja & V. T. Yarmishko (eds.), *Aerial pollution in Kola Peninsula: Proc. Intern. Workshop, April 14–16, 1992, St.Petersburg*, p. 370. Apatity.

Kozlov, M. V. & J. Jalava, 1994. Lepidoptera of the Kola Peninsula, Northwestern Russia. Entomologica Fennica 5 : 65–85.

Kryuchkov, V. V., 1993. Degradation of ecosystems around the "Severonikel" smelter complex. In: M. V. Kozlov, E. Haukioja & V. T. Yarmishko (eds.), *Aerial pollution in Kola Peninsula: Proc. Intern. Workshop, April 14–16, 1992, St.Petersburg*, pp. 35–46. Apatity.

Kryuchkov, V. V. & D. A. Eliseev, 1991. *Basic principles of the ecological expertise.* Kola Sci. Centre. 60 p. Apatity.

Kryuchkov, V. V. & T. D. Makarova, 1989. *Aerothechnogenic impact on the ecosystems of Kola North.* Kola Sci. Centre. 96. p. Apatity.

Nekrasova, T. N., 1938. Vegetation of alpine and subalpine zones of Tshuna-tundra. Trans. Lapland State Reserve, 1: 7-176.

Norin, B. N. & V. T. Yarmishko (eds.), 1990. *The influence of industrial air pollution on pine forests of Kola Peninsula.* Nauka, Leningrad, 195 pp. (in Russian)

Panichev, N. A., G. D. Kataev & E. B. Nosova, 1993. Contents of heavy metals in tissues of grey-sided voles (*Cletrionomys rufocanus*) in surroundings of Monchegorsk. In: M. V. Kozlov, E. Haukioja & V. T. Yarmishko (eds.), *Aerial pollution in Kola Peninsula: Proc. Intern. Workshop, April 14–16, 1992, St.Petersburg*, p. 210. Apatity.

Pozniakov, V. Ya. 1993. The "Severonikel" smelter complex: history of development. In: M. V. Kozlov, E. Haukioja & V. T. Yarmishko (eds.), *Aerial pollution in Kola Peninsula: Proc. Intern. Workshop, April 14–16, 1992, St.Petersburg*, pp. 16–19. Apatity.

Raitio, H., V. Nikonov & N. Lukina, 1993. Chemical composition of Scots pine needles in the industrial region of Monchegorsk. In: J. Derome, (ed.), *The Lapland Forest Damage Project: Russian-Finnish cooperative report*, pp. 23–30. The Finnish Forest Res. Inst., Rovaniemi.

Ramenskaja, M. L., 1983. *The analysis of the vegetation of the Murmansk region and Karelia.* Nauka, Leningrad, 215 pp. (in Russian).

Selikhovkin, A., 1993. Stressing agents in forests of the Kola Peninsula. In: M. V. Kozlov, E. Haukioja & V. T. Yarmishko (eds.), *Aerial pollution in Kola Peninsula: Proc. Intern. Workshop, April 14–16, 1992, St.Petersburg*, pp. 47–52. Apatity.

Tchekhin, L. P., 1993. Monitoring of industrial pollution on the Kola Peninsula. In: M. V. Kozlov, E. Haukioja & V. T. Yarmishko (eds.), *Aerial pollution in Kola Peninsula: Proc. Intern. Workshop, April 14–16, 1992, St.Petersburg*, pp. 112–115. Apatity.

Tikkanen, E., M. Varmola & T. Katermaa (eds.), 1992. *Symposium on the state of the environment and environmental monitoring in the Northern Fennoscandia and the Kola Peninsula, October 6-8, 1992, Rovaniemi, Finland. Extended Abstr. (Arctic Centre Publ. 4)*, pp. 226-231.

Tsvetkov, V. F. (ed.), 1982. *Development of methods of the reforestation in the zone of emissions of the "Severonikel" plant.* Unpublished manuscript of Arkhangelsk Inst. of Forest and Forest Chemistry, Arkhangelsk, 121 pp. (in Russian)

Tsvetkov, V. F., 1993. Threshold levels of air pollution and forest decline in surroundings of "Severonikel" smelter complex. In: M. V. Kozlov, E. Haukioja & V. T. Yarmishko (eds.), *Aerial pollution in Kola Peninsula: Proc. Intern. Workshop, April 14–16, 1992, St.Petersburg*, pp. 397–401. Apatity.

Zvereva, E. L., 1993. Population densities of some flies in an air pollution gradient near Monchegorsk. In: M. V. Kozlov, E. Haukioja & V. T. Yarmishko (eds.), *Aerial pollution in Kola Peninsula: Proc. Intern. Workshop, April 14–16, 1992, St.Petersburg*, pp. 371–373. Apatity.

Effects of environmental pollutants on the cold-hardiness of Arctic and boreal ectothermic animals

K. E. Zachariassen[1,2] & R. Lundheim[1]

[1]*Department of Ecotoxicology, ALLFORSK, Gryta 2, 7010 Trondheim*
[2]*Department of Zoology, The University of Trondheim, 7055 Dragvoll*

Keywords: melting point, osmolality, supercooling point, thermal hysteresis

1. Introduction

Substantial amounts of chemicals are released to the environment from industry, agriculture and other sources. These chemicals pollute the local areas where they are released, but due to airborne transport over long distances, they also cause global pollution. Large amounts of heavy metals and organic compounds are precipitated in Arctic and boreal areas, far away from their sources. The chemicals enter the food chain via plants, and they subsequently occur in all trophic levels of the ecosystem. The pollutants may interfere with biological processes in different ways and cause injury both by direct toxic actions, and indirectly, by interfering with processes which are important to organismal survival under extreme conditions.

Winter survival of Arctic and boreal ectothermic organisms depends on various physiological and biochemical cold-hardening mechanisms. Pollutants may interfere with these mechanisms and thus cause organisms to die from freezing injury. This would not be observed in ordinary toxicological tests, but may nevertheless be of great ecological importance.

The present paper deals with the relationship between environmental pollutants and the cold-hardiness of ectothermic organisms. The paper first presents a review of cold-hardening mechanisms, and then a review of the mechanisms of action of various environmental pollutants. Finally the documented and theoretical interactions between pollutants and cold-hardiness are outlined.

1.1 Definitions

Melting point. The melting point of a solution is the temperature at which the last tiny ice crystal disappears when a frozen sample of the solution is slowly thawed. The melting point of distilled water is 0°C and is depressed by increasing solute concentration. The term melting point is synonymous with the term equilibrium freezing point.

Osmolality. When solutes are dissolved in water, proportional changes take place in the so-called colligative properties of the system. The colligative properties comprise a number of physical parameters such as vapour pressure, melting point and boiling point. The changes are approximately proportional to the solute concentration expressed by molal units. However, for compounds which dissociate when dissolved, corrections must be made for the dissociation factor because the changes depend on the

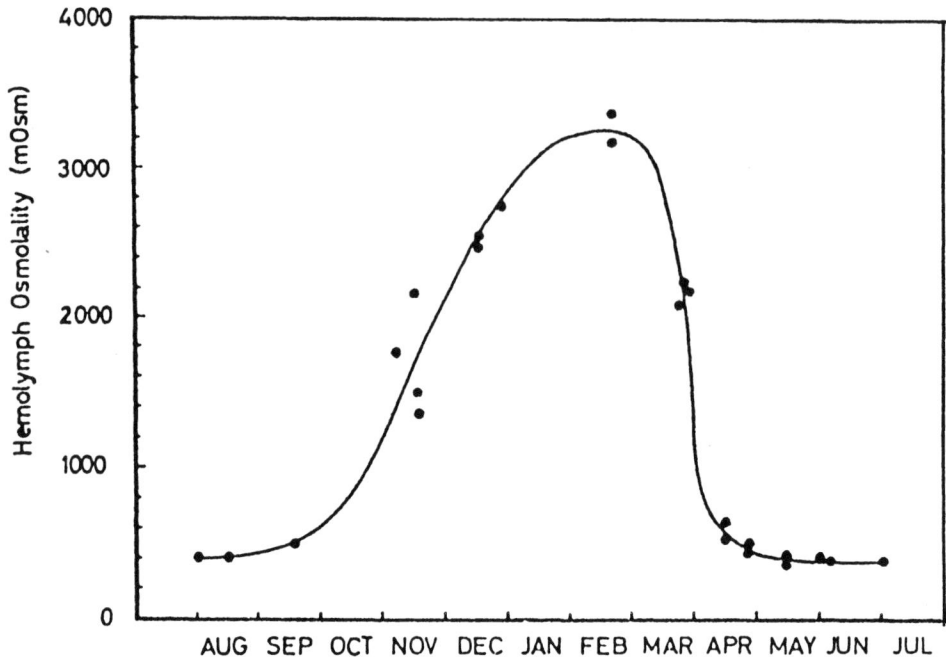

Fig. 1. Body fluid osmolality of *R. inquisitor* longhorn beetles throughout the year as it is brought to vary by the accumulation/breakdown of glycerol.

concentration of dissolved particles. To express the mole-based particle concentration corresponding to the measurable physical changes, the concept osmolality has been introduced. 1 Osm is the osmolality of a solution with a melting point of −1.86°C. It should be emphasized that the osmolality concept is not a genuine chemical concentration measure, but a measure which is precisely related only to the physical (colligative) properties of the solution.

Supercooling point. When a small sample of a solution is cooled, it will not freeze when the melting point is reached, but maintain the liquid state down to substantially lower temperatures. When a solution maintains the liquid state when cooled below the melting point, it is said to be supercooled, and the temperature at which spontaneous freezing occurs is termed the supercooling point. The supercooling point depends on solute concentration, sample volume and time.

Thermal hysteresis. The body fluids of polar fish and insects that overwinter in a supercooled state contain antifreeze peptides, which prevent the growth of ice crystals when the body fluid is cooled. The growth inhibition is limited, in that sufficient cooling will cause sudden rapid ice growth. The separation of the melting point, and the temperature of ice growth is called thermal hysteresis, and the temperature at which this ice growth occurs is termed the hysteresis freezing point (HFP). The peptides responsible for the hysteresis are often referred to as thermal hysteresis factors (THFs).

2. Mechanisms of cold hardiness

Normal, active animals and plants will freeze and die if cooled to temperatures ranging from −7 to −12°C (Zachariassen, 1985). This implies that poikilothermic animals in Arctic and boreal regions may freeze and die in the winter. The animals may, in principle, use two strategies for survival. They may seek to avoid freezing by attaining a high supercooling capacity for their body fluids, or they may develop tolerance to freezing, *i.e.* to ice formation of the body fluids. Insects hibernating above snow level are among the best known groups of cold-hardy animals, but special mechanisms for cold-hardiness have been found in a wide variety of ectothermic animals: nematodes (Namatov, 1969), tardigrades (Pigon & Weglarski, 1955), littoral molluscs (Kanwisher, 1959) and crustaceans (Cook & Lewis, 1971), polar teleost fish (DeVries & Wohlschlag, 1969), amphibians (Schmid, 1982; Storey & Storey, 1989) and reptiles (Pauktis *et al.*, 1989).

Hibernating insects use both strategies for cold-hardiness, and both strategies involve the accumulation of polyols to high concentrations in the body fluids (Fig. 1, Sømme, 1964). The changes in supercooling points accompanying the two strategies are illustrated in Fig. 2. This shows the supercooling points of the organisms as a function of body fluid osmolality, as it is brought to vary by the accumulation of polyols.

Samples of solutions with the same volume and osmolality as active summer insects freeze at about −20°C (Block & Young, 1979). The fact that summer insects freeze at temperatures as high as about −10°C, implies that insects contain an ice nucleating component which promotes freezing at a relatively high temperature (Zachariassen, 1985). The ice nucleators may be located in the intestine (Sømme & Conradi-Larsen, 1977) or in the intracellular fluid (Zachariassen, 1985). Insects that seek to avoid freezing remove the ice nucleators from their body fluids in the fall, and thus depress their supercooling points to − 20°C with no change in osmolality (Fig. 2).

The accumulation of polyols, particularly glycerol, starts in October, and may cause the body fluid osmolality to reach several thousand mOsm by midwinter (Fig. 1). Glycerol production takes place at the expense of glycogen stores, via a metabolic side pathway from anaerobic glycolysis, and the entire glycogen store may be mobilized and transferred to polyols (Ohyama & Asahina, 1972). Other polyols produced by hibernating insects are sorbitol (Sømme, 1967), mannitol (Sømme, 1969), ethylene glycol (Gehrken, 1984) and threitol (Miller & Smith, 1975).

The polyol accumulation leads to a further depression of the supercooling points. In freeze-avoiding insects, the supercooling point depression is more than twice the corresponding melting point depression (Sømme, 1964; Zachariassen, 1985). This may bring the midwinter supercooling points down to, or even below −30°C, implying that these insects may survive cold exposure down to this temperature range.

The supercooled state is physically metastable, and supercooled systems will sooner or later freeze spontaneously. Insects avoid spontaneous freezing by producing antifreeze peptides (THFs) which appear to stabilize the supercooled state (Zachariassen & Husby, 1982). In this way, insects can avoid a lethal freezing over their entire supercooling range throughout the cold season. THFs also appear to prevent freezing by inoculation of external ice from the hibernaculum (Gehrken 1992).

Freeze-tolerant species have supercooling points in the same range as the freeze-avoiding insects in the summer, but elevate their supercooling points in the early fall (Fig. 2). The increase in organismal supercooling points coincides with the production of potent ice nucleators in the haemolymph, the function of which appears to be to ensure that ice nucleation takes place in the extracellular compartment (Zachariassen &

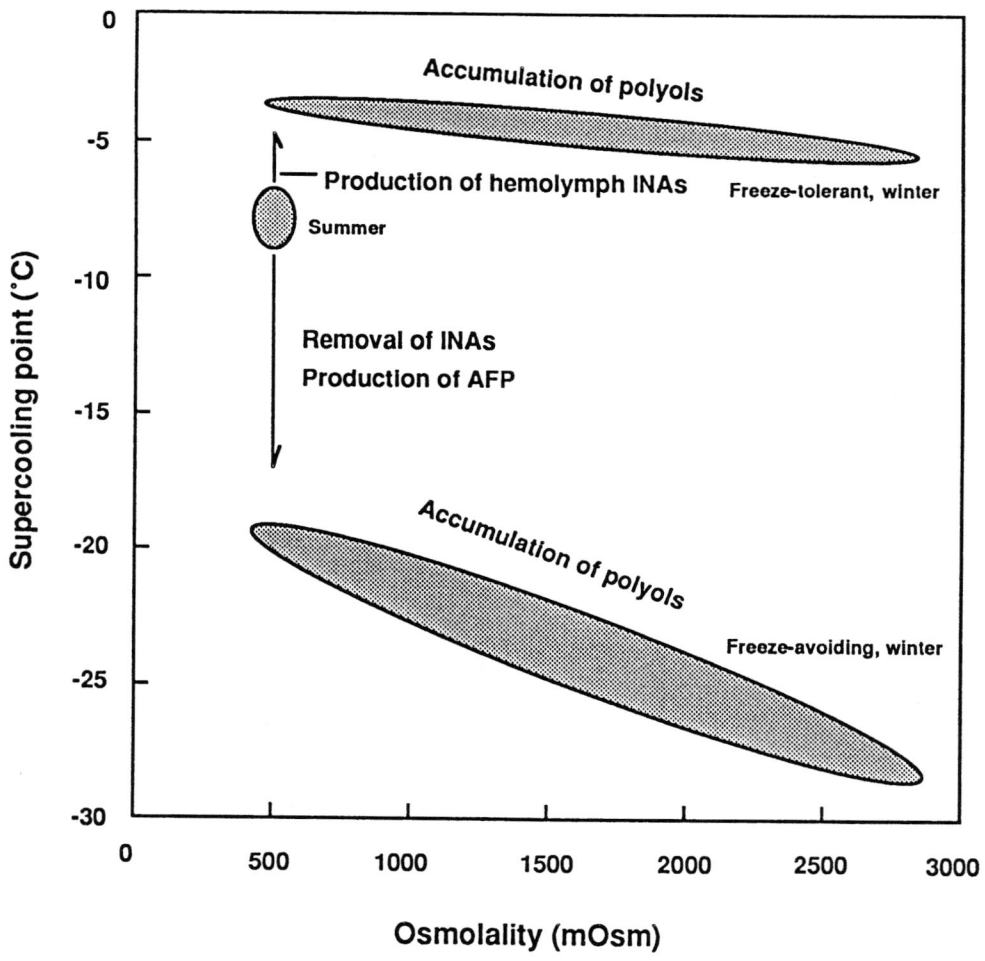

Fig. 2. Supercooling point (shaded areas) of freeze–tolerant and freeze–avoiding insects in summer and winter, plotted as a function of body fluid osmolality. The corresponding changes in levels of ice nucleating agents (INAs) and antifreeze petides (AFPs) are indicated. (Redrawn from Zachariassen 1980).

Hammel, 1976). The ice nucleators of insects are proteins or lipoproteins with a molecular weight of 75 to 800 kDa (Duman *et al.*, 1991).

The physiological function of extracellular ice nucleation is illustrated in Fig. 3.. In the absence of extracellular ice nucleators, ice nucleation is likely to take place in the intracellular or intestinal compartments. Ice formation will cause an increase in the osmolality of the fluid fraction of these compartments and thus lead to an osmotic influx of water. The cell or intestine will subsequently swell, and since cells and intestine are closed compartments with a limited swelling capacity, they may eventually rupture and become damaged (Zachariassen & Hammel, 1976).

The formation of extracellular ice, at a temperature higher than the temperatures of intracellular or intra-intestinal nucleation, will cause the osmolality of the extracellular fluid fraction to rise and thus lead to an osmotic outflux of water from the closed compartments. These compartments will subsequently shrink, until they reach osmotic

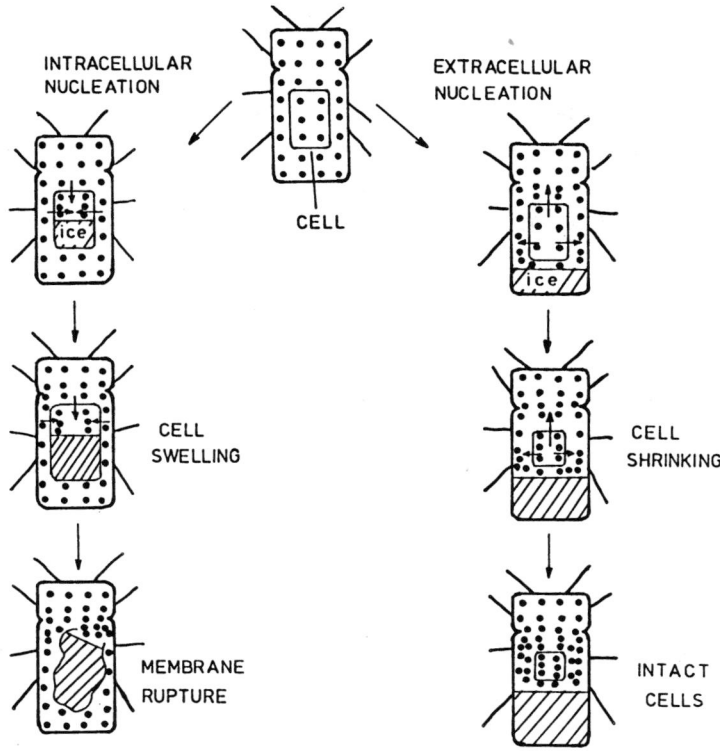

Fig. 3. Events involved in intracellular and extracellular freezing (From Zachariassen 1992).

equilibrium with the extracellular fluid (Kanwisher, 1959). In this state, neither the intracellular nor the extracellular compartments are supercooled, and another ice nucleation event is thus prevented. Closed compartments have a greater tolerance for shrinking than for swelling, and the cells may shrink to osmotic equilibrium without suffering any injury.

However, the tolerance for shrinking is not unlimited, and the cells may eventually shrink to a volume where the outer cell membrane rests on a matrix of intracellular structures (Meryman, 1971). In this state, further cooling and increase in extracellular osmolality cannot be compensated for by outflux of water from the cells, and the osmotic pressure may build up until the membrane ruptures. There may also be an increase in salt concentration of the fluid fraction to injurious levels, *i.e.* to levels where protein structures become irreversibly altered by lyotropic effects (Lovelock, 1953).

These injuries are counteracted by the accumulation of glycerol. Glycerol accumulation leads to a reduction in the amount of ice formed at all subfreezing temperatures, and thus displaces the formation of injurious salt concentration to lower temperatures (Lovelock, 1953). Since glycerol can penetrate cell membranes as it becomes concentrated in the extracellular fluid, it will also counteract the osmotic shrinking of cells and intestine, and displace the occurrence of critical volumes to lower temperatures (Meryman, 1971). Glycerol is also found to stabilize proteins during the stress imposed by low temperature and freezing (Gekko & Timasheff, 1981).

Freezing tolerance appears to require a moderate freezing rate (Miller, 1969;

Worland et al., 1992). A low cooling and freezing rate is secured if ice nucleation takes place at a high temperature, i.e. that the temperature gradient between the freezing system and the environment is low. This will reduce the freezing induced osmotic stress across the cell membranes (Zachariassen, 1992) and allow polyols to enter the cells before they become critically dehydrated.

The accumulation of glycerol provides a substantial cryoprotection, and insects with glycerol concentrations of some thousand mmolal may tolerate temperatures down to −50°C (Zachariassen, 1985; van der Laak, 1982). The changes in polyols, ice nucleators and antifreeze peptides in insects are controlled by endocrine mechanisms, involving hormones such as the juvenile hormone, ecdysterone and adipokinetic hormone (Zachariassen & Lundheim, 1992).

Marine littoral organisms such as *Mytilus edulis*, *Littorina* spp., *Patella* sp. and cirripeds are tolerant to freezing (Kanwisher, 1959; Aarset 1982). Similar to insects, *Mytilus* and *Littorina* have potent ice nucleators in the haemolymph in the winter (Aunaas, 1982a & 1982b). The ice nucleators apparently serve the same function as those found in insects, i.e. to secure a protective extracellular freezing at a high subzero temperature.

Littoral organisms have not been found to accumulate polyols, but cryoprotection is accomplished by intracellular free amino acids, which may reach levels of several hundred mmolal. With high intracellular amino acid concentrations, mussels may tolerate freezing to about −20°C (Murphy & Pierce, 1975).

Polar fish live in seawater which may be in contact with ice throughout the year. The temperature therefore stays at −2°C, which is the melting point of full salinity seawater. Since fish are ectothermic organisms, this is also the body temperature of the fish. However, the body fluids of the fish have a lower solute concentration than sea water, and thus the fish remain permanently supercooled by about 1.2°C. In ice laden waters, the fish may also frequently come into direct physical contact with ice. This could cause a lethal freezing by inoculation through the skin. Polar fish are thus in the danger of suffering a lethal freezing (Scholander et al., 1957).

Fish are protected against freezing by the presence of antifreeze proteins and glycoproteins in their body fluids. In most species, the antifreeze proteins are present throughout the year, but some species display seasonal variations in the antifreeze levels (DeVries, 1982).

Some species of fish have also been known to accumulate moderate amounts of sugars (Umminger, 1970) or glycerol (Raymond, 1992) in their body fluids. These substances probably protect the fish by making their body fluids less hyposmotic to seawater and thus less supercooled. This will also reduce the likelihood of lethal freezing.

3. Mechanisms of action of environmental pollutants

Environmental pollutants comprise a wide variety of chemical substances. The various categories of pollutants act through different mechanisms and may affect different parts of biological functional systems. The present chapter reviews the most important toxicological features of heavy metals, poly chlorinated biphenyls (PCBs), polycyclic aromatic hydrocarbons (PAHs), insecticides and dioxines.

Heavy metals or trace metals make up an important part of the environmental pollutants, both as locally released substances and as air transported substances, which may pollute wide areas. As pointed out by Nieboer & Richardson (1980), metals act by several mechanisms. They may bind to charged groups of proteins, including active sites of enzymes, or other charged groups which are functionally essential. They may replace the proper metal co-factor (zinc, iron, etc.) which is required to give enzymes and other proteins their optimal functional structure. They may also cause oxidation of cellular components, for example, by reduction of Cu^{2+} to Cu^+. Furthermore, they may form lipid soluble organometallic ions, which may penetrate cell membranes and thus interfere with intracellular structures (Connell & Miller, 1984). Enzyme deactivation may lead to inhibition of metabolic pathways and ionic pumps, with subsequent failures in ATP production, ionic pumping and osmoregulation (Staurnes et al., 1984; Aunaas et al., 1991; Børseth et al., 1992).

Metals have also been shown to cause a reduction in the total protein content of organisms. Cadmium has, for example, been found to induce a shift in metabolism, from utilization of lipids to utilization of proteins in the crustacean *Mysidopsis bahi* (Carr et al., 1985) and in chironomids (Rathore et al., 1979). The latter authors found that a short period of cadmium exposure led to a 30% reduction in total protein content in Chironomid larvae. Metals may also bind to DNA and thus cause mutations, and to lipids and cause membrane injury (Gebhart, 1984; Alberts et al., 1989).

Animals seek to inactivate toxic metals by binding them to metallothioneins or other specialized metal binding proteins (Hopkin, 1989). The capacity to deactivate metals varies from species to species, partly because the ability to produce metal binding proteins varies between species (Hopkin, 1989), but probably also because different proteins have a preference for different metals (Suzuki et al., 1989). Thus, the capacity of organisms to bind and detoxify metals is genetically determined and is likely to reflect the natural load of toxic metals experienced by each species.

Polycyclic biphenyls (PCBs) are soluble in fat and will accumulate in the adipose tissue of hibernating animals. PCBs are likely to be released to the circulatory system when the fat stores are consumed in winter and spring. In addition to being carcinogenic, PCBs induce chromosomal alterations and may affect reproduction (Peakall et al., 1972). They have also been found to influence hormonally controlled processes such as egg shell production in birds and the estrus cycle in mice (Connell & Miller, 1984).

Dioxins influence several enzyme systems (Connell & Miller, 1984) and are also probably carcinogenic (Kociba et al., 1978). Vos (1978) showed that dioxins have a negative effect on lymphocytes and alpha-, beta- and gamma-globulins, thus leading to an increased vulnerability to disease.

Some organic chemicals such as oil components, formaldehyde and benzene may have acute toxic effects. The acute toxicity may be accomplished by enzyme inhibition, with subsequent inhibition of ATP production and active transport of ions and free amino acids (Aunaas et al., 1991; Zachariassen et al., 1991).

PAHs have been shown to be carcinogenic and embryotoxic (Hoffman & Gay, 1981). They may also influence growth, development and feeding rates of animals (Neff, 1979).

The majority of the most widely used insecticides work by affecting the function of the nervous system (Connell & Miller, 1984). DDT and related compounds inhibit axonal transmission, probably by keeping sodium channels open and thereby interfering with the ion movement across the cell membrane (Corbett, 1974). Organophosphorous and carbamate insecticides inhibit acetylcholine esterase (DeBruin, 1976). Nicotine and cartap block the acetylcholine receptors.

4. Environmental pollution and cold-hardiness

Pollutants may influence mechanisms of low temperature tolerance, but there is also a possibility that the mechanisms of cold-hardiness will enhance the toxic effects of pollutants. However, the interaction of cold/cold-hardening mechanisms on one hand and environmental pollutants on the other has been the object of very few studies. Nevertheless, the knowledge of cold-hardening mechanisms and the general mechanisms of action of pollutants allows some speculation to be made regarding the interaction of cold and pollutants in causing lethal effects.

4.1 *Effect of pollution on cold-hardiness*

4.1.1 *Direct effects on components important to cold-hardiness*
Since ice nucleating agents and antifreeze peptides have many hydrophillic or negatively charged groups which are important to their function (Duman *et al.*, 1984), their activity may be reduced by heavy metals because metals may bind to the negative groups and thus reduce the availability of negative groups. This would reduce the capacity of ice nucleating agents to organize water molecules in an ice-like manner and thus reduce the nucleating potential of the nucleators. A reduction in the ability of extracellular ice nucleators to establish a protective extracellular freezing would reduce the cold-hardiness of freeze-tolerant insects, which depend on extracellular freezing at a high temperature for winter survival.

The effect of metals on ice nucleator activity was investigated in a series of experiments, in which samples of biological ice nucleators were added to solutions of $CdCl_2$ (100 mmolal), $CuCl_2$ (100 mmolal) and $PbCl_2$ (36 mmolal, *i.e.* saturated). The volumes of the native nucleator samples were adjusted so that the nucleators were diluted 1:100 in the various metal solutions. Previous experiments have revealed that when samples of a nucleator-rich biological fluid undergo isovolumetric dilution, the supercooling points remain nearly constant over a wide range of nucleator concentrations, but then drop steeply when the samples become diluted beyond a factor of 100 to 10000 (Zachariassen & Hammel, 1988). The supercooling points were determined on 20 µl samples. The dilutions used, implied that the results were within the steep part of the dilution curve of the nucleators employed, *i.e.* even a moderate change in the content of active nucleators would be expressed as a difference in supercooling points. The results revealed no effect compared to the supercooling points of control samples in which the nucleator samples were added to 100 mmolal solutions of NaCl. Thus, it appears that the metals did not have any inhibitory effect on the activity of these nucleators.

As in the case of ice nucleators, THFs and glycopeptides have charged groups which are essential for their function, and which might be blocked by metals. Metal interaction of this kind would make freeze-avoiding organisms less likely to avoid spontaneous or inoculative freezing of their body fluids. However, no data is available on the direct effect of metals on the activity of antifreeze peptides.

4.1.2 *Effects on processes involved in cold-hardiness*
Seasonal cold-hardening depends on production of proteinaceous substances such as extracellular ice nucleating agents and antifreeze peptides. A great number of toxic agents are known to inhibit protein synthesis (Connel & Miller, 1984), and these agents are likely to reduce or block the production of these cryoprotective proteins, and thus reduce or eliminate their cryoprotective action.

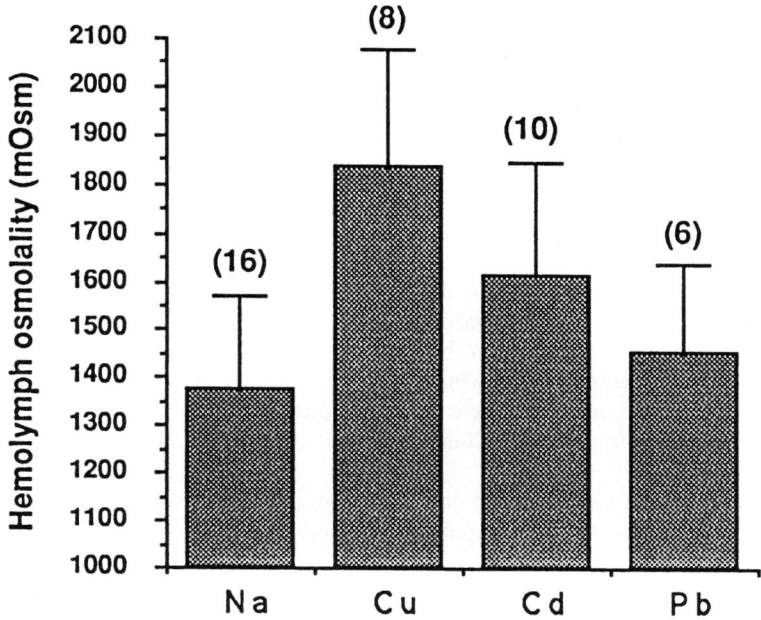

Fig. 4. Haemolymph osmolality of *Rhagium inquisitor* beetles exposed to 100 mmolal solutions of chlorides of different metals.

Reduced protein synthesis may also reduce the formation of enzymes involved in the production of cryoprotective sugars and polyols. This would reduce the capacity of hibernating insects to produce cryoprotectants and thus cause a reduced cold-hardiness in hibernating insects. However, no experimental data is available at this point.

Cryoprotectant levels may also be affected by environmental pollutants which interfere directly with enzyme activity. The effect of various metals on the reduction of glycerol levels in *Rhagium inquisitor* longhorn beetles in the spring is shown in Fig. 4. The beetles were taken from their winter hibernaculae in February and kept for seven days in constant darkness at 2°C in the laboratory. 1 µl of haemolymph was removed from the beetles with a microsyringe, whereafter the beetles were given an injection of 1µl samples of chloride salts solutions of lead (36 mmolal, saturated), cadmium (100 mmolal) or copper (100 mmolal). The injection of metals caused a marked reduction in glycerol breakdown after 10 days at +5° C compared to the control group, injected with 1µl of 100 mmolal solution of sodium chloride. Polyol production is catalyzed by the same enzymes as those catalyzing the breakdown, and if the same inhibitory effect exists for polyol production, the metals may also reduce the glycerol levels of hibernating insects. This would reduce the cold-hardiness of hibernating insects accordingly.

The cellular accumulation of free amino acids that act as cryoprotectants in marine littoral molluscs takes place by means of a sodium-coupled symport mechanism (Pequignat, 1973). Toxic agents which reduce the sodium gradient or inhibit the sodium-dependent symport mechanism directly, might lead to a reduced intracellular concentration of free amino acids and thus to reduced cold-hardiness. A number of organic pollutants and trace metals (oil, formaldehyde, benzene, mercury) have been shown to cause a reduction in the transmembrane sodium gradient of the adductor muscle of *Mytilus* (Børseth *et al.*, 1992), and may thus affect the cold-hardiness of littoral invertebrates. However, Aarset & Zachariassen (1983) did not find such pollutants (oil and oil dispersants) reduced the intracellular levels of FAAs or the cold-hardiness of *Mytilus*. Heavy metals and organic agents may also interfere with the hormonal control of the cold hardening mechanisms, but no data exist about such effects.

4.2 Effects of cold and cold adaptations on toxicity

Temperature has strong influence on the rates of biological processes. Whilst chemical reactions and diffusion are influenced in proportion to the absolute temperature (effects on solvent viscosity not considered), biological processes vary as described by the Q_{10} relationship. Q_{10} is the factor by which the rate changes over a temperature span of 10°C. Most enzymatic processes have Q_{10}-values between two and three, *i.e.* their rate will more than double when the temperature increases by 10°C (Schmidt-Nielsen, 1990).

The toxic effects of chemicals on ectothermic organisms are likely to vary in a manner that depends on the Q_{10} of the processes involved. For agents which have a direct toxic effect on organisms, the toxicity is likely to increase with temperature according to a Q_{10} relationship, *i.e.* the time it takes to develop toxic effects will be reduced by about half if the temperature is elevated by 10°C. For crude oil, the aromatic compounds are assumed to have a direct toxic effect, and Denstad (1988) found that the toxic effect on *Mytilus* was more pronounced at 14°C than at 5°C. Also, formaldehyde is assumed to have a direct toxic effect on *Mytilus*, and Tessem (1993) found that this effect has a Q_{10} value of about three.

This relationship is also likely to exist when the toxic effects are due to the enzymatic transformation of a non-toxic compound to a toxic compound in the organism. In this case, there will be a Q_{10} relationship for the formation of the toxic agent, as well as for the toxic interaction of that agent with the organism.

In cases where the toxic agent undergoes an enzymatic detoxification, the inactivation of the toxic agent will display a Q_{10} relationship, *i.e.* the de-activation will take place progressively faster as the temperature increases. In this case, the increasing direct toxic effect at rising temperatures will be counteracted by an increasing rate of detoxification. The net effect of temperature on toxicity will, in this case, depend on the relative values of the toxicity and the detoxifying mechanisms.

Membrane permeability to sodium is increased by the insecticide DDT (Corbett, 1974), leading to an intoxication of the nervous system. The transmembrane sodium influx associated with this elevated permeability is probably only moderately affected by reduced temperature, whereas the compensatory active Na/K ATP-ase dependent extrusion of sodium from the cells displays a Q_{10} relationship. Thus, passive influx is likely to remain high as the temperature drops, while there is a substantial reduction in the compensatory sodium extrusion. This implies that a pollutant-induced disturbance of the ionic balance is likely to be expressed at low temperatures.

A special situation exists in freeze-tolerant animals. Freezing of body fluids implies that solvent water is removed as ice, and as pointed out by Aarset & Zachariassen (1983), this will cause an increase in the concentrations of all dissolved substances, including pollutants. Aarset & Zachariassen (1983) found that freezing of *Mytilus* increased the toxic effect of oil and oil dispersants, apparently because of the freezing-induced increase in the concentrations of toxic components in their body fluids.

While freeze-tolerant organisms display increases in toxic agents due to the freezing out of solvent water, organisms that seek to avoid freezing may suffer a substantial evaporative water loss during winter. Hibernating insects may have values of relative water content as low as 25 percent (Richards et al., 1987; Ring, 1982). Unless the insects possess efficient mechanisms to keep the levels of toxic agents low, there will be a corresponding increase in the concentration of environmental pollutants dissolved in their body fluids, i.e. toxic effects may be expressed in the spring when the loss of solvent water is most pronounced.

5. Summary

Interaction between environmental pollutants and low temperature tolerance may take place via a number of mechanisms. The direct experimental documentation for interactions of this kind is very limited, but general knowledge about the mechanisms of action of environmental pollutants on one side and the mechanisms involved in cold-hardening on the other, suggests that this relationship may be of great importance. The importance of this relationship is further emphasized by the great biomass and the great ecological significance of the organisms which depend on seasonal cold-hardening for survival in Arctic and boreal regions.

Reference List

Aarset, A. V., 1982. Freezing tolerance in intertidal invertebrates (a review). Comp. Biochem. Physiol., 73A: 571–580.
Aarset, A. V. & K. E. Zachariassen, 1983. Synergistic effects of an oil dispersant and low temperature on the freezing tolerance and solute concentrations of the blue mussel (*Mytilus edulis* L.). Polar Research, 1: 223–229.
Alberts, B., D. Bay, W. Lewis, M. Raff, K. Roberts & J. Watson (ed.), 1989. The molecular genetics of cancer. In: *Molecular Biology of the Cell*. Garland Publishing Company Inc. New York. pp. 1203–1216.
Aunaas, T., 1982a. Ice nucleating agents in the haemolymph of intertidal molluscs tolerant to freezing. Experientia, 38: 1456–1457.
Aunaas, T., 1982b. Nucleating agents in the haemolymph of intertidal invertebrates tolerant to freezing. Cryo-Letters, 3: 287.
Aunaas, T., S. Einarson, T. Southon & K. E. Zachariassen, 1991. The effects of organic and inorganic pollutants on intracellular phosphate compounds in blue mussels (*Mytilus edulis*). Comp. Biochem. Physiol., 100C: 89–93.
Block, W. & S. R. Young, 1979. Measurements of supercooling points in small arthropods and water droplets. Cryo-Lett., 1: 85–91.
Børseth, J. F., T. Aunaas, S. Einarson, T. Nordtug, A. J. Olsen & K. E. Zachariassen, 1992. Pollutant-induced depression of the transmembrane sodium gradient in muscles of mussels. J. exp. Biol., 169: 1–18.
Carr, R. S., J. W. Williams, F. I. Saska, R. S. Buhol & J. M. Neff, 1985. Bioenergetic alterations correlated with growth, fecundity and body burdens of cadmium in mysids (*Mysidopsis bahia*). Environ. Toxicol. Chem., 4: 181–.
Connell, D. W. & G. J. Miller, 1984. *Chemistry and Ecotoxicology of Pollution*. John Wiley & Sons, New York. pp. 312–313.
Cook, P. A. & A. H. Lewis, 1971. Acquisition and loss of cold tolerance in adult barnacles (*Balanus balanoides*) kept under laboratory conditions. Mar. Biol., 9: 26–30.
Corbett, J. R., 1974. *The Biochemical Mode of Action of Pesticides*. Academic Press, London. 330 pp.
DeBruin, A., 1976. *Biochemical Toxicology of Environmental Agents*. Elsevier, Amsterdam 1544 pp.
Denstad, J.-P. 1988. Effects of temperature on physiological responses of blue mussels, *Mytilus edulis*, to water soluble fractions of oils and chemicals. In: K. E. Zachariassen (ed.), *Biological Effects of Chemical Treatment of Oil spills at Sea*. Report from the BECTOS-program 1985-1988. pp. 185–193. Tapir Trykk, Trondheim.
DeVries, A., 1982. Biological antifreeze agents in coldwater fishes. Comp. Biochem. Physiol., 73A: 627–640.
DeVries, A. & D. E. Wohlschlag, 1969. Freezing resistance in some Antarctic fish. Science, 163: 1074–1075.

Duman, J. G., J. P. Morris & F. J. Castellino, 1984. Purification and composition of an ice nucleating protein from queens of the hornet, *Vespula maculata*. J. Comp. Physiol., 154: 79–83.
Duman, J. G., D. W. Wu, L. Xu, D. Tursman & T. M. Olsen, 1991. Adaptations of insects to subzero temperatures. The Quarterly Review of Biology, 66: 387–409.
Gebhart, E., 1984. Mutagenität, Karzinogenität, Teratogenität. In: E. Merian (ed.), *Metalle in der Umwelt*. pp. 237–247. Verlag Chemie.
Gehrken, U., 1984. Winter survival of an adult bark beetle *Ips acuminatus* Gyll. J. Insect Physiol., 30: 421–429.
Gehrken, U., 1992. Inoculative freezing and thermal hysteresis in the adult beetles *Ips acuminatus* and *Rhagium inquisitor*. J. insect Physiol., 38: 519–524.
Gekko, K. & S. N. Timasheff, 1981. Mechanism of protein stabilization by glycerol: Preferential hydration in glycerol-water mixtures. Biochemistry, 20: 4667–4676.
Hoffman, D. J., & M. L. Gay, 1981. Embryotoxic effects of benzo[a]pyrene, chrysene, and 7,12-dimethyl benz[a]anthracene in petroleum hydrocarbon mixtures in mallard ducks. J. Toxicol. Environ. Health., 7: 775–787.
Hopkin, S. P., 1989. *Ecophysiology of Metals in Terrestrial Invertebrates*. Elsevier Applied Science Publishers Ltd., Essex. 366 pp.
Kanwisher, J. W. 1959. Histology and metabolism of frozen intertidal animals. Biol. Bull., 116: 258–264.
Kociba, R. I., D. G. Keyes & J. E. Beyer, 1978. Results of a two year chronic toxicity and oncogenicity study of 2,3,7,8,-tetrachlorodibenzo-*p*-dioxin in rats. Toxicol. Appl. Pharmacol., 46:279–303
Laak, S. van der, 1982. Physiological adaptations to low temperature in freezing tolerant *Phyllodecta laticollis* beetles. Comp. Biochem. Physiol., 73A: 613–620.
Lovelock, J. E., 1953. The mechanism of the cryoprotective effect of glycerol against freezing and thawing. Biochim. Biophys. Acta, 11: 28–36.
Meryman, H. T., 1971. Osmotic stress as a mechanism of freezing injury. Cryobiology, 8: 489–500.
Miller, L. K., 1969. Freezing tolerance in an adult insect. Science, 166: 105–106.
Miller, L. K. & J. S. Smith, 1975. Production of threitol and sorbitol by an adult insect: association with freezing tolerance. Nature, London, 258: 519–520.
Murphy, D. & S. K. Pierce, 1975. The physiological basis for changes in the freezing tolerance of intertidal molluscs. Response to subfreezing temperature and the influence of salinity and temperature acclimation. J. Exp. Zool., 193: 313–322.
Namatov, T., 1969. Resistance of the nematode *Anguina tritici* and *Rhabditis* sp. to low and extremely low temperatures. Dokl. Akad. Nauk SSSR, 185: 1382–1385.
Neff, J. M., 1979. *Polycyclic Aromatic Hydrocarbons in the Aquatic Environment*. Applied Science Publishers, London. 262 pp.
Nieboer, E. & D. H. S. Richardson, 1980. The replacement of the nondescript term "heavy metals" by a biologically and chemically significant classification of metal ions. Envir. Pollut. B, 1: 3–26.
Ohyama, Y. & E. Asahina, 1972. Frost resistance in adult insects. J. Insect Physiol., 18: 267–282.
Pauktis, G. L., R. D. Shuman & F. J. Janzen, 1989. Supercooling and freeze tolerance in hatching painted turtles (*Chrysemys picta*). Can. J. Zool., 67: 1082–1084.
Peakall, D.B., J. L. Lincer & S. E. Bloom, 1972. Embryonic mortality and chromosomal alterations caused by aroclor 1254 in ring doves. Environ. Health. Perspec., 1: 103–104.
Pequignat, E., 1973. A kinetic and autoradiographic study of the direct assimilation of amino acids and glucose by organs of the mussels *Mytilus edulis*. Mar. Biol., 19: 227–244
Pigon, A. & B. Weglarski, 1955. Anabiosis in Tardigrada. Metabolism and humidity. Bull. Acad. pol. Sci. Cl. II Ser. Sci. Biol., 3: 31–34.
Rathore, R. S., P. K. Sangui & H. Swarup, 1979. Toxicity of cadmium chloride and lead nitrate to *Chironomus tentants* larvae. Environm. Pollut. 18: 173–177.
Raymond, J. 1992. Glycerol is a colligative antifreeze in some northern fishes. J. exp. Zool., 262: 347–352.
Richards, J., M. J. Kelleher, K. B. Storey, 1987. Strategies of freeze avoidance of larvae of the goldenrod gall moth *Epiblema scudderiana*: Winter profiles of a natural population. J. Insect Physiol., 33: 443–450.
Ring, R., 1982. Freezing tolerant insects with low supercooling points. Comp. Biochem. Physiol., 73A: 605–612.
Schmid, W. D., 1982. Survival of frogs in low temperature. Science, 215: 697–698.
Schmidt-Nielsen, K. 1990. *Animal Physiology: Adaptation and Environment*. Cambridge University press, New York. 602 pp.
Scholander, P. F., L. Vandam, J. W. Kanwisher, H. T. Hammel & M. S. Gordon., 1957. Supercooling and osmoregulation in Arctic fish. J. Cell. Comp. Physiol., 49: 5–24.
Sømme, L., 1964. Effects of glycerol on cold-hardiness in insects. Can. J. Zool., 42: 87–101.
Sømme, L., 1967. The effects of temperature and anoxia in haemolymph composition and supercooling in three overwintering insects. J. Insect Physiol., 13: 805–814.
Sømme, L., 1969. Mannitol and glycerol in overwintering aphid eggs. Nor. Entomol. Tidsskr., 16: 107–111.

Sømme, L. & E.-M. Conradi-Larsen, 1977. Cold-hardiness of collembolans and oribatid mites from windswept mountain ridges. Oikos, 29: 118–126.

Staurnes, M., T. Sigholt & O. B. Reite, 1984. Reduced carbonic anhydrase and Na/K-ATPase activity in gills of salmonids exposed to aluminium-containing acid water. Experientia, 40: 226–227.

Storey, K. B. & J. M. Storey, 1989. Freeze tolerance and freeze avoidance in ectotherms. In: C. H. Wang (ed.), *Advances in Comparative and Environmental Physiology 4. Animal Adaptation to Cold.* Springer-Verlag. Berlin Heidelberg. pp.51–82.

Suzuki, K. T., H. Sunaga, S. Hatakeyama, Y. Sum & T. Suzuki., 1989. Differential binding of cadmium and copper to the same protein in a heavy metal tolerant species of mayfly (*Baetis thermicus*) larvae. Comp. Biochem. Physiol., 94C: 99–103.

Tessem, P. M., 1993. The significance of temperature on the effects of formaldehyde on physiological parameters of blue mussels, *Mytilus edulis*. Abstract from Int. Nord. Symp. on Chemicals in the Arctic-Boreal Environment, Helsinki, May 12–14, 1993. p. 84.

Umminger, B., 1970. Carbohydrate metabolism in fish at subzero temperatures. J. Exp. Zool., 173: 159–174.

Vos, J. G., 1978. 2,3,7,8-Tetrachlorodibenzo-p-Dioxin: Effects and Mechanisms. In: C. Ramel (Ed.) *Chlorinate Phenoxy Acids and their Dioxins,* Swedish Natural Science Research Council, Stockholm, p. 165–176.

Worland, R., W. Block & P. Rothery, 1992. Survival of sub-zero temperatures by two South Georgian beetles (Coleoptera, Perimylopidae). Polar Biol., 11: 607–613.

Zachariassen, K. E., 1980. The role of polyols and nucleating agents in cold-hardy beetles. J. Comp. Physiol., 140: 227–234.

Zachariassen, K. E., 1985. Physiology of cold-tolerance in insects. Physiol. Rev., 65: 799–832.

Zachariassen, K. E., 1992. Ice nucleating agents in cold-hardy insects. In: G. N. Somero, C. B. Osmond & C. L. Bolis (eds), *Water and Life*. pp. 261–281. Springer-Verlag, Berlin Heidelberg.

Zachariassen, K. E. & H. T. Hammel, 1976. Ice nucleating agents in the haemolymph of insects tolerant to freezing. Nature, 262: 285–287.

Zachariassen, K. E. & H. T. Hammel, 1988. The effect of ice nucleating agents on ice nucleating activity. Cryobiology, 25: 143–147.

Zachariassen, K. E. & J. A. Husby, 1982. Antifreeze effect of thermal hysteresis agents protects highly supercooled insects. Nature, 298: 865–867.

Zachariassen, K. E. & R. Lundheim, 1992. The endocrine control of insect cold hardiness. Zool. Jahrb., 96: 183–196.

Zachariassen, K. E., T. Aunaas, J. F. Børseth, S. Einarson, T. Nordtug, A. Olsen & G. Skjærvø, 1991. Physiological parameters in ecotoxicology. Comp. Biochem. Physiol., 100C: 77–79.

Mechanisms of thermal acclimation in relation to ecotoxicology

K. Y. H. Lagerspetz

*Laboratory of Animal Physiology, Department of Biology,
University of Turku, FIN-20 500, Turku, Finland*

Keywords: thermal ecotoxicology, temperature, acclimatization, membrane fluidity, enzymes

1. Introduction

In the northern cool (boreal) region the yearly variation of air temperature generally exceeds 30 degrees, and in some localities even 60 degrees. It is important to note that these ranges extend across the freezing point of water. Even in natural waters, the seasonal temperature variation is often 20 degrees, and the surface of the waters generally freezes in the winter.

The survival of organisms in such a climate presupposes special wintering strategies. Some of these are linked with the life cycles of the organisms, and some others with their behavioral and physiological adjustments. Such adaptive adjustments are called seasonal acclimatizations. They are individual and reversible and repeatable during the life of the individual. Some of these may be triggered, in part, for instance by variations in the availability of food, but also prolonged changes in the environmental temperature suffice to produce thermal acclimation phenomena. These are an important part of the seasonal acclimatization and the wintering of animals. Experimental studies on thermal acclimation have given us much knowledge about its cellular and molecular mechanisms.

In the following, I will first define some basic terms and concepts used in thermal physiology, then give some examples of seasonal and thermal acclimation phenomena and discuss their mechanisms, and finally point out some of the possibly important but virtually unstudied interactions of environmental pollutants with temperature.

2. Definitions

Body temperature (T_b) is the internal temperature of animals, measured from their body core (*e.g.* from their alimentary tract or from their tissues). Ambient temperature (T_a) is the actual environmental temperature of the animal, measured near its body.

In poikilothermic animals the body temperature varies with the ambient temperature, in homeothermic animals T_b is more or less constant in spite of large variations in T_a. Typical mammals and birds are homeotherms. Some reptiles are homeotherms within narrower season dependent limits. Generally all other animals are poikilothermic.

Most animals are ectothermic, which means that their T_b depends mainly on external heat sources. The tissues of mammals and birds, and the muscles of some large active fish like tuna, produce so much heat energy, that the body temperature of these animals is mainly (or in tuna, locally) dependent on internal heat production. These animals are called endothermic.

Stenothermic species live in conditions with a small local and temporal variation of Ta. Eurythermic species have a larger ecological temperature range. Accordingly, stenothermic species show evolutionary adaptations to a narrow range of environmental temperature and the eurytherms for living at more variable temperatures.

Like the other distinctions made in thermal physiology, the stenothermy-eurythermy distinction is relative and indicates only positions in a continuous dimension. This is clear when we compare the ecological temperature relations in some families of fish (Elliott, 1981). Coregonids (*e.g.* whitefish) and salmonids are relatively stenothermic cold water fish. They tolerate temperatures between the freezing point and about 25 °C, and are most active between about 5 and 17 °C. For instance most cyprinids are relatively eurythermic, and tolerate temperatures from 0 to about 35 °C, and are most active between 8 and 28 °C.

The difference between stenothermic and eurythermic species is well demonstrated by a comparison of the temperature ranges for normal egg development. These are below 10 °C for coregonids, below 20 °C for salmonids, and extend from about 5 to 30 °C in most cyprinids. The larger the ranges, the more eurythermic we call the species in question.

3. Thermal acclimation

An important observation is that the eurythermic species show a better ability for thermal acclimation than the stenothermic species. Thermal acclimation is individual, reversible, phenotypic adaptation to a prolonged change in ambient temperature. Examples of the ability for thermal acclimation in two isopod crustaceans are given in Table 1. Asellus aquaticus is a common eurythermic fresh water isopod which lives on shallow bottom and on the plants of ponds, lakes and rivers and also along the coasts in the Baltic sea. When turned on its back, it usually in a few seconds returns to the upright position. This righting reflex is lost at high temperatures. The temperature at which it is lost, is called the critical thermal maximum (CTMax) of this animal. It is critical for the animal, because if the righting reflex is permanently lost, the animal will not survive. CTMax values are shifted upwards by keeping Asellus at higher temperatures. Thus in animals which have been kept at 10 °C, CTMax is about 31 °C, and in those kept for two-three weeks at 20 °C, CTMax is about 35 °C.

Saduria entomon is a stenothermic cold water isopod living in the bottom at depths below 10 m in the Baltic. Its close relatives live in the Arctic seas. Its CTMax is about 5 °C lower than in Asellus, and it is not significantly changed by keeping the animals at 14 °C instead of 4 °C, *i.e.* by acclimation to a temperature 10 °C higher. Thus, stenothermic cold water animals may show much less thermal acclimation ability and in some functions no thermal acclimation at all. This is important when assessing the possible effects of long-term climate changes.

Critical thermal maximum is a behavioral measure but its thermal acclimation is obviously based on physiological changes in the functions of certain nerve cells which mediate the righting reflex. The same holds also for other behavioral adaptations to temperature and wintering, some examples of which are given below.

The most rapid reaction to a change of ambient temperature is seen in the avoidance

Table 1. Effect of thermal acclimation on the critical thermal maximum temperature (CTMax) in a eurythermic isopod *Asellus aquaticus* (Lagerspetz & Bowler, 1993) and in a stenothermic cold water isopod *Sadura entomon* (Kivivuori and Lagerspetz, 1990). AT – acclimation temperature. Acclimation time is given in brackets.

AT (°C)	CTMax (°C)	P
A. aquaticus		
10	30.6 ± 0.3	
20 (2-3 wks)	34.9 ± 0.5	< 0.001
S. entomon		
4	26.4 ± 0.5	
14 (2 wks)	27.3 ± 0.6	n.s.
14 (4 months)	26.0 ± 0.4	n.s.

of extreme temperatures. This phenomenon occurs in all animals from protozoans to humans. Adaptations for wintering include for instance search for frost-free wintering places and resistance to starvation.

4. Seasonal acclimatization

The avoidance of starvation in winter is made possible by the accumulation of reserve foodstuffs and by a decrease of the rates of metabolic and neural functions. An example of the first is provided by the seasonal variation in the glycogen content of the liver in the grass frog (Rana temporaria), shown in Figure 1. The relative content of glycogen is highest and lasts for the longest period in frogs from the most northern latitude, and lowest and least lasting in frogs from the southernmost latitude. This shows the adaptive significance of reserve foodstuffs in this species which breeds early in the spring and without previous feeding. The accumulation and use of reserve foodstuffs obviously depends on changes in the production of hormones (Lagerspetz, 1977). It is a case of seasonal acclimatization.

An example of the relative similarity of the effects of seasonal acclimatization and experimental thermal acclimation is provided by the study of Meyer and Hegmann (1971). The conduction velocity in the sciatic nerve of the frog Rana pipiens is increased at intermediate and high temperatures in the summer, and similarly by an acclimation of the frogs to a higher temperature during both seasons. The faster conduction of nervous impulses in the motor nerves is obviously adaptive for the generally higher activity in the summer and at higher temperatures.

If the rates of metabolic and neural functions are slowed down in the cold, this contributes to the saving of energy during wintering but puts the animals in torpor. On the other hand, these rates in some cases do become faster during cold acclimation. This again tends to keep up the metabolism and reactivity of the animal even during temporary cold exposures, which may occur in nature during weather changes also in summer.

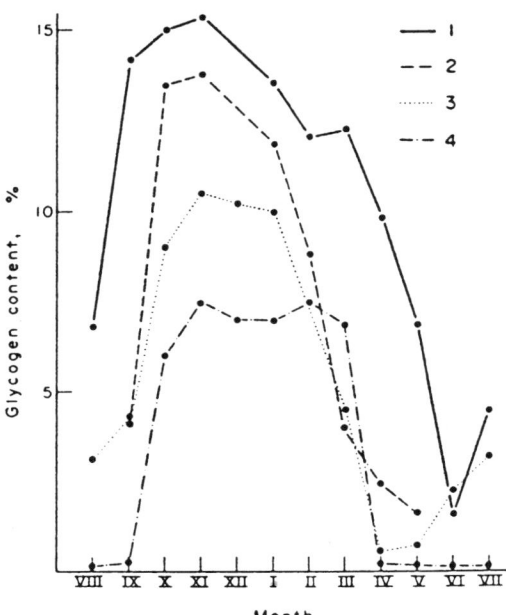

Fig. 1. Seasonal variation in glycogen content (%) of the liver in mature frog *Rana temporaria* at different latitudes. 1. Pasanen & Koskela (1974), 64–65°N; 2. Kato (1910), 54–55°N; 3. Smith (1950), 53–54°N; 4. Hong *et al.* (1968); 37–38°N. From Pasanen & Koskela (1974), with permission.

5. Cellular mechanisms of thermal acclimation

5.1 *Two types of mechanisms*

The thermal acclimation of functions of nerve and other cells may depend on two cellular mechanisms. One of them is the homeoviscous adaptation of the fluidity of the lipid matrix of cell membranes. When poikilothermic animals are transferred from the cold to a higher temperature, the thermodynamic ordering of the lipid matrix of their cell membranes decreases, and the membranes become more fluid, just like butter taken from the fridge to room temperature. However if the exposure to this new temperature is prolonged, the matrix gradually becomes more ordered and less fluid again. However, if such animals or cells are then returned to cold, the lipid matrix of the cell membranes becomes first more ordered and rigid, and then gradually more fluid again. Cell membrane fluidity first decreases in cold, but then shows a compensatory increase. This homeoviscous adaptation tends to keep the membrane fluidity at a more or less constant level in spite of changes in the ambient temperature (Cossins & Bowler, 1987).

An example of membrane alteration is seen in Figure 2. It shows the membrane fluidity in brain cell membranes of rainbow trout acclimated to 5 °C and 21 °C. Higher fluorescence polarization values indicate a higher orderliness and a lower fluidity of the membranes. When membrane fluidity is measured in animals acclimated to 21 °C at that temperature (A), the fluidity is higher (the polarization lower) than when measured at 5 °C (B). During an exposure of the fish to a few weeks at 5 °C, the membrane

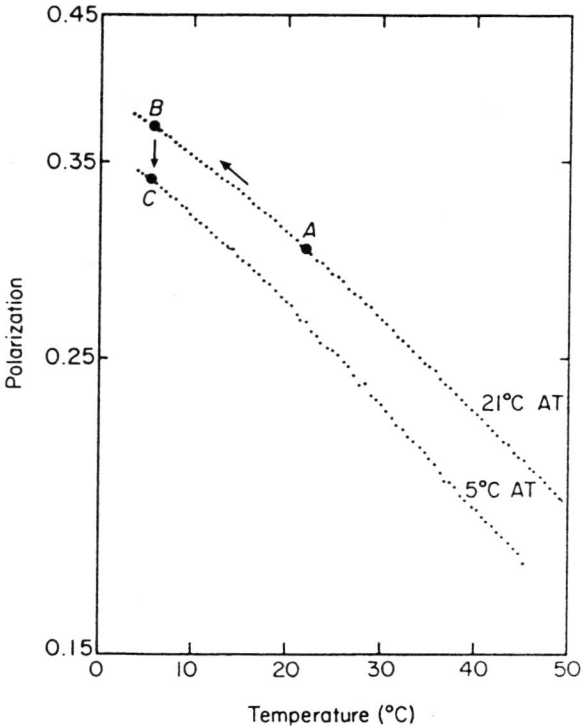

Fig. 2. Homeoviscous adaptation of brain membrane preparations isolated from trout acclimated to 5 °C and 21 °C (AT). The results are from fluorescence polarization measurements with a lipophilic fluorescent probe molecule. Higher values of polarization indicate a lower fluidity of the membranes. Sudden transfer of fish from 21 °C to 5 °C results in substantial decrease in fluidity (point A to point B). During a few weeks at 5 °C the fluidity is increased (B to C), but not to the same level as originally at 21 °C (A). Thermal acclimation causes a partial compensation of the immediate effects of cold. Modified from Cossins & Bowler (1987), with permission.

fluidity measured at that temperature is adjusted upwards (the polarization downwards) to the point C. However, the adjustment observed in cold-acclimated animals is not sufficient to completely compensate for the immediate effects of cooling.

Homeoviscous adaptation is known to affect the passive flux of ions across the membrane, and it may also affect the ability for conformational changes in proteins embedded in the lipid matrix. These include transport proteins, ion channels and receptor molecules.

The homeoviscous adaptation is based on changes in the synthesis of fatty acid saturates and other enzymes which affect the lipid and sterol constituents of the membrane (Thompson, 1983; Cossins & Bowler, 1987).

Besides homeoviscous adaptation, another molecular mechanism of thermal adaptation is the temperature induced change in the synthesis of cytoplasmic and membrane proteins.

In their extensive review Hazel & Prosser (1974) gave lists of 57 enzymes which had been studied in different tissues of different ectothermic animals, mostly in fishes but also in amphibians, crustaceans, insects, molluscs and in other invertebrates. The activities of 40 of these enzymes show partial thermal compensation; their activity when measured at intermediate temperatures is increased by the acclimation to cold.

The remaining 17 enzymes show reverse compensation or no compensation; their activities are either decreased by acclimation to cold or not affected by it.

Partial thermal compensation is typical for many glycolytic, hexose monophosphate shunt, citric acid cycle and electron transport enzymes. Those enzymes which show reverse thermal compensation of their activity are the peroxisomal and lysosomal enzymes and the enzymes of nitrogen catabolism. This means that the activities of such enzymes which decompose intermediate products of catabolism are decreased during prolonged stay of the animals in cold, in which usually less catabolic intermediates are produced. In some cases thermal acclimation is known to affect the relative activities and the synthesis of different isozymes.

5.2 Species-specificity of the mechanisms

Homeoviscous adaptation of membranes and changes in the synthesis of proteins not directly involved in membrane lipid production and desaturation, may occur independently from each other and be species specific. This is shown in the example from the recent work of Schwarzbaum *et al.* (1991, 1992).

The roach (Rutilus rutilus) is a relatively eurythermic cyprinid which lives often in the same alpine lakes in Austrian Tyrol as the Arctic char (Salvelinus alpinus), a stenothermic cold water salmonid. Schwarzbaum *et al.* (1991, 1992) studied the effects of temperature acclimation on the membranes of kidney cells in these two fish species. The acclimation temperatures were generally 5 °C and 20 °C for the roach and 5 °C and 15 °C for the Arctic char. The acclimation time was at least 3 weeks. The two species responded differently to the cold acclimation. In the kidney cell membranes of the roach an activation of the Na+-K+ pump occurred, as shown by the increase of ouabain-binding, ouabain-sensitive tissue oxygen consumption, specific Na++K+-ATPase activity and thermolability of this enzyme. Such differences were not found between the cold- and warm-acclimated Arctic char. In this fish a homeoviscous adaptation of the cell membranes occurred that increased the membrane fluidity in cold-acclimated animals and, linked with this, a decrease in the passive K+ efflux from the kidney cells. Such changes did not occur in the roach cell membranes or, as in the case of homeoviscous adaptation, were small.

Because of the different temperature coefficients of the active transport and the passive diffusion of ions across cell membranes, low temperatures (as well as high temperatures) may cause an imbalance between the pump and leak of potassium ions. Two different strategies may be used in cold acclimation to avoid this: 1) an increase of ion pumping, like in the roach, or 2) a decrease of ion leak, as in the Arctic char.

We do not know if these species differences in the mechanisms of thermal acclimation are connected with the relative eurythermy and stenothermy of the two species, or perhaps are rather family specific, contrasting cyprinids and salmonids. However, such differences should be taken into account also when studying the effects of different types of environmental pollutants, like enzyme inhibitors (*e.g.*, some heavy metal ions which inhibit ATPase enzymes) and detergents (which perturb lipid membranes) also at low temperatures. Their overall effects may be different in different species.

Both neural and kidney cell functions depend on membrane permeability and ion transport. Changes in the fluidity of cell membranes and in protein synthesis seem to be the mechanisms of these acclimation phenomena. The fluidity of cell membrane is changed again by differential synthesis of enzymes affecting the saturation degree of the fatty acids of membrane lipids and by other enzymes acting on phospholipid synthesis.

In conclusion, the basic mechanism of thermal acclimation seems to be altered gene expression (protein synthesis) triggered directly (in each cell), or indirectly (through mediation by the nervous and endocrine systems) by a prolonged change of Ta.

6. Interactions of temperature, thermal acclimation and toxic chemicals

Temperature and thermal acclimation have profound effects on cell membranes. On the other hand, the contacts of organisms with their chemical environment are mediated by cell membranes or across them. This happens when the chemicals affect sensory receptors, or enter the body through the skin or through respiratory or alimentary epithelia. Therefore cell membranes are one place at which the effects of temperature and chemicals may interact.

As we have seen, temperature and thermal acclimation have definite effects on the activity and synthesis of many enzyme proteins. As enzymes have often been found to be the targets of toxic chemicals, significant interactions between temperature, thermal acclimation status, and effects of chemicals are to be expected.

On the basis of what we know separately about thermal physiology and the effects of ecotoxic compounds, the combinations of such studies would be theoretically rewarding and practically important. For lack of knowledge we cannot at present answer such questions as "Is this chemical more toxic or less toxic in boreal than in temperate conditions?" "How does a rise of the environmental temperature by a few degrees affect the biotransformation of this chemical?" The review by Cairns *et al.* (1975) usefully summarized our meagre knowledge at that time, but we have not advanced much since. Dave (1986) wrote in his review of toxicity testing procedures on fish: "Because of the non-predictable effect of temperature on toxicity, the need for simplified methods which can be applied systematically on a great number chemicals is obvious. Since temperature and toxicity have much in common, models of temperature tolerance may be applicable" (Dave, 1986). Bourdeau *et al.* (1989) considered in their symposium-summary the current understanding of the interrelations between toxicity and temperature to be inadequate. "It may be worthwhile to carry out a series of experiments over a range of temperatures – using a range of species more representative of different ecosystems" (Bourdeau *et al.*, 1989).

The toxicity of most chemicals studied either increases with increasing temperature, is not affected by temperature, or is lowest at intermediate temperatures (see Cairns *et al.*, 1975; Dave, 1986). An exception is DDT which is most toxic to fish and water fleas at low temperatures (for references see Cairns *et al.*, 1975). DDT seems to disrupt the osmotic regulation in fish by an inhibition of membrane (Na^++K^+)-ATPase (Janicki & Kinter, 1971). The (Na^++K^+)-ATPase is known to be sensitive to changes in the fluidity of membrane lipid matrix (*e.g.* Cossins, 1983; Schwarzbaum *et al.*, 1992). A decrease in temperature would decrease the activity of this enzyme directly and by decreasing the membrane fluidity. The homeoviscous adaptation of membranes would in cold be compensatory for the activity of (Na^++K^+)-ATPase (Schwarzbaum *et al.*, 1992).

Mg^{2+} is a necessary cofactor for the activity of this enzyme. On the other hand, Mg^{2+}-ions at high concentrations decrease membrane fluidity (*e.g.* Newton *et al.*, 1978). Mg^{2+}-ions are also more toxic for Daphnia magna at low than at high temperatures (my own observations). The higher toxicity of some substances at lower temperatures may therefore be connected with the decrease of cell membrane fluidity in cold. In extreme cold, the membrane functions, passive permeability and active ion transport are impaired. Freezing increases the sensitivity of Mytilus edulis to oil dispersant, and this dispersant also reduces the ability of mussels to survive freezing (Aarset & Zachariassen, 1983).

On the other hand, the increase of membrane fluidity at high temperatures is a factor which may affect the tolerance of animals to toxic compounds. The toxicity of cadmium to mussels (Veldhuizen-Tsoerkan et al., 1991) and to polyps (Theede et al., 1979) is increased at ecologically high temperatures, and then especially at low salinities in which the ion regulation becomes critical for the organisms (Theede et al., 1979).

In conclusion, important interactions of environmental temperature and environmental contaminants certainly occur on the cell plasma membrane level. They may explain why organisms are more sensitive to ecotoxic compounds at extreme temperatures, and why sublethal concentrations of ecotoxic compounds affect the tolerance of organisms to extreme temperatures. Membrane changes may also at least in part explain why the thermal acclimation status affects the toxicity (Hodson & Sprague, 1975; Kovacs & Leduc, 1982).

Three types of proteins not bound to the plasma membrane of cells have been shown to be important in the defence of animals to environmental contaminants. Mixed-function oxygenase enzymes (MFO), which generally have P-450 cytochromes as the terminal oxidase of the enzyme chain, detoxicate many organic compounds (Payne et al., 1987). Metallothioneins (MT) are important in the detoxication of heavy metal ions (Kägi & Schäffer, 1988). Stress proteins (SP) were originally called heat shock proteins (HSP) because of the rapid induction of their synthesis in many animals and cultured animal cells after an increase in the ambient temperature. The synthesis of SP is also induced by many prospective environmental contaminant chemicals (Sanders, 1993). There are some studies on the effects of environmental temperature, seasonal acclimatization and thermal acclimation on the synthesis and functions of these types of proteins.

Seasonal differences in MFO activity have often been demonstrated (e.g., Koivusaari et al., 1981; Lindström-Seppä, 1985; Payne et al., 1987). Low temperatures may extend the period of time for the induction of MFO in rainbow trout (Andersson & Koivusaari, 1985), but the final MFO activity shows an ideal temperature compensation (acclimatization) during water cooling in autumn (Koivusaari et al., 1981). Obviously MFO enzyme induction can be used as a sensitive biological monitoring index during the whole year (Payne et al., 1987). However, the temperature dependence of accumulation and biotransformation of ecotoxic compounds varies according to the thermal ecology and physiology of the organism.

As both the heat shock and the acute exposure of animals to toxic chemicals may induce synthesis of SP (Sanders, 1993), synergistic effects on heat tolerance and SP synthesis could be expected. However, exposure to arsenic or to selenium decrease the heat tolerance (measured as CTMax) in fish (Paladino & Spotila, 1978; Watenpaugh & Beitinger, 1985). Acute, but not chronic exposure to cadmium induces SP synthesis in the gills of Mytilus edulis. On the other hand, chronic exposure to cadmium enhances the subsequent effect of a heat shock on the synthesis of SP (Veldhuizen-Tsoerkan et al., 1990, 1991). The effects of changes in environmental temperature and in contaminant concentrations certainly interact in the induction of SP synthesis, but this area of research is in its beginning (Sanders, 1993).

Interactions between environmental temperature and chemicals are evident at the levels of cell membranes and protein syntheses, but such interactions are reflected also in the physiological regulation of the energy metabolism of the animals. Ecotoxic compounds may interfere with or inhibit thermal acclimation and seasonal acclimatization. Thermal acclimation of the oxygen consumption of the bluegill (Lepomis macrochirus) is inhibited by exposure to copper (Felts & Heath, 1984). In addition, the recovery from ecotoxic effects may be slow and perhaps impossible at low temperatures (because of the metabolic costs).

Apparently there are very few systematic studies covering the ecotoxicology of any organism throughout its whole ecological temperature range. Such work would be important for elucidating the effects of contaminants on organisms in the Arctic-boreal environment.

7. Summary

Many animals living in boreal climate show seasonal acclimatization in nature and a capacity for thermal acclimation in laboratory conditions. Relatively eurythermic species have usually a better capacity for thermal acclimation than the more stenothermic species. Thermal acclimation is obviously a part of seasonal acclimatization.

The two best known mechanisms of thermal acclimation are the homeoviscous adaptation of cell membranes and differential enzyme synthesis. Both depend on changes in gene expression triggered by prolonged changes in environmental temperature. These effects may be direct effects on cells or mediated by the nervous and endocrine systems. Different types of mechanisms occur in different species.

The structure and function of cell membranes and the synthesis and activity of enzymes are profoundly affected by temperature and thermal acclimation. Cell membranes and enzymes are also important targets for the actions of ecotoxic compounds. Therefore studies on the interactions of ecotoxic chemicals and temperature covering the whole ecological temperature range of animals are needed. Such work would be especially important for the studies on pollution effects in Arctic-boreal environments.

Acknowledgements

A part of this article has been written during the holding of a personal grant from the Academy of Finland.

References

Aarset, A. V. & K. E. Zachariassen, 1983. Synergistic effects of an oil dispersant and low temperature on the freezing tolerance and solute concentrations of the blue mussel (Mytilus edulis L.). Polar Research, n.s. 1: 223-229.
Andersson, T. & U. Koivusaari, 1985. Influence of environmental temperature on the induction of xenobiotic metabolism by ß-naphthoflavone in rainbow trout, Salmo gairdneri. Toxic. appl. Pharmac. 80: 45-50.
Bourdeau, P., J. A. Haines, W. Klein & C. R. Krishna Murti, 1989. Introduction, conclusions, and recommendations. In: P. Bourdeau, J. A. Haines, W. Klein & C. R. Krishna Murti (eds.), Ecotoxicology and Climate. SCOPE 38: 3-11. John Wiley & Sons, Chichester.
Cairns, J. Jr., A. G. Heath & B. C. Parker, 1975. The effects of temperature upon the toxicity of chemicals to aquatic organisms. Hydrobiologia 47: 135-171.
Cossins, A. R., 1983. The adaptation of membrane structure and function to changes in temperature. In: A.R. Cossins & P. Sheterline (eds.), Cellular Acclimatisation to Environmental Change. Pp. 3-32. Cambridge Univ. Press, Cambridge.

Cossins, A. R. & K. Bowler, 1987. Temperature Biology of Animals. Chapman and Hall, London. 337 pp.
Dave, G., 1986. Toxicity testing procedures. In: S Nilsson & S. Holmgren (eds.), Fish Physiology: Recent Advances. Pp. 170-195. Croom Helm, London.
Elliott, J. M., 1981. Some aspects of thermal stress on freshwater teleosts. In: A. D. Pickering (ed.), Stress and Fish. Pp. 209-245. Academic Press, London.
Felts, P. A. & A. G. Heath, 1984. Interactions of temperature and sublethal environmental copper exposure on the energy metabolism of bluegill, Lepomis macrochirus Rafinesque. J. Fish Biol. 25: 445-454.
Hazel, J. R. & C. L. Prosser, 1974. Molecular mechanisms of temperature compensation in poikilotherms. Physiol. Rev. 54: 620-677.
Hodson, P. V. & J. B. Sprague, 1975. Temperature-induced changes in acute toxicity of zinc to Atlantic Salmon (Salmo salar). J. Fish. Res. Board Canada, 32: 1-10.
Hong, S. K., C. S. Park, Y. S. Park, & J. K. Kim, 1968. Seasonal changes of antidiuretic hormone action on sodium transport across frog skin. Am. J. Physiol. 215: 439-443.
Janicki, R. H. & W. B. Kinter, 1971. DDT: Disrupted osmoregulatory events in the intestine of the eel Anguilla rostrata adapted to seawater. Science 173: 1146-1148.
Kato, K., 1910. Über das Verhalten des Glykogenes im Eierstocke der Frösche zu den verschiedenen Jahreszeiten. Pflügers Arch. ges. Physiol. 132: 545-579.
Kivivuori, L. & K. Y. H. Lagerspetz, 1990. Thermal resistance and behaviour of the isopod Saduria entomon (L.). Ann. Zool. Fennici 27: 287-290.
Koivusaari, U., M. Harri & O. Hänninen, 1981. Seasonal variation of hepatic biotransformation in female and male rainbow trout (Salmo gairdneri). Comp. Biochem. Physiol. 70C: 149-157.
Kovacs, T. G. & G. Leduc, 1982. Acute toxicity of cyanide to rainbow trout (Salmo gairdneri) acclimated at different temperatures. Can. J. Fish. 39: 1426-1429.
Kägi, J. H. R. & A. Schäffer, 1988. Biochemistry of metallothionein. Biochemistry 27: 8509-8515.
Lagerspetz, K. Y. H., 1977. Interactions of season and temperature acclimation in the control of metabolism in Amphibia. J. therm. Biol. 2: 223-231.
Lagerspetz, K. Y. H. & K. Bowler, 1993. Variation in heat tolerance in individual Asellus aquaticus during thermal acclimation. J. therm. Biol. 18: 137-143.
Lindström-Seppä, P., 1985. Seasonal variation of the xenobiotic metabolizing enzyme activities in the liver of male and female vendace (Coregonus albula L.) Aquat. Toxicol. 6: 323-331.
Meyer, J. R. & J. P. Hegmann, 1971. Environmental modification of sciatic nerve conduction velocity in Rana pipiens. Amer. J. Physiol. 220: 1383-1387.
Newton, C., W. Pangborn, S. Nir & D. Papahadjopoulos, 1978. Specificity of Ca^{2+} and Mg^{2+} binding to phosphatidylserine vesicles and resultant phase changes of bilayer membrane structure. Biochim. Biophys. Acta 506: 281-287.
Paladino, F. V. & J. R. Spotila, 1978. Effect of arsenic on the thermal tolerance of newly hatched muskellunge fry (Esox masquinongy). J. therm. Biol. 3: 223-227.
Pasanen, S. & P. Koskela, 1974. Seasonal and age variation in the metabolism of the common frog, Rana temporaria L. in northern Finland. Comp. Biochem. Physiol. 47A: 635-654.
Payne, J. F., L. L. Fancey, A. D. Rahimtula & E. L. Porter, 1987. Review and perspective on the use of mixed-function oxygenase enzymes in biological monitoring. Comp. Biochem. Physiol. 86C: 233-245.
Sanders, B. M., 1993. Stress proteins in aquatic organisms: an environmental perspective. Crit. Rev. Toxicol. 23: 49-75.
Schwarzbaum, P. J., W. Wieser & H. Niederstätter, 1991. Contrasting effects of temperature acclimation on mechanisms of ionic regulation in a eurythermic and a stenothermic species of freshwater fish (Rutilus rutilus and Salvelinus alpinus). Comp. Biochem. Physiol. 98A: 483-489.
Schwarzbaum, P. J., W. Wieser & A. R. Cossins, 1992. Species-specific responses of membranes and the $Na^{++}K^+$ pump to temperature change in the kidney of two species of freshwater fish, roach (Rutilus rutilus) and Arctic char (Salvelinus alpinus). Physiol. Zool. 65: 17-34.
Smith, C. L., 1950. Seasonal changes in blood sugar, fat body and liver glycogen and gonad in the common frog, Rana temporaria. J. exp. Biol. 26: 412-429.
Theede, H., N. Scholz & H. Fischer, 1979. Temperature and salinity effects on the acute toxicity of cadmium to Laomedea loveni (Hydrozoa). Mar. Ecol. Prog. Ser. 1: 13-19.
Thompson, G. A. Jr., 1983. Mechanisms of homeoviscous adaptation in membranes. In: A.R. Cossins & P. Sheterline (eds.), Cellular Acclimatisation to Environmental Change. Pp. 33-53. Cambridge Univ. Press, Cambridge.
Veldhuizen-Tsoerkan, M. B., D. A. Holwerda, C. A. van der Mast & D. I. Zandee, 1990. Effects of cadmium exposure and heat shock on protein synthesis in gill tissue of the sea mussel Mytilus edulis L. Comp. Biochem. Physiol. 96C: 419-426.
Veldhuizen-Tsoerkan, M. B., D. A. Holwerda, C. A. van der Mast & D. I. Zandee, 1991. Synthesis of stress proteins under normal and heat shock conditions in gill tissue of sea mussels (Mytilus edulis) after chronic exposure to cadmium. Comp. Biochem. Physiol. 100C: 699-706.
Watenpaugh, D. E. & T. L. Beitinger, 1985. Se exposure and temperature tolerance of fathead minnows, Pimephales promelas. J. therm. Biol. 10: 83-86.

The role of natural organic material on the fate and toxicity of xenobiotics in aquatic environment

J. Kukkonen

University of Joensuu, Department of Biology, P.O. Box 111, FIN-80101 Joensuu, Finland

Keywords: Dissolved Organic Material, Humus, Organic Pollutants, Binding, Bioavailability

1. Introduction

Boreal inland waters have a high dissolved organic material (DOM) content. The pool of DOM in natural waters consists of various organic molecules. While some of these molecules have a defined chemical structure, most of the DOM in natural waters has no readily identifiable structure, and the members of this heterogeneous group of organic macromolecules are referred to as humic substances. Since Berzelius (1806, 1833; according to Thurman, 1985) first isolated DOM from natural water, noting that the organic acids present were similar to those isolated from soil, and Aschan (1908, 1932; according to Thurman, 1985) examined the colour-causing substances in six Finnish lakes and rivers, research on dissolved humic substances has been largely dominated by attempts to characterize these substances chemically (Waksman, 1936; Shapiro, 1957; Gjessing, 1976; Thurman, 1985). These studies have been influenced by research on organic matter in soil (Schnitzer & Khan 1978; Stevenson, 1982). Although there is clearly an evolution of structural models of humic substances due to the efforts of many scientists, as well as continuous development of more sensitive and accurate methods in chemistry (Hayes et al., 1989), no definite structure has been shown. However, some interesting hypothetic models of humic molecules have been developed (Wershaw, 1986; Ziechmann, 1988), and these models can also be used to assess the environmental effects of humic substances and other dissolved macromolecules.

DOM is an important environmental factor in natural waters. It affects the chemistry and ecology of aquatic habitats and, more importantly, the fate of organic pollutants. Visser (1964) reported that sodium humate lowers the surface tension of water considerably, and thus is capable of solubilizing otherwise insoluble organic xenobiotics. Wershaw et al., (1969) showed that soluble salts of humic acid increased DDT solubility twenty-fold. Since these studies, a number of investigations have demonstrated that DOM can bind many kinds of hydrophobic organic pollutants that load aquatic environments (Ogner & Schnitzer, 1970; Hassett & Anderson, 1979; Gjessing & Bergling, 1981; Carter & Suffet, 1982; Carlberg & Martinsen 1982; Landrum et al., 1984; Hassett & Millicic, 1985; McCarthy & Jimenez, 1985a; Sithole & Guy 1985; Chiou et al., 1986; Lara & Ernst, 1989; Lee & Farmer, 1989). The magnitude of the binding, expressed as a partition coefficient, K_{oc}, is linearly related to the hydrophobicity of the contaminant (McCarthy & Jimenez, 1985a; Chiou et al., 1986; Lara & Ernst, 1989).

Further, the adsorption of organic xenobiotics (like PCB congeners) on humic acid is high compared to adsorption of the same congeners onto inorganic particle surfaces. This has been explained as being due to a combination of the large surface area and the number of functional groups present in humic acids (Haque & Schmedding, 1976).

Several of the studies cited in this review used Aldrich humic acid preparation as a source of DOM, even though this acid has a greater affinity for hydrophobic xenobiotics than natural aquatic DOM does (Carter & Suffet, 1982; Landrum et al., 1985; McCarthy et al., 1989; Servos & Muir, 1989; Eadie et al., 1990; Jota & Hassett, 1991). Nor are commercially available humic materials fully characteristic of aquatic DOM; they are more like the less-polar, high-molecular weight soil humic acids, and the quality of the material varies from batch to batch (Malcolm & MacCarthy, 1986; Grasso et al., 1990). In spite of this, such studies do provide a useful reference, although caution should be used when data obtained with commercially available humic acids are extrapolated to natural aquatic environments.

2. Binding of xenobiotics to natural DOM

Several mechanisms have been postulated for the sorption of organic xenobiotics to DOM. Two or more mechanisms may occur simultaneously, depending on the chemical characteristics of the xenobiotic and the DOM (Choudhry, 1983). The mechanisms most likely involved are: van der Waals forces, hydrophobic bonding, hydrogen bonding, charge transfer, ion exchange and ligand exchange (Choudhry, 1983).

The search for the structure in humic substances has continued since the development of chemistry allowed progress to be made in characterizing organic material (see for example Hayes et al., 1989). No definitive structure has been shown, although several models have been developed. In order to explain the DOM-xenobiotic interactions and other chemical reactions of DOM on the molecular level, one possible approach is the structural model of humic substances proposed by Wershaw (1986, 1989). Humic substances are composed of plant- and animal-derived components which are partly degraded, or have been degraded and resynthesized by micro-organisms or chemical reactions. These subcomponents, which can be called monomers, are held together in ordered, membrane-like or micelle-like aggregated structures by weak interactions, such as hydrogen bonding, pi bonding, van der Waals interactions or hydrophobic forces. The micelles have hydrophobic interiors and hydrophillic exterior surfaces with a variety of functional groups. In fact, the model is basically analogous to the tertiary structure of proteins. Many of the important physical-chemical properties of DOM are merely functions of the structure of the micelle or membrane rather than of the individual monomers of the aggregate. The effects of increasing pH on the micelle may, for example, result in ionization of functional groups and loosening (or even breaking), of bonds, which can be seen in the smaller molecular size of the DOM.

Several methods have been introduced to determine the organic xenobiotics binding to DOM. These methods include: liquid-liquid extraction procedures (Gjessing & Bergling, 1981; Johnsen, 1987), equilibrium dialysis (Carter & Suffet, 1982), reverse-phase separation method (Landrum et al., 1984), ultrafiltration (Means & Wijayaratne, 1982), size exclusion chromatography (Hassett & Milicic, 1985), measurement of apparent water solubility enhancement by discreet solubility measurements (Chiou et al., 1987) or generation columns (Caron & Suffet, 1989), fluorescence quenching (Gauthier et al., 1986) and headspace methods (Yin & Hasset, 1986). All of these methods mentioned have limitations. For example, for benzo(a)pyrene, the differences

2.1 Effects of chemical composition of DOM on binding

In waters from different sources, the affinity of organic matter for binding a given xenobiotic compound appears to differ (Carter & Suffet, 1982; Landrum et al., 1984; Morehead et al., 1986; Kukkonen & Oikari, 1991). The underlying causes of this difference in the binding affinity of different waters for organic contaminants is not fully understood. This lack of understanding hampers most attempts to describe and predict the importance of natural organic matter in the transport and fate of organic pollutants in aquatic systems. One approach which elucidates the source of variability between natural waters is to examine relationships between the chemical and structural properties of DOM and its capacity to bind organic contaminants. It is possible to fractionate DOM, either chemically into more similar fractions or by molecular size, and subsequently determine whether there are underlying similarities in the binding affinities of functionally similar subcomponents of the total DOM. Non-ionic macroporous sorbents such as the Amberlite XAD resins have been used to fractionate DOM into subcomponents based on hydrophobicity and the charge of the molecules (Leenheer & Huffman, 1979). For example, Gauthier et al., (1987) reported that for pyrene, the binding coefficients in natural waters correlated with the aromaticity of DOM; and McCarthy et al., (1989) have related the association coefficient of benzo(a)pyrene (BaP) with DOM to aromaticity, molecular size and hydrophobic acid content of DOM.

DOM in one natural bog water (DOC = 49.9 mg l^{-1}, Hyde County, NC, USA) was fractionated by XAD-8 resin and the affinity of the fractions obtained was studied with selected model xenobiotics and compared to the filtered unfractionated water. The hydrophobic fractions of natural DOM in the bog water had the greatest affinity for binding BaP. The hydrophobic acids (HbA) fraction had a significantly higher organic carbon normalized partition coefficient (K_{oc}) than the hydrophobic neutrals (HbN) fraction. On the other hand, the hydrophillic (Hl) fraction had much lower affinity for binding BaP (Kukkonen et al., 1990). Amy and Liu (1990) have shown similar type of results for phenanthrene in the XAD-8 fractions of natural DOM. Further, the K_{oc} of xenobiotics to DOM in the total water appears to reflect the sum of the binding affinities of the individual fractions with little indication of interactive effects between the fractions (Kukkonen et al., 1990).

The relative binding affinities of the DOC fractions for 2,2',5,5'-, 3,3',4,4'-, 2,2',4,4'-tetrachlorobiphenyls and 2,2',4,4',5,5'-hexachlorobiphenyl exhibited a different pattern from that observed for BaP (Kukkonen et al., , 1990). In particular, the HbN fraction had the highest affinity for binding all PCBs. The different affinity for binding to isolated XAD-8 fractions of DOM may be due to differences in the electron densities of DOM fractions and model compounds. BaP is an electron-rich compound, and has a tendency to donate its electrons via charge transfer mechanism to electron deficient compounds, which HbA fraction could be. Conversely, PCB's are electrophiles and may be attracted to the richer electron densities of HbN material (Kukkonen et al., 1990).

Using a series of natural surface waters from Eastern Finland, the observed partition coefficients were related to some chemical parameters of DOM by Kukkonen and Oikari (1991). There was a strong positive correlation between the K_{oc} values of BaP and the portion of hydrophobic acids (HbA) in the natural waters and a strong negative correlation between the K_{oc} values of BaP and the hydrogen to carbon ratio of DOM. The absorptivity at 270 nm (ABS_{270}) being related to the aromaticity of DOM (Gauthier et al., 1987; Traina et al., 1990) gave the best correlation with the K_{oc} values for BaP. The dominating importance of HbA fraction on the binding hydrophobic xenobiotics, especially BaP, was documented. The BaP results agree with those reported by McCarthy et al., (1989), who showed a strong correlation between binding of BaP to DOM and ABS_{270} and HbA content of DOM in series of six ground and five surface water samples and Aldrich humic acid preparation. Also, Gauthier et al., (1987), revealed a good correlation between both aromatic content and ABS_{270} of humic materials extracted from soils and sediments, and the K_{oc} values of pyrene to these samples. Taken together, these studies suggest that, between different water sources, observed differences in affinities of BaP and other organic pollutants are, at least in part, explained by different proportions of hydrophobic acids and the aromaticity of DOM.

2.2 *Effects of molecular size of DOM on binding*

Besides the fractions obtained by using XAD-8 resin, the binding of BaP to different molecular weight fractions of DOM in the natural water from Hyde County, NC, was measured. This data strongly suggests the importance of high molecular weight or colloidal material in binding of BaP. When cutting off 10 percent of DOC (the molecules > ~100 kilodalton) the K_{oc} value decreases about 45 percent. Further, the binding capacity of molecules ~5 kilodalton and smaller, which constitutes 35 percent of DOC of the water sample, have very low K_{oc} values (Kukkonen, 1991a). This study is in accordance with Amy and Liu (1990) who fractionated natural DOM <5 kilodalton and >5 kilodalton molecular weight fractions and noticed that binding of phenanthrene was higher to >5 kilodalton fraction. Atrazine, diazinon and lindane are shown to have highest affinity to the high molecular weight (>300 kilodalton) humic acids (Saint-Fort & Visser, 1988). Also, in the studies where the DOM-xenobiotic interaction was evaluated by liquid chromatography (Whitehouse, 1985) or gel chromatography (Hassett & Anderson, 1979; Wijayaratne & Means, 1984), it was concluded that the high molecular weight DOM will bind most of the studied compounds. Similarly, Jota and Hassett (1991) fractionated Aldrich humic acid by ultrafiltration and found that it was mostly the high molecular weight fraction (>300 kilodalton) that binds 2,2',5,5'-tetrachlorobiphenyl. On the other hand, they pointed out that the small molecular weight fraction (<1 kilodalton) of Aldrich humic acid preparation also has a measurable partition coefficient value, as measured with a gas purging technique. This may cause some underestimation in K_{oc}'s measured by equilibrium dialysis, because of the molecular cut-off 1 kilodalton of the dialysis tubing (Jota & Hassett, 1991). However, the possible leaching of small molecular weight DOM cannot significantly affect the measured partitioning, because the binding capacity of this material is, although measurable, low (1.6×10^4) compared to the binding capacity of larger molecular weight fractions (1.04×10^5). Furthermore, the DOM concentration outside the dialysis membrane will always be very low (dilution from ~5 ml to 180 ml), and therefore the percentages of the DOM-bound xenobiotic outside the bag is low in any case. For example, if we assume a humic water sample (5 ml, DOC ~50 mgC l^{-1}), having molecules 10 per cent smaller than 1 kilodalton sealed in the dialysis tubing and immersed in 180 ml of clean water. If these small molecules distribute equally in the waters, we get DOC concentration 0.135 mgCl^{-1} for these molecules. Now, if these small molecules

2.3 Effects of pH on the binding

Besides high DOM concentration, boreal waters have often naturally low pH. This may have its own effect on the fate of xenobiotics. The chemical properties of humic material change as a function of pH. For example, the colour, apparent molecular size and molecular weight are reported to change as the pH changes (Gjessing, 1976; Chen & Schnitzer, 1976; De Haan et al., 1983; Thurman, 1985). The average size of humus molecules decreases with increasing pH. This can be explained as arising from the opening of hydrogen bonding or van der Waals interactions, but all explanations are based on the hypothetical molecular structure of humic substances. Based on surface pressure and viscosity measurements, Ghosh and Schnitzer (1980) suggested a model for the molecular conformation of humic substances. In their model, macromolecular conformations of humic and fulvic acids vary with the changes in the solution properties: a spherocolloidal structure exists when the sample concentrations are high or the pH is medium or low, or when appreciable amounts of neutral electrolytes are present.

The partition coefficient (K_{oc}) of BaP to DOM in one Finnish humic lake was 18×10^4, 14×10^4 and 8.4×10^4 at pH 3.5, 6.5 and 8.5, respectively. The K_{oc}'s of pentachlorophenol (PCP) in the same lake water were 9.8×10^3 and 1×10^3 at pH 3.5 and 6.5, respectively, and at pH 8.5 it was not possible measure any K_{oc} value with equilibrium dialysis method (Kukkonen, 1991b). Because BaP is a neutral compound, it can be assumed that the illustrated effect of pH on the K_{oc} value is due to effects on the structure or functional groups of DOM. With increase in pH, the degree of protonation of the DOM micelle carboxylic acid groups decreases and apparently, the polarity of the surface increases and it becomes less attractive to BaP molecules. In the case of PCP, the most important factor is the ionization of PCP itself with increasing pH (pK_a = 4.7). An ionized form of PCP is much more water soluble than the free acid compound and it may be assumed that the ionized form does not have interaction with natural DOM.

The conclusions regarding the inverse relationship between K_{oc} and pH described here for BaP is much like the results for DDT (Carter & Suffet, 1982), 2,2',5,5'-tetrachlorobiphenyl (Jota & Hassett, 1991), lindane [hexachlorocyclohexane] (Tramonti et al., 1986; Saint-Fort & Visser, 1988), tetracycline (Sithole & Guy, 1987), atrazine [2-chloro-4-ethylamino-6-isopropyl amino-s-triazine] (Haniff et al., 1985; Saint-Fort & Visser, 1988), methylene blue, diquat [1,1'-ethylen-2,2'-bipyridylium-cation], paraquat [1,1'-dimethyl-4,4'-bipyridyliumcation] (Best et al., 1972; Khan, 1973), benzidine, o-toluidine and azobenzene (Means & Wijayaratne, 1989). Traina et al., (1989) reported no pH effect on the binding of naphthalene to DOM, evidently because naphthalene does not have much interaction with DOM (Chiou et al., 1986; Kukkonen et al., 1990). For PCP, increased adsorption by the sediment, with decreasing pH, was observed in an aquatic microcosms study (Fisher, 1990).

Guy et al., (1980) and Narine and Guy (1982) reported quite opposite data for methylene blue, diquat and paraquat: binding increased with increasing pH. Similarly, the binding of napropamide [2-(α-naphthoxy)-N,N-diethylpropionamide] (Lee & Farmer, 1989), atrazine and linuron [3-(3,4-dichlorophenyl)-1-methoxy-1-methylurea] (Means & Wijayaratne, 1982; Zhang et al., 1990) is reported to increase with increasing pH. The later results may be explained by the ionization of the compound studied, because some of the compounds in the list form a cation in low pH and the interaction between DOM and an ionized compound can be weaker than the interaction between an unionized compound and 'ionized' DOM at higher pH. On the other hand, complexation of inorganic metal cations with dissolved humic material also increases with increasing pH and this is explained by the ionic interactions (O'Shea & Mancy, 1978; Saar & Weber, 1979; Dempsey & O'Melia, 1983).

In their study with four secondary amines, Sithole and Guy (1985) concluded that the ion exchange is the main adsorption mechanism between amines and humic acid. They explained the increasing binding to humic acid with an increase in pH by viewing humic acid as weak acid cation exchanger: at low pH protons compete for sites and at high ionic strength the sodium ions compete with the amine cations. This mechanism may work for weak organic bases, but it does not explain the results shown for neutral compounds or organic acids. For neutral compounds, like lindane and chloranil [2,3,5,6-tetrachloro-p-benzoquinone], it is suggested that the formation of electron donor-acceptor complexes between donor sites of fulvic acid with the lindane as acceptor (Tramonti et al., 1986; Melcer et al., 1989).

Taken altogether, the pH effects on the binding of organic contaminants to DOM has been studied widely but the obtained data is so scattered that it is difficult to draw any conclusion about the mechanisms involved. A possible explanation for this seemingly anomalous or even contradictory behaviour may be found in the common practise of using isolated aquatic DOM fractions or base-extracted humic materials from soils in the experiments. At present, it is not known how the humic material is affected by the extraction and cleanup procedures used in many studies. The structure of DOM is not definitively known and, for instance, the hypothetical micelle structure may be rather fragile and easily disrupted by the extractions or other chemical treatments. The best way to study the effects of aquatic DOM is to use the fresh sample as it is without any further treatment or extractions, except filtration to remove the particles. If any isolation or fractionation procedure is used, besides the obtained fractions, the unfractionated water samples should also be used in the experiments to see possible effects of the isolation procedures.

3. Bioavailability of xenobiotics in the presence of DOM

The bioavailability of organic xenobiotics in natural waters is decreased by DOM- (Leversee et al., 1983; Carlberg et al., 1986; Servos & Muir, 1989; Kukkonen et al., 1989), and the magnitude of the decrease is related to the extent of the binding between the contaminant and the organic matter (Landrum et al., 1985; McCarthy & Jimenez, 1985b; McCarthy et al., 1985; Black & McCarthy, 1988). Thus, the ability to predict the role of organic macromolecules in the accumulation and toxicity of hydrophobic organic contaminants in aquatic environments is dependent on, and limited by, the differences in the binding affinities of DOM in natural waters.

Accumulation of model compounds by *Daphnia magna* is reduced by increasing concentration of DOM. The median DOM concentration in Finnish lakes (TOC=12 mg l^{-1}, colour 100 mg Pt l^{-1}, Forsius et al., 1990) is high enough to significantly reduce the bioavailability of highly lipophilic compounds. The inverse relationship between water DOC concentration and bioconcentration factors (BCF) for xenobiotics in animals is a logarithmic type, both for series of natural waters from a boreal area (Kukkonen & Oikari, 1991) and for diluted DOM series of one humic water (Kukkonen et al., 1989, 1990) or fulvic acid preparation (Kukkonen et al., 1989). Logarithmic relationship between accumulation of xenobiotic and Aldrich humic acid concentrations is reported by McCarthy et al., (1985) and Day (1991). Accumulation of

1,3,6,8-tetrachlorodibenzo-*p*-dioxin expressed as uptake rate constant was also reduced by increasing DOC concentration (DOC ~3-30 mgC l^{-1}) in a series of lake and sediment interstitial waters (Servos & Muir, 1989).

As an exception, accumulation of naphthalene in natural waters having a low DOM concentration (< 4 mgC l^{-1}) revealed 2–3 times higher BCF values than in the DOM-free control water (Kukkonen & Oikari, 1991). Similar type of effect of DOM, but slighter, for naphthalene accumulation was detected by diluting a natural water sample down to 5 mgC l^{-1} (Kukkonen *et al.*, 1990). The increased bioavailability of methylcholanthrene to *D. magna* in Aldrich humic acid containing water is reported by Leversee *et al.*, (1983), but another study by McCarthy *et al.*, (1985) showed opposite results for this compound.

3.1 *Effects of chemical composition of DOM on the accumulation of xenobiotics*

The effect of different XAD-8 fractions on the bioavailability of xenobiotics has also been studied. Similar to binding capacities, the HbA fraction had the strongest effect on the bioavailability of BaP (Kukkonen *et al.*, 1990). The HbN and Hl fractions also reduced the bioaccumulation of BaP compared to the control, but the effect was not as large as for the HbA fraction. Approximately five times more organic carbon from the Hl fraction was required to reduce the BCF to the same extent as observed for the HbA fraction or the DOM in total water. The capacity of the HbN fraction was intermediate to the two other fractions (Kukkonen *et al.*, 1990).

The BCF values for model xenobiotics in a series of natural waters correlated with the percentage of hydrophobic acids, absorptivity at 270 nm and hydrogen to carbon ratio (Kukkonen & Oikari, 1991). This result indicates that, in addition to the total DOM concentration, the quality of DOM also affects the bioavailability of organic xenobiotics. The K_{oc}'s of naphthalene and dehydroabietic acid (DHAA) are too low to be measured by the equilibrium dialysis technique (Kukkonen & Oikari, 1991). Nevertheless, for lipophilic xenobiotics, the bioaccumulation of naphthalene and DHAA in this series of surface waters also appeared to correlate with chemical parameters, such as aromaticity, of DOM (Kukkonen & Oikari, 1991).

3.2 *Effects of pH on the accumulation of organic xenobiotics in the presence of DOM*

The pH of the water strongly affects the toxicity and accumulation of organic weak acids, such as pentachlorophenol (PCP) and dehydroabietic acid (DHAA) (McLeay *et al.*, 1979a, 1979b; Saarikoski & Viluksela, 1981; Spehar *et al.*, 1985; Fisher & Wadleigh, 1986; Saarikoski *et al.*, 1986; Stehly & Hayton, 1990; Fisher, 1991). The pH also affects the toxicity and accumulation of neutral compounds, although this effect is not as obvious as in the case of acids (Fisher, 1985; Fisher & Lohner, 1986).

The accumulation of PCP and DHAA by *D. magna* and *Heptagenia fuscogrisea* was clearly pH dependent both in DOM-free control water and in natural water containing aquatic DOM (Kukkonen, 1991b). DOM reduced the accumulation of PCP into *H. fuscogrisea* significantly at the pH range from 4.5 to 7.5. However, at pH 3.5, the difference between humus and control water was not significant, although a fifteen percent lower BCF value was noted in humic water than in control water. At pH 8.5 there were no differences between humic and control treatments. This accumulation data agrees with the K_{oc} values and also with the percentages of the unionized PCP at different pHs showing that the unionized form of PCP both accumulates better in animals and binds better to the humic material than the ionized form. This is also in line with the literature stating that the unionized form of PCP is more available and environmentally toxic than ionized form (Saarikoski & Viluksela, 1981, 1982; Spehar et al., 1985; Fisher & Wadleigh, 1986; Fisher, 1990).

4. Effects of DOM on the acute toxicity of xenobiotics

Humic substances have been shown to be toxic to *D. pulex* at high DOC concentrations at pH 5 (Petersen & Persson, 1987). At pH 7 there was no difference between humus waters and reference. The result was explained in that DOM contains a small fraction of phenolic or tannin type of compounds, which may become bioavailable to organisms in acidic conditions.

By binding the xenobiotics, DOM affects the physiochemical state of these compounds. In most papers reviewed in this presentation, DOM decreases the bioavailability of xenobiotics. This leads to the assumption that the acute toxicity of xenobiotics may decrease too. In some studies, the presence of DOM has lowered the acute toxicity of synthetic pyrethroids (Day, 1991), cationic polymers (Cary et al., 1987; Goodrich et al., 1991), anthracene (Oris et al., 1990), diazinon, 4-chloroanilin and 4-nitrophenol (Steinberg et al., 1992; Lee et al., 1993) to aquatic organisms and the acute toxicity of fenvalerate, permethrin, azinophosmethyl, chlorpyrifos and carbofuran to bacteria (Benson & Long, 1991; Ortego & Benson, 1992). On the other hand, DOM does not have any effect on the acute toxicity of tetrachloroguaiacol, tetrachlorocymene, 1,2,4trichlorobenzene, methylparathion, diazinon, o-toluidine and 2,6-dimethylaniline (Kukkonen & Oikari, 1987; Oikari et al., 1992; Steinberg et al., 1992) or, DOM can decrease the acute toxicity of diazinon and 4-chloroanilin to D. magna and have no effect on the toxicity of these compounds to fish (Lee et al., 1993). However, acute toxicity may also be enhanced in the presence of DOM. This has been shown, for example, with fenvalerate, lindane, dehydroabietic acid and with chlorinated phenols and anilines (Stewart, 1984; Kukkonen & Oikari, 1987; Oikari et al., 1992; Steinberg et al., 1992).

The enhanced toxicity of several compounds in the presence of DOM is an interesting phenomenon which needs more research. Currently, changes in the toxicity of the various tested chemicals, caused by DOM, are without firm theoretical background. Some possible explanations exist: First, DOM molecules may interact with cell membranes of the test organisms and thereby alter the accumulation rate of the compound. Second, DOM mediates chemical (photo induced?) changes in the structure of the xenobiotics, resulting in daughter products that are more toxic to the organisms than the original compounds. Third, addition of a new chemical in humic water displaces others, *e.g.* heavy metals, which naturally associate with DOM. (Presented results, however, do not provide strong support for this idea). Fourth, natural waters tested containing DOM may differ from the 'standard' water used, by some physico-chemical characteristics, *e.g.* water hardness or concentrations of some essential ions, so that these unknown factors modify the toxicity of the compounds but not the DOM.

5. Environmental implications

DOM in fresh waters binds hydrophobic organic xenobiotics. The quantity of binding can be positively related both to the hydrophobicity or aromaticity of DOM and lipophilicity (K_{ow}) of the xenobiotic chemical. The charge on the surface of a DOM micelle, a functional unit of aquatic DOM, may be modified by environmental pH. As the water pH is lowered, the charge density decreases, leading to a higher hydrophobicity of the micelle. Therefore, the hydrophobic neutral xenobiotics will be more likely to associate with micelles in acidic environments. Also, with decreasing pH, the associations between DOM and weak organic acids will increase more steeply than that of neutral compounds because of the effects of pH on both DOM and the pollutant. This aspect can be of paramount importance when evaluating the possible effects of the acidification of aquatic and soil systems on the fate and transport of potentially hazardous organic contaminants.

The observed bioaccumulation of model compounds in waters containing DOM can be compared to the predicted BCF values. This prediction is based on the assumption that xenobiotics bound to DOM are unavailable for uptake, mainly because the xenobiotic-DOM complex may be too large to penetrate the biomembranes. Accordingly, bioaccumulation in water containing DOM is assumed to be proportional to the fraction of the contaminant that is freely dissolved. The present data on the effects of natural DOM suggests that this assumption is valid in short term exposures. Reviewed data supports the conclusion that little or none of the contaminant associated with DOM is available to animals, like *D. magna*, in the short-term (<48 h) experiments (McCarthy et al., 1985; Kukkonen et al., 1989, 1990; Kukkonen & Oikari, 1991). This is in accordance with the results obtained with *Pontoporeia hoyi* (Landrum et al., 1985) and rainbow trout (Black & McCarthy, 1988) for BaP as well as on rainbow trout for 2,2',5,5'-tetrachlorobiphenyl accumulation (Black & McCarthy, 1988). This kind of effect was also reported for DOM on the bioavailability of 1,3,6,8-tetrachlorodibenzo-*p*-dioxin to *Crangonyx laurentianus* (Amphipoda) (Servos & Muir, 1989).

The binding of BaP or 2,3,7,8-tetrachlorodibenzo-*p*-dioxin to DOM is shown to be reversible to a large extent (McCarthy & Jimenez 1985a; Kukkonen, 1992). The fact that a xenobiotic molecule, once bound, can be freed again if the concentration of free molecules in the water column decreases for any reason, must have important ecotoxicological consequences in aquatic ecosystems. In short term exposures in the laboratory, or maybe also close to the point sources of xenobiotics, natural DOM effectively reduces the bioavailability of organic xenobiotics, possibly leading, for instance, to reduced chronic toxicity. On the other hand, the reversibility of the binding of xenobiotics to macromolecules may cause the biota, on larger areas of lake system than expected, to be exposed to low levels of xenobiotics, which would, in the absence of carrier macromolecules, have only limited spatial distribution.

Finally, the plausibility of strong binding of lipophilic xenobiotic chemicals to DOM may also have important analytical consequences when trying to analyze the trace amounts of lipophilic xenobiotics in natural water samples containing DOM. The normal liquid-liquid extraction with organic solvent without any other treatments (hydrolysis, etc.) do not necessarily extract out all the compounds, but only the free fraction. There is a need to measure the mass balance in this kind of system, to be sure that the total amount of compound is extracted and analyzed. To a certain extent, this can be assessed with radiolabelled tracers. On the other hand, addition of an internal standard to water samples may also minimize this potential source of inaccuracy.

6. Summary

Interaction of hydrophobic contaminants with natural DOM plays a major role in contaminant distribution and bioavailability in the aquatic environment. For non-polar organic compounds, the binding affinity generally relates directly to the hydrophobicity of the contaminant. This is measured as the octanol-water partition coefficient or the inverse of its aqueous solubility and the organic content of a particulate sorbent such as sediment particles. However, the hydrophobicity of xenobiotics is not the only factor affecting the binding.

DOM fractions obtained by XAD-8 resin have different affinities for PAHs and PCBs. The different DOM fractions also reduce the uptake and accumulation of model compounds by organisms in proportion to their binding capacity, thus showing that the bound xenobiotic is not bioavailable to the organisms in short term exposures. Reviewed results show that both the quantity and the quality of DOM in natural inland waters affect the bioavailability of organic xenobiotics.

An important phenomenon related to the environmental transport and fate of xenobiotics is the reversibility of the binding between DOM and xenobiotics. This means that the biota may be exposed to low levels of xenobiotics in larger areas than expected, because without carrier macromolecules xenobiotics might have only a limited spatial distribution.

The effects of DOM on the toxicity of less hydrophobic, and in many cases, polar compounds, are not well known. However, DOM can both increase and decrease the toxicity of some metals and organic xenobiotics when compared to DOM-free water. Currently, there exists no firm theoretical background for the phenomena of enhanced toxicity. More research on mechanisms involved are required to understand the system and to get this information usable for regulatory purposes.

References

Amy, G. L. & H. Liu, 1990. PAH binding to natural organic matter (NOM): A comparison of NOM fractions and analytical methods. Presented before the Division of Environmental Chemistry, 199th American Chemical Society National Meeting, Boston 22–27, 1990, Preprints Extended Abstracts 30(1): 243–246.
Aschan, O., 1908. Soluble humus material of Northern fresh waters. Journal fuer Praktische Chemie, 77: 172–189.
Aschan, O., 1932. Om vattenhumus och dess medverkan vid sjomalmsbildningen. Arkiv Kemi, Mineralogie. Geologie, 15: 1–143.
Benson, W. H. & S. F. Long, 1991. Evaluation of humic-pesticide interactions on the acute toxicity of selected organophosphate and carbamate insecticides. Ecotoxicol. Environ. Safety, 21: 301–307.
Best, J. A., J. B. Weber & S. B. Weed, 1972. Competitive adsorption of diaquat, paraquat and Ca^{2+} on organic matter and exchange resins. Soil Sci., 114: 444–450.
Black, M. C. & J. F. McCarthy, 1988. Dissolved organic macromolecules reduce the uptake of hydrophobic organic contaminants by the gills of rainbow trout (*Salmo gairdneri*). Environ. Toxicol. Chem., 7: 593–600.
Carlberg, G. E. & K. Martinsen, 1982. Adsorption/complexation of organic micropollutants to aquatic humus. Sci. Total Environ., 25: 245–254.
Carlberg, G. E., K. Martinsen, A. Kringstad, E. T. Gjessing, M. Grande, T. Källqvist & J. U. Skåre, 1986. Influence of aquatic humus on the bioavailability of chlorinated micropollutants in atlantic salmon. Arch. Environ. Contam. Toxicol., 15: 543–548.
Carter, C. W. & I. H. Suffet, 1982. Binding of DDT to dissolved humic materials. Environ. Sci. Technol., 16: 735–740.
Caron, G. & I. H. Suffet, 1989. Binding of nonpolar pollutants to dissolved organic carbon: Environmental fate modelling. In: I. H. Suffet & P. MacCarthy (eds.), *Aquatic Humic Substances: Influence on Fate and Treatment of Pollutants*. Advances in Chemistry Series 219: 117–130. American Chemical Society. Washington, DC.
Cary, G. A., J. A. McMahon & W. J. Kuc, 1987. The effect of suspended solids and naturally occurring dissolved organics in reducing the acute toxicities of cationic polyelectrolytes to aquatic organisms. Environ. Toxicol. Chem., 6: 469–474.
Chen, Y. & M. Schnitzer, 1976. Viscosity measurements on soil humic substances. Soil Sci. Soc. Am. J., 40: 866–872.
Chiou, C. T., R. L. Malcolm, T. I. Brinton & D. E. Kile, 1986. Water solubility enhancement of some organic pollutants and pesticides by dissolved humic and fulvic acids. Environ. Sci. Technol., 20: 502–508.

Chiou, C. T., D. E. Kile, T. I. Brinton, R. L. Malcolm, J. A. Leenheer & P. MacCarthy, 1987. A comparison of water solubility enhancements of organic solutes by aquatic humic materials and commercial humic acids. Environ. Sci. Technol., 21: 1231–1234.

Choudhry, G. G., 1983. Humic substances. Part III: Sorptive interactions with environmental chemicals. Toxicol. Environ. Chem., 6: 127–171.

Day, K. E., 1991. Effects of dissolved organic carbon on accumulation and acute toxicity of fenvalerate, deltamethrin and cyhalothrin to Daphnia magna (Straus). Environ. Toxicol. Chem., 10: 91–101.

De Haan, H., G. Werlemark & T. De Boer, 1983. Effect of pH on molecular weight and size of fulvic acids in drainage water from peaty grassland in NW Netherlands. Plant and Soil, 75: 63–73.

Dempsey, B. A. & C. R. O'Melia, 1983. Proton and calcium complexation of four fulvic acid fractions. In: R. F. Christman & E. T. Gjessing (eds.), *Aquatic and Terrestrial Humic Substances.* pp. 239–273. Ann Arbor Science, Ann Arbor, MI.

Eadie, B. J., N. R. Morehead & P. F. Landrum, 1990. Three-phase partitioning of hydrophobic organic compounds in great lakes waters. Chemosphere, 20: 161–178.

Fisher, S. W., 1985. Effects of pH on the toxicity and uptake of [^{14}C]lindane in the midge, *Chironomus riparius*. Ecotoxicol. Environ. Safety, 10: 202–208.

Fisher, S. W., 1990. The pH dependent accumulation of PCP in aquatic microcosms with sediment. Aquat. Toxicol., 18: 199–218.

Fisher, S. W., 1991. Changes in the toxicity of three pesticides as a function of environmental pH and temperature. Bull. Environ. Contam. Toxicol., 46: 197–202.

Fisher, S. W. & T. W. Lohner, 1986. Changes in the aqueous behaviour of parathionunder varying conditions of pH. Arch. Environ. Contam. Toxicol., 16: 79–84.

Fisher, S. W. & R. W. Wadleigh, 1986. Effects of pH on the acute toxicity and uptake of [^{14}C]pentachlorophenol in the midge, *Chironomus riparius*. Ecotoxicol. Environ. Safety, 11: 1–8.

Forsius, M., J. Kämäri, P. Kortelainen, J. Mannio, M. Verta & K. Kinnunen, 1990. Statistical lake survey in Finland: Regional estimates of lake acidification. In: P. Kauppi, P. Anttila & K. Kenttämies (eds.), *Acidification in Finland.* pp. 759–780. Springer-Verlag. Berlin.

Gauthier, T. D., E. C. Shane, W. F. Guerin, W. R. Seitz & C. L. Grant, 1986. Fluorescence quenching method for determining equilibrium constants for polycyclic aromatic hydrocarbon binding to dissolved humic materials. Environ. Sci. Technol., 20: 1162–1166.

Gauthier, T. D., W. R. Seitz & C. L. Grant, 1987. Effects of structural and compositional variations of dissolved humic materials on pyrene K_{oc} values. Environ. Sci. Technol., 21: 243–248.

Ghosh, K. & M. Schnitzer, 1980. Macromolecular structures of humic substances. Soil Sci., 129: 266–276.

Gjessing, E. T., 1976. *Physical and Chemical Characteristics of Aquatic Humus.* Ann Arbor Science Publishers Inc., Ann Arbor, MI.

Gjessing, E. T. & L. Bergling, 1981. Adsorption of PAH to aquatic humus. Arch. Hydrobiol., 92: 24–30.

Goodrich, M. S., L. H. Dulak, M. A. Friedman & J. J. Lech, 1991. Acute and long-term toxicity of water-soluble cationic polymers to rainbow trout (*Oncorhynchus mykiss*) and the modification of toxicity by humic acid. Environ. Toxicol. Chem., 10: 509–515.

Grasso, D., Y.-P. Chin & W. J. Weber, 1990. Structural and behavioral characteristics of a commercial humic acid and natural dissolved aquatic organic matter. Chemosphere, 21: 1181–1197.

Guy, R. D., D. R. Narine & S. DeSilva, 1980. Organocation speciation. I. A comparison of the interactions of methylene blue and paraquat with bentonite and humic acid. Can. J. Chem., 58: 547–554.

Haniff, M. I., R. H. Zienius, C. H. Langford & D. S. Gamble, 1985. The solution phase complexing of atrazine by fulvic acid: Equilibria at 25°C. J. Environ. Sci. Health, B20: 215–262.

Haque, R. & D. W. Schmedding, 1976. Studies on the adsorption of selected polychlorinated biphenyl isomers on several surfaces. J. Environ. Sci. Health, B11: 129–137.

Hassett, J. P. & M. A. Anderson, 1979. Association of hydrophobic organic compounds with dissolved organic matter in aquatic systems. Environ. Sci. Technol., 13: 1526–1529.

Hassett, J. P. & E. Milicic, 1985. Determination of equilibrium and rate constants for binding of polychlorinated biphenyl congener by dissolved humic substances. Environ. Sci. Technol., 19: 638–643.

Hayes, M. H. B., P. MacCarthy, R. L. Malcolm & R. S. Swift (eds.), 1989. *Humic Substances II. In Search of Structure.* John Wiley and Sons, Guildford, Great Britain.

Johnsen, S., 1987. Interactions between polycyclic aromatic hydrocarbons and natural aquatic humic substances. Sci. Total Environ., 67: 269–278.

Jota, M. A. & J. P. Hassett, 1991. Effects of environmental variables on binding of a PCB congener by dissolved humic substances. Environ. Toxicol. Chem., 10: 483–491.

Khan, S. U., 1973. Interaction of humic substances with bipyridylium herbicides. Can. J. Soil. Sci., 53: 199–204.

Kukkonen, J., 1991a. Effects of dissolved organic material in fresh waters on the binding and bioavailability of organic pollutants. PhD Thesis. University of Joensuu, Joensuu, Finland. 39 pp.

Kukkonen, J., 1991b. Effects of pH and natural humic substances on the accumulation of organic pollutants into two freshwater invertebrates. In: B. Allard, H. Borén & A. Grimvall (eds.), *Humic Substances in the*

Aquatic and Terrestrial Environment. Proceedings of an International Symposium, Linköping, Sweden, August 21–23, 1989. Lecture Notes in Earth Sciences 33: 413–422.

Kukkonen, J., 1992. Effects of lignin and chlorolignin in pulp mill effluents on the binding and bioavailability of hydrophobic organic pollutants. Water Res., 26: 1523–1532.

Kukkonen, J. & A. Oikari, 1987. Effects of aquatic humus on accumulation and acute toxicity of some organic micropollutants. Sci. Total Environ., 62: 399–402.

Kukkonen, J. & A. Oikari, 1991. Bioavailability of organic pollutants in boreal waters with varying levels of dissolved organic material. Water Res., 25: 455–463.

Kukkonen, J. & J. Pellinen, 1994. Binding of organic xenobiotics to dissolved organic macromolecules: comparison of analytical methods. Sci. Total Environ., 152: 19–29.

Kukkonen, J., A. Oikari, S. Johnsen & E. Gjessing, 1989. Effects of humus concentrations on benzo(a)pyrene accumulation from water to *Daphnia magna*: Comparison of natural waters and standard preparations. Sci. Total Environ., 79: 197–207.

Kukkonen, J., J. F. McCarthy & A. Oikari, 1990. Effects of XAD-8 fractions of dissolved organic carbon on the sorption and bioavailability of organic micropollutants. Arch. Environ. Contam. Toxicol., 19: 551–557.

Landrum, P. F., S. R. Nihart, B. J. Eadie & W. S. Gardner, 1984. Reverse phase separation method for determining pollutant binding to Aldrich humic acid and dissolved organic carbon of natural waters. Environ. Sci. Technol., 18: 187–192.

Landrum, P. F., M. D. Reinhold, S. R. Nihart & B. J. Eadie, 1985. Predicting the bioavailability of organic xenobiotics to *Pontoporeia hoyi* in the presence of humic and fulvic materials and natural dissolved organic matter. Environ. Toxicol. Chem., 4: 459–467.

Lara, R. & W. Ernst, 1989. Interaction between polychlorinated biphenyls and marine humic substances. Determination of association coefficients. Chemosphere, 19: 1655–1664.

Lee, D.-Y. & W. J. Farmer, 1989. Dissolved organic matter interaction with napropamide and four other nonionic pesticides. J. Environ. Quality, 18: 468–474.

Lee, S. K., D. Freitag, C. Steinberg, A. Kettrup & Y. H. Kim, 1993. Effects of dissolved humic materials on acute toxicity of some organic chemicals to aquatic organisms. Water Res., 27: 199–204.

Leenheer, J. A. & E. W. D. Huffman, 1979. Analytical method for dissolved organic carbon fractionation. Water-Resour. Invest. U.S. Geol. Sur. 79–4: 1–16.

Leversee, G. J., P. F. Landrum, J. P. Giesy & T. Fannin, 1983. Humic acids reduce bioaccumulation of some polycyclic aromatic hydrocarbons. Can. J. Fish. Aquat. Sci., 40(Suppl. 2): 63–69.

Malcolm, R. L. & P. MacCarthy, 1986. Limitations in the use of commercial humic acids in water and soil research. Environ. Sci. Technol., 20: 904–911.

McCarthy, J. F. & B. D. Jimenez, 1985a. Interactions between polycyclic aromatic hydrocarbons and dissolved humic material: binding and dissociation. Environ. Sci. Technol., 19: 1072–1076.

McCarthy, J. F. & B. D. Jimenez, 1985b. Reduction in bioavailability to bluegills of polycyclic aromatic hydrocarbons bound to dissolved humic material. Environ. Toxicol. Chem., 4: 511–521.

McCarthy, J. F., B. D. Jimenez & T. Barbee, 1985. Effect of dissolved humic material on accumulation of polycyclic aromatic hydrocarbons: structure-activity relationship. Aquat. Toxicol., 7: 15–24.

McCarthy, J. F., L. E. Roberson & L. W. Burris, 1989. Association of benzo(a)pyrene with dissolved organic matter: prediction of K_{dom} from structural and chemical properties of the organic matter. Chemosphere, 19: 1911–1920.

McLeay, D. J., C. C. Walden & J. R. Munro, 1979a. Influence of dilution water on the toxicity of kraft pulp and paper mill effluent, including mechanisms of effect. Water Res., 13: 151–158.

McLeay, D. J., C. C. Walden & J. R. Munro, 1979b. Effect of pH on toxicity of kraft pulp and paper mill effluent to salmonid fish in fresh and seawater. Water Res., 13: 249–254.

Means, J. C. & R. Wijayaratne, 1982. Role of natural colloids in the transport of hydrophobic pollutants. Science, 215: 968–970.

Means, J. C. & R. D. Wijayaratne, 1989. Sorption of benzidine, toluidine and azobenzene on colloidal organic matter. In: I.H. Suffet & P. MacCarthy (eds.), *Aquatic Humic Substances: Influence on Fate and Treatment of Pollutants.* Advances in Chemistry Series 219: 209–222. American Chemical Society. Washington, DC.

Melcer, M. E., M. S. Zalewski, J. P. Hassett & M. A. Brisk, 1989. Charge-transfer interaction between dissolved humic materials and chloranil. In: I. H. Suffet & P. MacCarthy (eds.), *Aquatic Humic Substances: Influence on Fate and Treatment of Pollutants.* Advances in Chemistry Series 219: 173–183. American Chemical Society. Washington, DC.

Morehead, N. R., B. J. Eadie, B. Lake, P. F. Landrum & D. Berner, 1986. The sorption of PAH onto dissolved organic matter in Lake Michigan waters. Chemosphere, 15: 403–412.

Narine, D. R. & R. D. Guy, 1982. Binding of diquat and paraquat to humic acid in aquatic environments. Soil. Sci., 133: 356–363.

Ogner, G. & M. Schnitzer, 1970. Humic substances: fulvic acid-dialkyl phthalate complexes and their role in pollution. Science, 170: 317–318.

Oikari, A., J. Kukkonen & V. Virtanen, 1992. Acute toxicity of chemicals to *Daphnia magna* in humic waters. Sci. Total Environ., 117/118: 367–377.

Oris, J. T., A. T. Hall & J. D. Tylka, 1990. Humic acids reduce the photo-induced toxicity of anthracene to fish and daphnia. Environ. Toxicol. Chem., 9: 575–583.
Ortego, L. S. & W. H. Benson, 1992. Effects of dissolved humic material on the toxicity of selected pyrethroid insecticides. Environ. Toxicol. Chem., 11: 261–265.
O'Shea, T. A. & K. H. Mancy, 1978. The effect of pH and hardness metal ions on the competitive interaction between trace metal ions and inorganic and organic complexing agents found in natural waters. Water Res., 12: 703–711.
Petersen, R. C. Jr. & U. Persson, 1987. Comparison of the biological effects of humic materials under acidified conditions. Sci. Total Environ., 62: 387–398.
Saar, R. A. & J. H. Weber, 1979. Complexation of cadmium(II) with water- and soil-derived fulvic acids: effect of pH and fulvic acid concentration. Can. J. Chem., 57: 1263–1268.
Saarikoski, J. & M. Viluksela, 1981. Influence of pH on the toxicity of substituted phenols to fish. Arch. Environ. Contam. Toxicol., 10: 747–753.
Saarikoski, J. & M. Viluksela, 1982. Relation between physicochemical properties of phenols and their toxicity and accumulation in fish. Ecotoxicol. Environ. Safety, 6: 501–512.
Saarikoski, J., R. Lindström, M. Tyynelä & M. Viluksela, 1986. Factors affecting the absorption of phenolics and carboxylic acids in the guppy (*Poecilia reticulata*). Ecotoxicol. Environ. Safety, 11: 158–173.
Saint-Fort, R. & S. A. Visser, 1988. Study of the interactions between atrazine, diazon and lindane with humic acids of various molecular weights. J. Environ. Sci. Health, A23: 613–624.
Schnitzer, M. & S. U. Khan, 1978. *Soil Organic Matter.* Elsevier, Amsterdam.
Servos, M. R. & D. C. G. Muir, 1989. The effects of dissolved organic matter from the Canadian Shield lakes on the bioavailability of 1,3,6,8-tetrachlorodibenzo-*p*-dioxin to the amphipod *Crangonyx laurentianus*. Environ. Toxicol. Chem., 8: 141–150.
Shapiro, J., 1957. Chemical and biological studies on the yellow organic acids of lake water. Limnol. Oceanogr., 2: 161–179.
Sithole, B. B. & R. D. Guy, 1985. Interactions of secondary amines with bentonite clay and humic materials in dilute aqueous systems. Environment International, 11: 499–504.
Sithole, B. B. & R. D. Guy, 1987. Models for tetracycline in aquatic environments. II. Interaction with humic substances. Water Air Soil Pollut., 32: 315–321.
Spehar, R. L., H. P., Nelson, M. J. Swanson & J. W. Renoos, 1985. Pentachlorophenol toxicity to amphipods and fathead minnows at different test pH values. Environ. Toxicol. Chem., 4: 389–397.
Steinberg, C. E. W., A. Sturm, J. Kelbel, S. K. Lee, N. Hertkorn, D. Freitag & A. A. Kettrup, 1992. Changes of acute toxicity of organic chemicals to *Daphnia magna* in the presence of dissolved humic material (DHM). Acta hydrochim. hydrobiol., 20: 326–332.
Stehly, G. R. & W. L. Hayton, 1990. Effect of pH on the accumulation kinetics of pentachlorophenol in goldfish. Arch. Environ. Contam. Toxicol., 19: 464–470.
Stevenson, F. J., 1982. *Humus Chemistry.* John Wiley and Sons, New York.
Stewart, A. J., 1984. Interactions between dissolved humic materials and organic toxicants. In: K.E. Cowser (ed.), *Synthetic Fossil Fuel Technologies. Results of Health and Environmental Studies.* pp. 505–521. Butterworths, Boston.
Thurman, E. M., 1985. *Organic Geochemistry of Natural Waters.* Martinus Nijhoff, Dordrecht.
Traina, S. J., D. A. Spontak & T. J. Logan, 1989. Effects of cations on complexation of naphthalene by water-soluble organic carbon. J. Environ. Quality, 18: 221–227.
Traina, S. J., J. Novak & N. E. Smeck, 1990. An ultraviolet absorbance method of estimating the percent aromatic carbon content of humic acids. J. Environ. Quality, 19: 151–153.
Tramonti, V., R. H. Zienius & D. S. Gamble, 1986. Solution phase interaction of lindane with fulvic acid: Effect of solution pH and ionic strength. Intern. J. Environ. Anal. Chem., 24: 203–212.
Visser, S. A., 1964. Oxidation reduction potentials and capillary activities of humic acids. Nature, 204: 581.
Waksman, S. A., 1936. *Humus. Origin, Chemical Composition and Importance in Nature.* Williams and Wilkins, Baltimore.
Wershaw, R. L., 1986. A new model for humic materials and their interactions with hydrophobic organic chemicals in soil-water or sediment-water systems. J. Contam. Hydrol., 1: 29–45.
Wershaw, R. L., 1989. Application of a membrane model to the sorptive interactions of humic substances. Environ. Health Perspect., 83: 191–203.
Wershaw, R. L., P. J. Burcar & M. C. Goldberg, 1969. Interactions of pesticides with natural organic matter. Environ. Sci. Technol., 3: 271–273.
Whitehouse, B., 1985. The effects of dissolved organic matter on the aqueous partitioning of polynuclear aromatic hydrocarbons. Estuar. Coast. Shelf Sci., 20: 393–402.
Wijayaratne, R. D. & J. C. Means, 1984. Affinity of hydrophobic pollutants for natural estuarine colloids in aquatic environments. Environ. Sci. Technol., 18: 121–123.
Yin, C. & J. P. Hasset, 1986. Gas partitioning approach for laboratory and field studies of mirex fugacity in water. Environ. Sci. Technol., 20: 1213–1217.

Zhang, H., C. E. Clapp, U. Mingelgrin, W. C. Koskinen & R. H. Dowdy, 1990. Complexation of atrazine by humic acid. Presented at Annual Meetings, Am. Soc. Agron., October 22, 1990, San Antonio, TX. Agron. Abstr. 1990: 261.

Ziechmann, W., 1988. Evolution of structural models from consideration of physical and chemical properties. In: F.H. Frimmel & R.F. Christman (eds.), *Humic Substances and Their Role in the Environment.* Dahlem Workshop Reports. Life Science Research Report 41: 113–132. John Wiley & Sons. New York.

Degradation of the halogenated organic fraction of biologically treated bleached kraft pulp mill effluents in Finnish lake water mesocosms

E. K. Saski,[1] K. Salonen,[2] A. Vähätalo[2] & M. Salkinoja-Salonen[1]

[1]*Department of Applied Chemistry and Microbiology, P.O. Box 27, FIN-00014 University of Helsinki, Finland*
[2]*Lammi Biological Station, University of Helsinki, FIN-16900 Lammi, Finland*

Keywords: waste water, mesocosm, organic halogen, molecular weight

1. Introduction

Activated sludge systems and aerated lagoons are used by the pulp industry to diminish the organic load of bleached pulp mill waste waters (Häggblom & Salkinoja-Salonen, 1991; Jokela *et al.*, 1993). Owing to widely applied secondary treatment, the residual discharge of the Finnish mills in 1992 was ca. 330 000 metric tons of chemical oxygen demand (COD_{Cr}) and 4900 metric tons of adsorbable organic halogen (AOX).

The fate of the organohalogens and other organic compounds contained by bleached kraft pulp mill effluents (BKME) has been actively investigated in waste water treatment plants (Zender *et al.*, 1994; Stuthridge & McFarlane, 1994; Jokela *et al.*, 1993). Much less is known on the subsequent events after the biologically purified waste water entered the recipient lake.

The aim of our study was to assess the post-treatment degradability of organic carbon and halogen of BKME in the freshwater environment. Laboratory microcosms and *in situ* mesocosms were used as the experimental tools. Waste waters were exposed to natural daylight, temperature and lake water biota in mesocosms to simulate the pelagic lake ecosystem. Both quantitative (AOX) and qualitative (molecular weight distribution) changes in the BKME halogens were measured during aging in natural Finnish lake waters over a year. The potential of mesocosm sediments, generated in course of the experiment, to adsorb and incorporate organic halogens of various molecular sizes from the water column was also studied.

2 Materials and methods

2.1 *The waste waters*

BKME was collected at the outlet of the biological treatment plants, and truck transported in 1 m³ plastic containers filled to the top and capped. Table 1 summarizes some features of the pulping technologies and waste water treatment processes of the mills studied. Further details are described elsewhere (Jokela *et al.*, 1993). Our Mill A is equal to AS2, and Mill C is equal AL2 in Table 1 of the paper by Jokela *et al.* (1993).

Table 1. Description of production and waste water treatment processes of the mills. BKME from Mill A contained the effluents from kraft pulping and bleaching, carboxy methyl cellulose production, debarking and the paper mill, and the sanitary effluents from the mills. BKME from Mill C contained the effluents from kraft pulping and bleaching, and the municipal sewage from the community of 2000 residents. Chemical treatments used in pulp bleaching; D = chlorine dioxide; E = alkaline extraction; O = oxygen; EO = alkaline extraction with oxygen; C = chlorine gas.

	Mill A		Mill C	
	May 1991	October 1992	May 1991	October 1992
Bleaching	D-EO-D-E-D	D-EO-D-E-D	O-C/D-EO-D-E-D	O-D-EO-D-E-D
Wood species	spruce and pine	spruce and pine	spruce and pine	spruce and pine
Debarking	wet	wet	dry	dry
Active Cl used kg (1000 kg)$^{-1}$ of pulp	86	74	70	65
Secondary treatment process	activated sludge	activated sludge	anaerobic-aerobic lagoon	anaerobic-aerobic lagoon
Volume of secondary treatment plant, m^3	34 700	34 700	650 000	650 000
Hydraulic retention time of the waste water	20 h	~18 h	~10 d	~10 d

Table 2. Experimental set-up of the mesocosms. Waste waters of both mills were collected from the discharge channels leading from the biological treatment plant to the recipient water. For the actual percent of dilution, see Fig. 3. The actual dilution was calculated from the measured conductivity before and after dilution.

	summer-to-summer	winter-to-winter
Experimental period	from Jun 3, 1991 to Aug 6, 1992	from Nov 4, 1992 to Mar 10, 1994
Waste waters used		
BKME from Mill A	May 24, 1991	October 20, 1992
BKME from Mill C	May 23, 1991	October 16, 1992
Dilution water		
clear lake water	+	+
humic lake water	+	+
Dilution of BKME into lake water (v/v)		
1:10	+	+
1:50	+	–
Filling volume of the enclosures	1.8 m^3	2.5 m^3

2.2 CO_2-evolution test

The CO_2-evolution test was run for 24 days in 1-litre microcosms according to the European Community procedure (Commission of the European Communities, 1992 equivalent to OECD, 1992) with a few modifications. The temperature used was 15 °C instead of 22 °C to simulate the conditions in an average Finnish lake in summer time. BKME from Mill A (October 1992, see Table 1) was tested, both in lake water (clear and humic) and inoculated (with the activated sludge from a municipal waste water treatment plant) phosphate buffer medium (2.8 mM, pH 7.4), in duplicate. 4 ml of sodium hypochlorite solution (11 % w/v of active chlorine) was added into a parallel microcosm to inactivate the microbes (abiotic control). Sodium acetate was used as a positive control in the lake water and in the inoculated buffer medium experiments.

2.3 Mesocosms

Two series of *in situ* mesocosm experiments were conducted to determine BKME degradability in the field, the set-up is presented in Table 2. Two lakes were used for the study; one with clear water (lake Valkea Mustajärvi), the other was highly humic (lake Mekkojärvi). Both lakes are located in Southern Finland (61°13'N, 25°08'E). The enclosures (cylindrical in the summer-to-summer run and cylindrical with a conical bottom in the winter-to-winter run) were made of polyethene (194 g m^{-2}), translucent for ambient light and black for dark controls. The dark enclosures were covered with black polyethene sheets, while the light enclosures were open to daylight. To monitor for adsorption of organic halogen to the polyethene walls of the enclosures, strips of the same polyethene were incubated inside the enclosures.

The enclosures were supported by wooden frames divided into eight subframes. The frame was placed in the lake, backed with styrofoam flotation aids and anchored to trees ashore. The enclosures were filled with BKME and lake water in ratios (v/v) of about 1:10 and 1:50, and aliquots of the BKMEs used were stored frozen (–20 °C) for later analyses.

Sediment was pumped as a slurry from the mesocosms using a centrifuge pump, measured for volume, and homogenized by mixing. Two litres were taken for analysis, and the remainder returned to the enclosure.

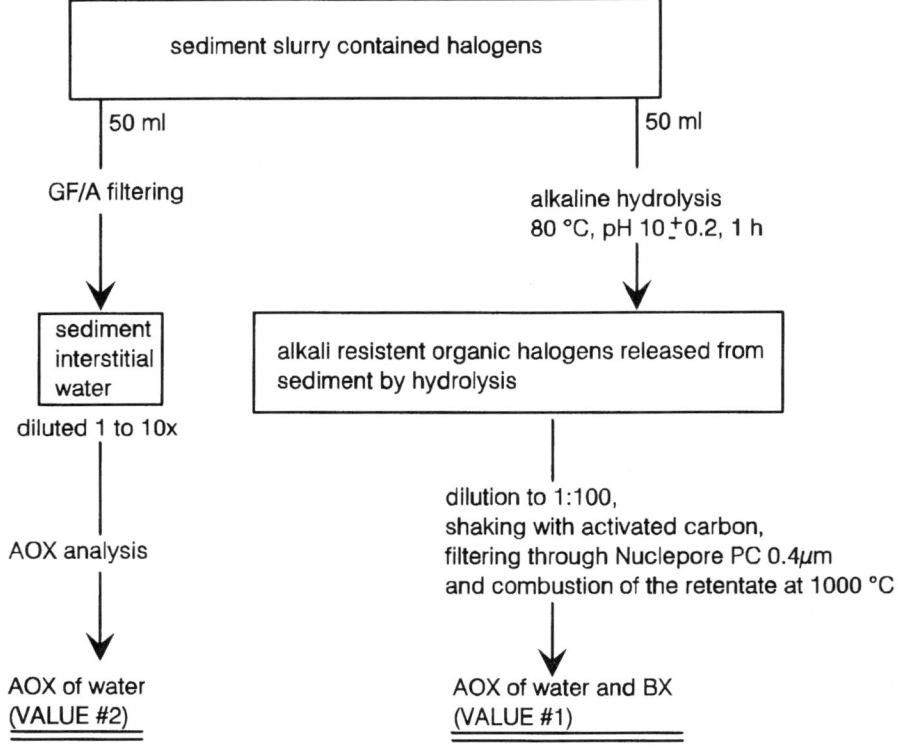

Fig. 1. Protocol for analysis of organic halogen in sediments formed in the mesocosms over about one year, BX = sediment bound halogen.

In the summer-to-summer experiment, correction factors were used from 275 days onwards. The factors for dilution by storm water based on conductivity were 2.33; 2.63; 1.45; 1.54 and 2.13 for the clear lake enclosures Mill C (2 %); Mill C (11 %) open, Mill C (11 %) dark; Mill A (10 %) open and Mill A (11 %) dark, respectively. The figures in the brackets indicate the concentrations (%, v/v) of BKME in the enclosures.

2.4 *Halogen analysis*

The adsorbable organic halogen (AOX) content of water was measured with a microcoulometric halogen analyzer (Euroglass) according to ISO standard 9562 (International Standardization Organization, 1989). To include the halogen retained by suspended solids (eg. microbes and algae), the AOX analysis of the mesocosm water column was performed without filtering the sample first. For the sediments, bound halogen (BX) was measured. The sample preparation protocol is presented in Figure 1. Whatman GF/A glass fibre filters were used. Active carbon for AOX assays was from Euroglass (Delft, NL). After filtration and/or hydrolysis as indicated in Fig. 1, the extracts were acidified (pH 2, conc. HNO_3) and analyzed for AOX. The bound halogen (BX) content of the sediment was calculated by subtracting the AOX content of the sediment interstitial water (= VALUE #2 in Fig. 1) from the sediment slurry contained halogens (VALUE #1 in Fig. 1).

Volatile organic halogen compounds (POX) were assayed by the halogen analyzer (Euroglass) after purging water samples of 100 ml (60 °C) with carrier gas (O_2,) according to the manufacturer's instruction.

For the extractable organic halogen (EOX) assay, samples of 10 ml of undiluted BKME or 300 ml of the mesocosm water column were freeze dried, and the residues extracted with 2 to 4 ml of tetrahydrofuran (THF) acidified (pH 2) with 15 µl (BKME) or 100 µl (mesocosm samples) of conc. HNO_3. After extraction on a shaker (0.5 h) and in a bath sonicator (0.5 h), the extracts were centrifuged and the supernatants (0.5 ml) filtered (0.45 µm Nylon Acrodisc 13) and evaporated to dryness in a stream of N_2. The residue was analyzed for halogen content in the Euroglass analyzer after direct combustion (1000 °C). An injection volume of 200 µl of the same extract (prior to evaporation) was used for molecular weight distribution analysis with high performance size exclusion chromatography (HPSEC) as described elsewhere (Jokela & Salkinoja-Salonen, 1992). A total of 25 fractions were collected. The calibration range was from 58 to 498 000 g mol^{-1}.

To determine the molecular weight distribution of organic halogen in sediment formed *de novo* in the mesocosms, the suspended solids of 50 ml of sediment slurry were collected on a glass fibre filter (Whatman GF/D). The filter was freeze dried, weighed (dry weight) and extracted in 30 ml of tetrahydrofuran spiked with 30 µl of conc. HNO_3. The subsequent steps were as described above for EOX analysis of water samples. A rotary evaporator was used to concentrate the sediment extracts from 30 ml to 3 ml.

Organic halogen adsorption (onto the polyethene walls of the enclosures and to adhering periphyton) was checked by two extraction methods. (1) For alkaline water extraction, ca. 100 cm^2 of the polyethene strip, colonized by periphyton, was submerged in dilute aqueous KOH (pH 10.5), heated at 80 °C, cooled, acidified to pH 2 (HNO_3) and processed for halogen analysis like the sediment slurry (see above). Another strip of polyethene from each enclosure was first washed with water to remove periphyton, and then treated as described above. (2) For ethanol extraction, ca. 200 cm^2 of the polyethene strip was washed and brushed to remove adhering periphyton, and then extracted with 300 ml of ethanol for 12 h in a Soxhlet apparatus (Kitunen & Salkinoja-Salonen, 1991). The ethanol extract was evaporated to about 20 ml and the halogen was determined in 1 ml according to the same protocol as the tetrahydrofuran extracts (see above).

2.5 Other analyses

The total (< 0.18 mm) organic carbon (TOC) content of the water samples were measured with Shimadzu TOC-5000 analyzer equipped with an auto sampler (ASI-5000) according to the protocol described elsewhere (Jokela & Salkinoja-Salonen, 1992). Carbon in freeze-dried sediment was measured using a Leco CHN-900 analyzer.

The sum of nitrite and nitrate was determined by the cadmium reduction method (USEPA no. 353.2). Total nitrogen (CEN/TC230/WG1/TG3 N31) and total phosphorus (USEPA no. 365.1) were analyzed after persulfate oxidation (Koroleff, 1979). Orthophosphate was determined according to the USEPA standard no. 365.1. Conductivity was measured using a Philips PW 9529 conductivity meter, and pH with an Orion 91-06 gel-filled combination electrode. Chloride was determined using an Orion 94-17B chloride electrode and an Orion 90-02 double junction reference electrode. Sulphate was analyzed by ion chromatography and metals by a plasma emission

Table 3. Characteristics of the BKMEs and lake waters used in the microcosm and mesocosm experiments. Most lake water results are an average of several measurements from two separate enclosures without BKME. Some parameters (indicated by *) were measured from a single (May 29) sample. nd = no data, BOD_7 = biological oxygen demand (7-day incubation), COD_{Cr} = chemical oxygen demand (dichromate oxidation).

	Lake water		BKME			
	Clear	Humic	Mill A May 24, 1991	Mill A Oct 20, 1992	Mill C May 23, 1991	Mill C Oct 16, 1992
NUTRIENTS						
$NO_2^- + NO_3^-$-N (mg m^{-3})	35	30	118	4	98	35
Total N (mg m^{-3})	350	560	3620	2000	5050	4500
PO_4^{3-}-P (mg m^{-3})	1	2	620	140	620	440
Total-P (mg m^{-3})	8	12	902	340	1040	880
SO_4^{2-} (g m^{-3})	6	10	499	nd	453	nd
Ca (g m^{-3})	2.5	4.6	101	77	128	48
Na (g m^{-3})	1.3	1.9	884	887	654	225
K (g m^{-3})	0.6	0.8	nd	83	nd	23
Cl (g m^{-3})	2	2	887	840	635	350
Si (g m^{-3})	0.1*	5.7*	7.1	nd	7.9	nd
TRACE ELEMENTS						
Zn (g m^{-3})	<0.1*	<0.1*	0.2	0.2	0.1	0.1
Fe (g m^{-3})	<0.1	0.4	0.5	nd	1.4	nd
Mg (g m^{-3})	0.6	1.0	nd	10	nd	7
Mn (g m^{-3})	<0.1*	<0.1*	1.7	nd	1.4	nd
Cu, Co and F (g m^{-3})	<0.1*	<0.1*	<0.1	<0.1	<0.1	<0.1
TOXIC ELEMENTS						
Pb (g m^{-3})	<0.5*	<0.5*	<0.5	nd	<0.5	nd
Cd (g m^{-3})	<0.01*	<0.01*	<0.01	nd	<0.01	nd
Al (g m^{-3})	<0.1*	0.3*	0.5	nd	3.4	nd
Ni, Cr and As (g m^{-3})	<0.1*	<0.1*	<0.1	nd	<0.1	nd
ORGANICS						
AOX (g m^{-3})	0.009	0.03	26	17	21	7
EOX (g m-3)	nd	nd	nd	16	nd	4
POX (g m-3)	nd	nd	nd	0,15	nd	0,07
TOC (g m^{-3})	6	21	380	300	280	190
OTHERS						
Alkalinity (mol m^{-3})	0.06	0.08	7.0	nd	6.4	nd
Acidity (mol m^{-3})	0.03*	0.05*	nd	nd	nd	nd
Colour (g Pt m^{-3})	17	280	3930	2460[a]	3230	1490[a]
pH	6.5	6.0	7.6	7.6	7.6	7.6
Conductivity (mS m^{-1})	3	4	380	460	300	280
BOD_7 (g m^{-3})[b]	nd	nd	40	30	30	50
COD_{Cr} (g m^{-3})[b]	nd	nd	910	660	670	630

[a] analyzed from frozen samples
[b] annual average of the mill, data from Finnish Environment Agency

spectrometer. The colour of the water was determined from filtered (Whatman GF/C) samples by spectrophotometer at 420 nm using platinum-cobalt standard solutions.

3. Results and discussion

Degradation of AOX and TOC of secondary treated BKME from two mills was studied in clear and humic lake waters. The chemical characteristics of the BKMEs are presented in Table 3.

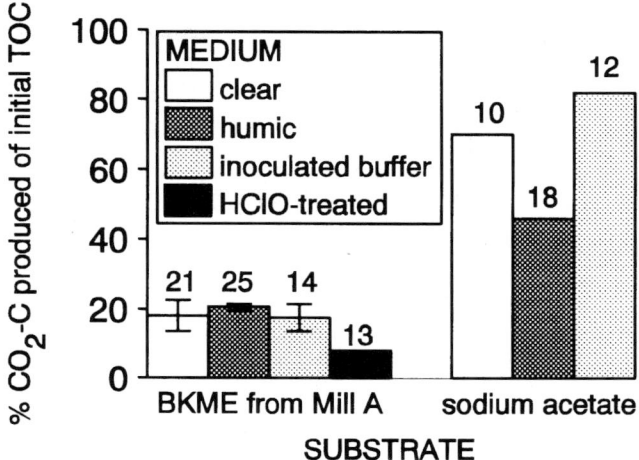

Fig. 2. Ability of clear and humic lake water and inoculated buffer medium to mineralize Mill A waste water (10 % v/v) and sodium acetate. Mineralization is expressed by the percentage of TOC–carbon converted to IC–carbon over 24 days at + 15 °C. Values of two replicates are joined by vertical lines. The waste water–buffer mixture was treated with hypochlorite to supply the abiotic control. Numbers above the bars are for TOC input (mg).

3.1 Biodegradability of BKME organic carbon in microcosms

The standardized set-up for measuring carbon dioxide evolution was used for the microcosm study in two different matrixes, non-modified lake water and sludge inoculated buffer medium. The results (Fig. 2) showed that the yield of carbon dioxide from BKME (Mill A) in 24 d was 18 to 21 percent of the TOC under experimental conditions where the yield of CO_2 from sodium acetate, the positive control, was 50 to 80 percent of TOC. The yield of carbon dioxide from BKME was 8 percent in hypochlorite treated parallel microcosms. This leaves 10 to 13 percent of the mineralization to be explained by biodegradation. This recalcitrance towards further biodegradation shows that the previous secondary treatment had effectively removed the easily biodegradable organic carbon.

The ratio of P:N:C in the clear lake water was 1:40:750, and 1:50:1800 in the humic lake water (Table 3). In lake water with 10 percent (v/v) of BKME the ratio of P:N:C was 1:10:800, and in the synthetic medium with 10 percent of BKME 1:0.002:0.1. The standard (OECD, 1992) medium used here greatly increased the ratio of the nutrients to the carbon, necessarily affecting biodegradation. Therefore, to evaluate organic matter degradability in natural environments, lake water is preferable to phosphate buffered, nutrient amended, synthetic medium. However, the results obtained with lake water microcosms showed that the microcosm pH increased in 24 d by 1 to 2 units. This pH instability was due to the low alkalinity of the lake waters (0.06 to 0.08 mM, Table 3) *vs.* the synthetic medium (2.8 mM). The large pH change may adversely affect biodegradation.

3.2 Degradability of BKME organic carbon and halogen in mesocosms

Because of the instability of the laboratory scale test system, units larger than micro-

Table 4. Conditions in the mesocosms over a one–year observation period. The parameters were measured every 0.6 to 19 weeks over a year. The extreme values and the period when such values were observed, including both summer–to–summer (s–s) and winter–to–winter (w–w) experiments (exp.), are displayed. Where the extreme value was observed twice during the experimental period, 1) and 2) are used to separate. The mesocosms were ice–covered (down to – 0.5 m) from October to April.

measurement (depth)				lake							
			clear				humic				
			BKME		when observed		BKME		when observed		
		none	Mill A	Mill C	months	exp.	none	Mill A	Mill C	months	exp.
temperature, °C (- 1.5 m)	min	2	3	3	Mar Nov-Mar	s-s w-w	3	3	3	Mar Nov-Mar	s-s w-w
	max	21	22	22	Jul Jun	s-s w-w	10	10	10	Sep Aug	s-s w-w
dissolved oxygen, g m^{-3} (- 1.5 m)	min	2	1	2	Mar Mar	s-s w-w	0	0	0	1) Aug-Sep 2) Mar-May Feb-Aug	s-s s-s w-w
	max	11	11	12	Jul-Oct 1) Jun-Aug 2) Nov	s-s w-w w-w	6	6	8	1) Jun-Jul 2) Oct Nov-Dec	s-s s-s w-w
pH (0 to - 1 m)	min	6.0	6.7	6.6	Mar Feb	s-s w-w	4.8	6.2	6.2	Mar Feb	s-s w-w
	max	7.4	7.5	8.3	Aug Jun	s-s w-w	6.1	7.2	7.6	Jun-Jul Jun	s-s w-w

cosms are needed for better simulation of the pelagic lake ecosystem. The results obtained with microcosms showed that no extensive biodegradation of BKME organic carbon occurred in 24 d. Hence, long-term biodegradation studies were required. We used outdoor enclosures, *i.e.* mesocosms placed in two different lakes, humic and clear, over four seasons.

The enclosures with humic lake water without BKME, maintained temperature, oxygen, and shallow salinity stratification (≤ 3 g Cl m^{-3}) during a year. The enclosures with clear lake water without BKME, maintained very shallow temperature and oxygen stratification, and no salinity gradient at any season (data not shown). The chloride concentration of the BKME holding mesocosms was 10 to 100 g m^{-3}.

After the input of 2 percent (v/v), 7 to 11 percent (v/v, summer-to-summer) and 11 to 13 percent (v/v, winter-to-winter) of BKME to the mesocosm, P:N:C ratios (average) were 1:20:700, 1:10:400, and 1:10:450, respectively.

Table 4 summarizes the range of physico-chemical conditions to which the mesocosm ecosystems were exposed during one year incubation. The mesocosms with BKME maintained temperature stratification similar to the mesocosms without BKME. They were aerated during the filling at the beginning of the experiments, but later on the concentration of oxygen (–1.5 m) in the humic lake enclosures both with and without BKME, decreased to zero in 50 and 100 days in the summer-to-summer and winter-to-winter experiments, respectively. Clear lake enclosures remained aerobic throughout the year. The pH remained higher by 0.3 to 1 units in the enclosures with BKME compared to the enclosures with lake water only.

The natural background of AOX was low in both lakes, 0.01 to 0.05 g of Cl m^{-3} (Fig. 3). BKME introduction of ca. 10 percent (v/v) into the mesocosms increased AOX concentration by 1 to 3 g Cl m^{-3}. After the incubation of either 275 or 393 d, 0.3 to 1.7 g Cl m^{-3} (20 to 60 %) of AOX became removed. Some (≤ 0.0034 g Cl m^{-3})

Degradation of the halogenated organic fraction 117

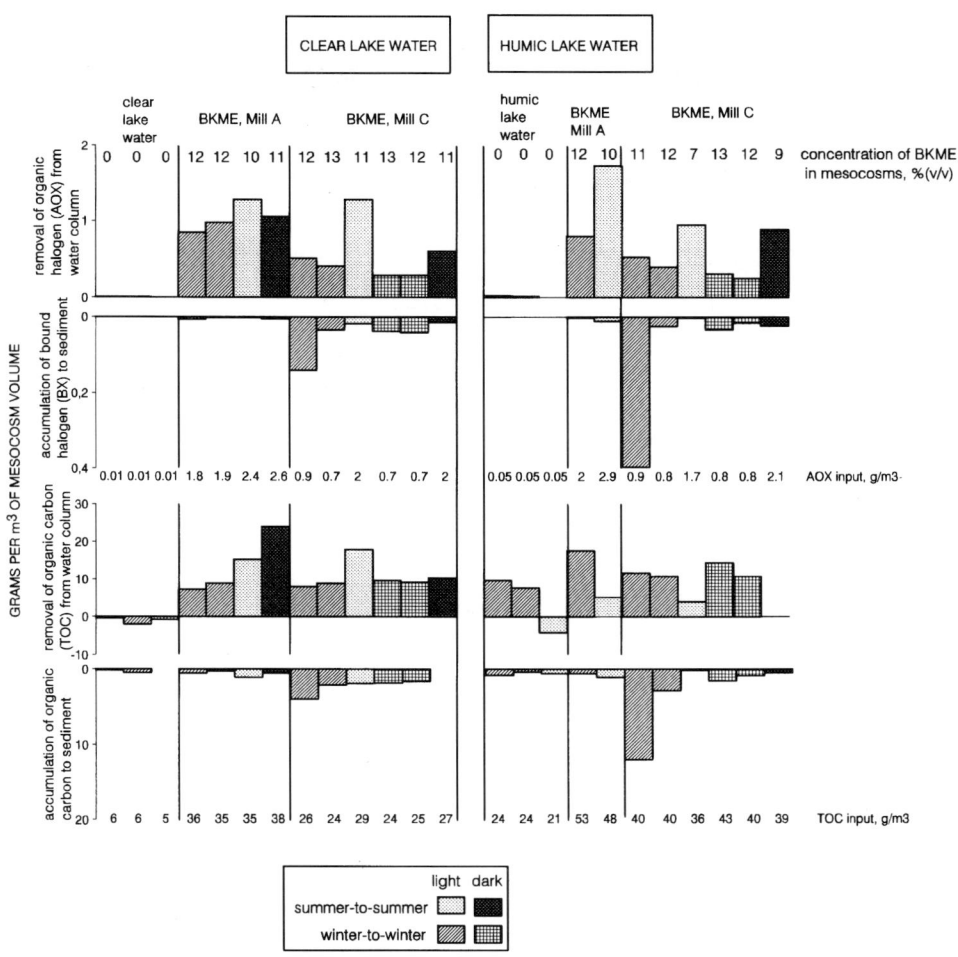

Fig. 3. Removal of organic halogen and TOC from the water column, and migration to sediment of the mesocosms over four seasons. The decrease of water column containing AOX and TOC from day 0 to day 275 (summer–to–summer), and to day 393 (winter–to–winter) is displayed. The migration of AOX and TOC to the sediment from day 0 to 275 (summer–to–summer), and to day 290 (winter–to–winter) is displayed. The volume of the mesocosms were 1.8 m^3 in the summer–to–summer, and 2.5 m^3 in the winter–to–winter experiment.

volatile organic halogen (POX) was found in the mesocosms after the introduction of the BKMEs. Most likely, the aeration in the secondary treatment plant had purged out POX from the BKMEs before they were introduced to the mesocosms. The organic carbon content of the clear lake mesocosms (less than 10 g C m^{-3}) was lower than that of the humic (ca. 20 g C m^{-3}) (Table 3). BKME introduction increased the carbon content by 20 to 30 g C m^{-3}. After the incubation of either 275 or 393 d, on the average 10 g C m^{-3} was removed from water.

On day zero, the mesocosms were filled with lake water and BKME, no sediment was added. But during the incubation period of either 275 or 290 d, sediment was formed *de novo*, 1 to 80 g of dry weight per mesocosm. Migration of organic carbon to the sediment was on the average 1.7 g C m^{-3}. Sedimentation explained in most cases only a minor proportion 0.2 to 14 percent (0.002 to 0.04 g Cl m^{-3}) of the organic halo-

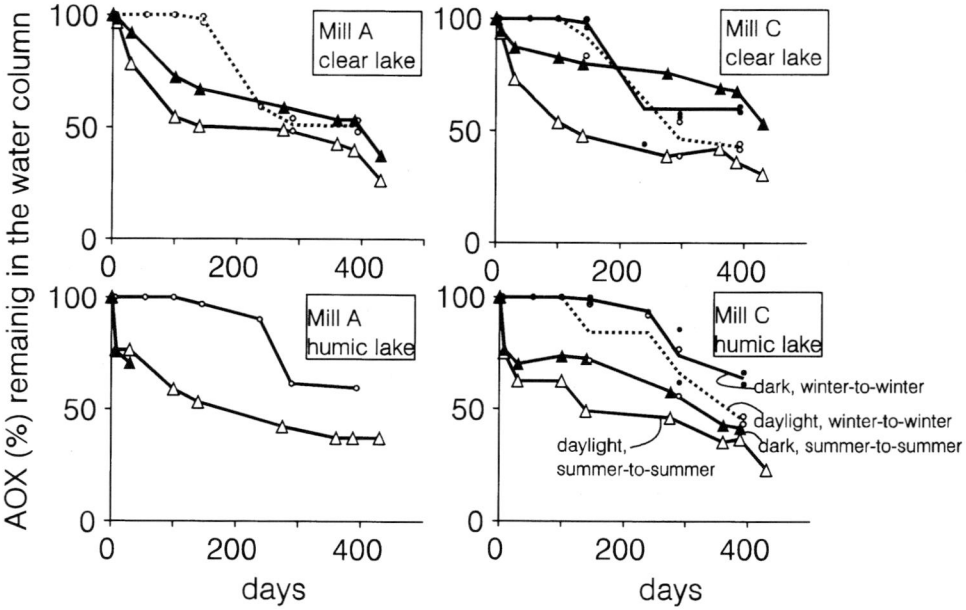

Fig. 4. Time course of the removal of AOX in the mesocosms from two BKMEs diluted 1:10 in two different lake waters. Day zero was November 4 in the winter–to–winter, and June 3 in the summer–to–summer experiment. For the mesocosms run in duplicate (Mill C, winter–to–winter), the line represents the average.

gen removal (Fig. 3). Two exceptions were found, one of Mill C daylight exposed mesocosms in the clear (27 %) and the other in the humic (75 %) lake. An explanation may lie in the elevated periphytic algal growth in these individual enclosures leading to an increase in sediment production. This illustrates that similar experimental conditions can lead to a very different response by the mesocosmic ecosystem.

To monitor the role of wall growth and physico-chemical adsorption on polyethene walls of the enclosures, polyethene strips were placed in the mesocosms on day 0, and examined after 275 d. It was found that 0.001 to 0.004 g Cl m^{-3} (2 to 7 mg Cl per mesocosm) was extractable to ethanol from polyethene. By alternative extraction into alkaline water (pH 10), even less AOX was extracted; from periphyton covered polyethene 0.0001 to 0.002 g Cl m^{-3} (0.1 to 3.1 mg of Cl per mesocosm) was measured, and from cleaned polyethene 0.0001 to 0.0005 g Cl m^{-3} (0.1 to 0.9 mg of Cl per mesocosm).

In conclusion, the above results of the fate of organic halogen in the mesocosms indicate that AOX removal from the water column including plankton and other suspended solids was due to degradation rather than to evaporation, sedimentation, or sorption to the walls of the enclosures. On the basis of the present results, we suggest that although organic halogen is known to accumulate into the sediment downstream from pulp mills to a concentration clearly higher than in the BKME-free reference site (*e.g.* Maatela *et al.*,1990), sediment is not the main sink for AOX. Our finding that a minor portion of AOX discharged by pulp mills migrated to sediment is consistent with a number of other studies (Grimvall *et al.*, 1991; Parker *et al.*, 1991; Bryant *et al.*, 1988).

If the extensive degradation of BKME organohalogen in the mesocosms was mainly due to biodegradation, a different degradation rate might be expected in the warm and cold seasons. Figure 4 shows the kinetics of AOX removal from the water column of the mesocosms during either 393 or 430 d. Initial degradation of AOX was faster when the BKMEs were introduced to the mesocosms in June as compared to the results of the experiment started in November. This was similar for both the different BKMEs and the different recipient lakes. Therefore, the warm and light season promoted degradation better than dark and cold (1 to 4 °C for the first five months in the winter-to-winter mesocosms). The final removal achieved in 390 d was similar (30 to 60 %, see Fig. 4) both in the summer-to-summer and winter-to-winter mesocosms. To monitor for the natural daylight as an agent promoting degradation of AOX, one of each pair of mesocosms was covered from daylight. In seven cases (with duplicates), light and dark counterparts of mesocosms both survived through all four seasons, and the results are recorded in Fig. 4. During the summer season of the (61°st latitude) summer-to-summer run, the majority of the annual solar irradiation entered the mesocosms during the first 120 days. During this period, the degradation of AOX was slower in the dark mesocosms compared to their light counterparts. The winter-to-winter mesocosms were started in early November, and were therefore frozen (down to -0.50 m in depth) for the first five months, and exposed to significant daylight only after day 200. The results in Fig. 4 show that the onset of the light season enhanced AOX degradation in the daylight exposed mesocosms more than in their darkened counterparts. During the 393 to 430 d exposure 30 to 60 percent of the AOX in BKME became degraded also in the dark mesocosms. This points to a significant role of biodegradation, and other reactions independent of daylight, in the removal of AOX.

Material balances shown in Fig. 3, indicated that accumulation to sediment was not the major sink for AOX from the water column. Irrespective of light exposure, 11 mesocosms out of 19 showed less than a 500-day half life for AOX. Grimvall *et al.* (1991) reported in Lake Vättern half-time of ≤ 2 years for AOX from a bleach-plant with no biological waste water treatment.

3.3 *BKME organic halogen: solubility, filterability and molecular weight distribution*

In order to find out whether the AOX persisting (Figs 3 and 4) after about a year was different from that introduced on day zero, molecular weight distributions of organic halogens were studied. The major part of organic halogen was soluble in tetrahydrofuran (Fig. 5). The amount of volatile materials (POX) was negligible on day 0 (< 1 % of AOX), and it did not increase during the incubation to any measurable level (< 0.004 g Cl m^{-3}) indicating that products of AOX and EOX degradation were not volatile.

Fig. 5 shows molecular weight distributions on day 0 from tetrahydrofuran extracts prepared from total BKME and from BKME from which suspended solids were removed by GF/C filtration. All tetrahydrofuran soluble halogen eluted similarly in the molecular weight range of 50 to 8000 g mol^{-1}, regardless of whether or not the suspended solids were included in the analysis. Hence, the sedimentation of solids is unlikely to bring any major change to the molecular weight distribution of the EOX in water column.

Next, we studied whether the aging of the BKMEs in the outdoor mesocosms brought any change in the molecular weight distribution. The results are shown for two clear water and two humic water mesocosms from November (393 days) in Figure 6. It shows that mesocosm incubation which had removed 40 to 60 percent (see Fig. 4) of the AOX from the water column, brought no major change to the molecular weight distribution of the tetrahydrofuran soluble halogen in the BKME (compare the WA-

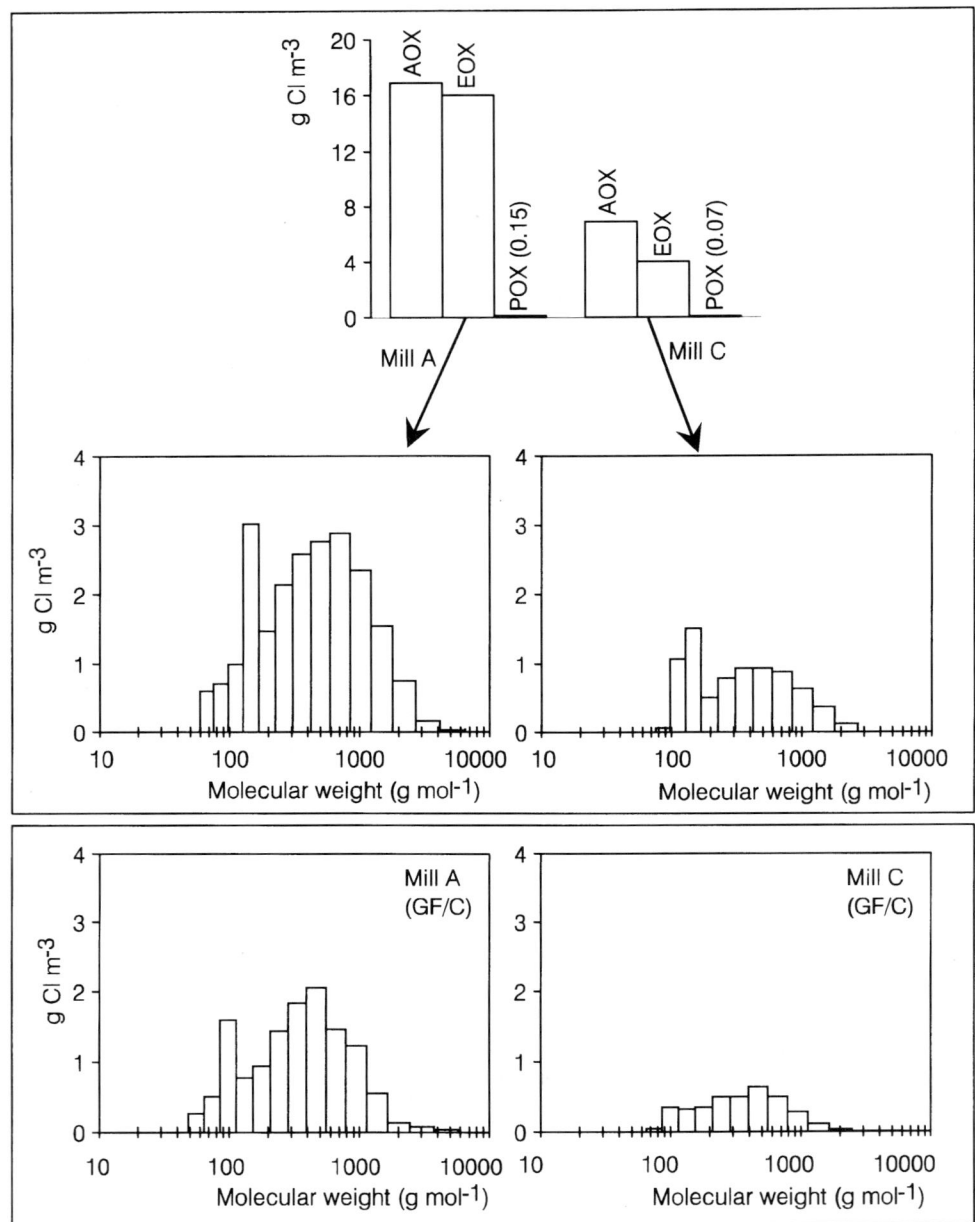

Fig. 5. The concentrations of adsorbable (AOX), tetrahydrofuran extractable (EOX) and volatile (POX) organic halogens, and the molecular size distributions of the tetrahydrofuran extracts of BKME. The total waste waters (BKME from Mill A and C, October 1992, upper panel) and the GF/C filtrates of the waste waters (lower panel) were extracted into tetrahydrofuran. Molecular weight distributions were analyzed by size exclusion chromatography with a calibration range from 58 to 498 000 g mol^{-1}.

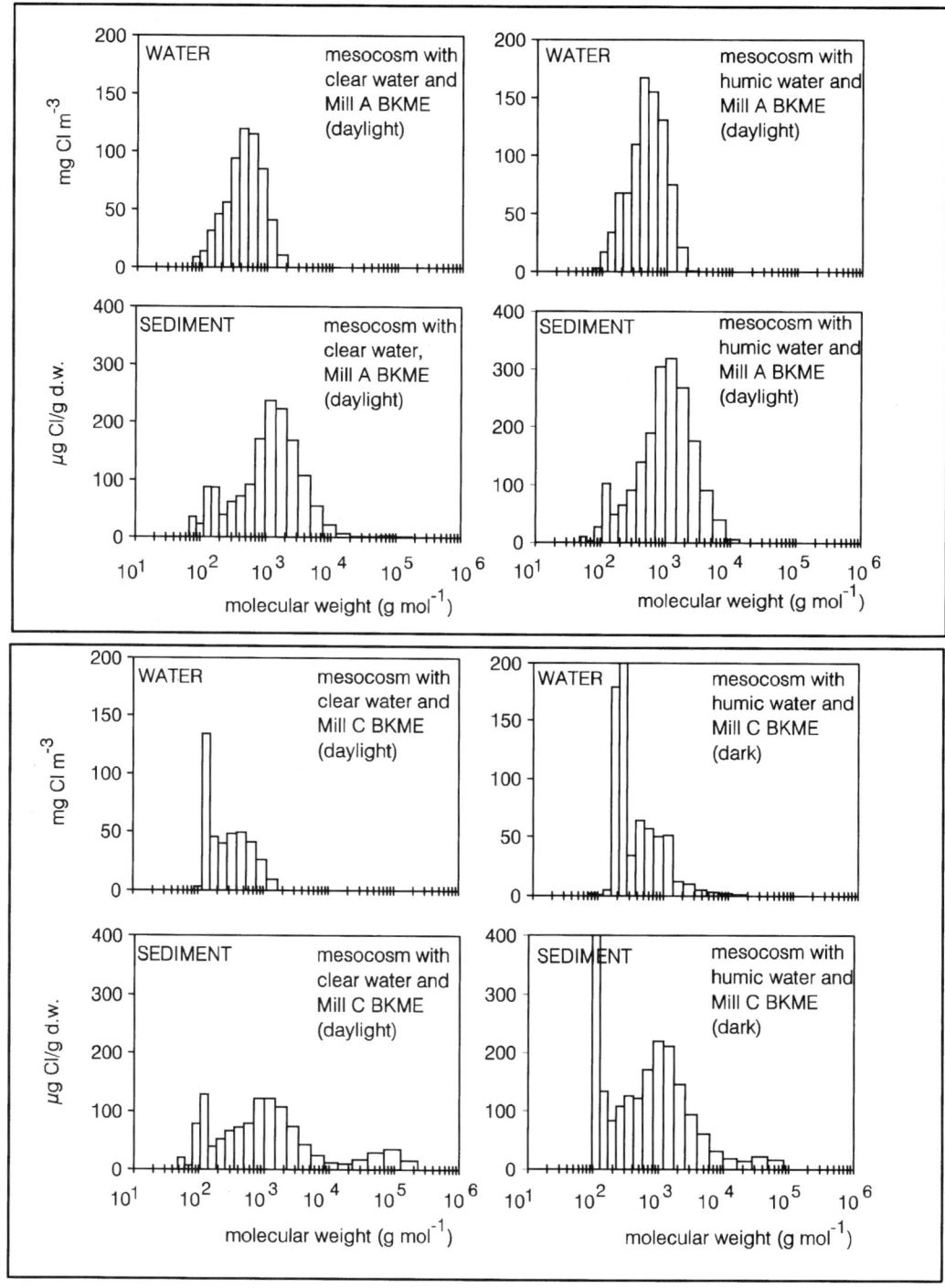

Fig. 6. Molecular weight distribution of tetrahydrofuran extractable organic halogen of water column and sediment of the mesocosms. Two mesocosms contained 12 percent (v/v) of BKME from Mill A (upper panel) and two contained 12 percent (v/v) of BKME from Mill C (lower panel). Water was sampled 393 d and sediment 290 d after day zero (November 5).

TER patterns of Fig. 6 to Fig. 5). Both large and small halogenated molecules, ranging from 10^2 to 10^4 g mol^{-1}, became removed rather similarly. We saw no sign of depolymerization analogous to that observed by O'Connor & Voss (1992) who showed that chlorinated phenolic compounds were released abiotically from high-molecular-mass fractions of E-stage liquor upon storage.

In order to find out whether any particular molecular size fraction preferably accumulated to sediment, also the sediments were extracted into tetrahydrofuran and analyzed for molecular size distribution. The results are shown in the SEDIMENT patterns of Fig. 6. They show that the major part of the sediment bound halogenated molecules were of larger size (average 1000 to 2000 g mol^{-1}) than those in the water column (100 to 500 g mol^{-1}) of the same mesocosm. We are not aware of any published reports about the effect of weathering on the molecular size distribution of organic halogens of BKME, under recipient lake conditions. But analogous phenomenon was observed in soil by Laine *et al.* (this volume) who found that chlorinated phenolic compounds accumulated from soil to earthworms as larger molecules than those found in the soil. Several authors have observed in the laboratory or aerated lagoon scale that the low-molecular-mass fraction of organic chlorine in bleaching effluent or lignin was biodegraded more efficiently than the high-molecular-mass fraction (Lindström & Mohamed, 1988; Fitzsimons *et al.*, 1990; Kern & Kirk, 1987). However, it may be difficult to extrapolate from laboratory results to the events taking place in the pelagic lake conditions.

In summary, the fate of organic halogen discharged with secondary treated total waste waters from bleached kraft pulp mills was studied. To this end, waste waters were incubated for one year *in situ*, in the lake mesocosms. The results indicate that organic halogen underwent extensive degradation, and increased in molecular weight when incorporated to sediment.

4. Summary

Degradability of total organic carbon (TOC) and organic halogen (AOX) of bleached kraft pulp mill effluents (BKME) was studied after the secondary treatment. Ten to 13 percent biodegradation (in 24 d) of TOC was observed in laboratory microcosms with BKME. To simulate the environmental fate of organic carbon and halogen in lake ecosystem, BKMEs were diluted to 2 to 13 percent (v/v) with lake water, and placed in outdoor mesocosms (ca. 2 m^3). During ca. one-year incubation, 20 to 60 percent of the AOX was removed from the water column of the mesocosms. From 0.3 to 3 percent of the removed AOX was recovered from the sediment formed *de novo* in the mesocosms, and less than 1 percent was found adsorbed to the mesocosm walls. Volatile halogenated metabolites were not formed to a measurable extent. The major portion of the AOX removed from the mesocosms by weathering, was thus explained by degradation. Natural daylight enhanced the degradation of AOX. Faster initial but similar final degradation of AOX was observed in one-year mesocosm incubation starting from June, compared to that starting from November. The molecular weight distribution of the organic halogen persisting in the water column of the mesocosm after one-year incubation, was similar to that of the BKME before incubation. The sediment formed *de novo* in the BKME mesocosms, contained organic halogenated molecules larger than those in the water column of the same mesocosm. The present study indicates that organic halogens discharged by bleached kraft mill waste water treatment plants, continued degrading in recipient mesocosms, and increased in molecular weight as a result of incorporation into sediment.

Abbreviations

AOX	= adsorbable organic halogen (activated carbon as adsorbent)
BKME	= bleached kraft pulp mill effluent
EOX	= extractable organic halogen (extracted to tetrahydrofuran)
POX	= volatile organic halogen purged at 60 °C
TOC	= total organic carbon (< 0.18 mm)

Acknowledgements

This work was financially supported by Maj and Tor Nessling Foundation, Ministry of the Environment and Academy of Finland. We thank Maarit Herranen, Carl Curtén and Ann-Christine Eriksson for cooperation.

References

Bryant, C. W., G. L. Amy, R. Neill & S. Ahmad, 1988. Partitioning of organic chlorine between bulk water and benthal interstitial water through a kraft mill aerated lagoon. Wat. Sci. Technol. 20(1): 73–79.
Commission of the European Communities, 1992. Determination of "ready" biodegradability. Official Journal of the European Communities L 383 A 35: 187–206.
Fitzsimons, R., M. Ek & K.-E. L. Eriksson, 1990. Anaerobic dechlorination/degradation of chlorinated organic compounds of different molecular masses in bleach plant effluents. Environ. Sci. Technol. 24(11): 1744–1748.
Grimvall, A., S. Jonsson, S. Karlsson & R. Sävenhed, 1991. Organohalogens in unpolluted waters and large bleach-plant recipients. In: *TAPPI Proc. Environ. Conf. Book 1*. pp. 147–154.
Häggblom, M. & M. Salkinoja-Salonen, 1991. Biodegradability of chlorinated organic compounds in pulp bleaching effluents. Wat. Sci. Tech., 24(3/4): 161–70.
International Standardization Organization, 1989. Water quality – Determination of Adsorbable Organic Halogens (AOX). ISO 9562, 8pp.
Jokela, J. K. & M. Salkinoja-Salonen, 1992. Molecular weight distributions of organic halogens in bleached kraft pulp mill effluents. Environ. Sci. Technol., 26(6):1190–1197.
Jokela, J. K., M. Laine, M. Ek & M. Salkinoja-Salonen, 1993. Effect of biological treatment on halogenated organics in bleached kraft pulp mill effluents studied by molecular weight distribution analysis. Environ. Sci. Technol., 27(3): 547–57.
Kern, H. W. & T. K. Kirk, 1987. Influence of molecular size and ligninase pretreatment on degradation of lignins by *Xanthomonas* sp. strain 99. Appl. Environ. Microbiol. 53(9): 2242–2246.
Kitunen, V. H. & M. S. Salkinoja-Salonen, 1991. Distribution of PCDD/PCDFs in PE-coated paperboards – an explanation for migration of PCDD/PCDFs from board products. Chemosphere, 23: 1561–1568.
Koroleff, F., 1979. Chemical analysis of sea water. Meri, 7: 1–60. (In Finnish).
Lindström, K. & M. Mohamed, 1988. Selective removal of chlorinated organics from kraft mill total effluents in aerated lagoons. Nordic Pulp and Paper Res. J., 1/1988, pp. 26–33.
Maatela, P., J. Paasivirta, J. Särkkä & R. Paukku, 1990. Organic chlorine compounds in lake sediments. II Organically bound chlorine. Chemosphere 21(12):1343–1354.
O´Connor, B. I. & R. H. Voss, 1992. A new perspective (sorption/desorption) on the question of chlorolignin degradation to chlorinated phenolics. Environ. Sci. Technol., 26: 556–560.
OECD (Organisation for Economic Cooperation and Development), 1992. Ready biodegradability. Carbon dioxide evolution. OECD test guidelines 301B.
Parker, W. J., E. R. Hall & G. J. Farquhar, 1991. Dechlorination of segregated kraft mill bleach plant effluents in high rate anaerobic reactors. In: *TAPPI Proc. Environ. Conf. Book 2*. pp. 787–795.
Stuthridge, T. R. & P. N. McFarlane, 1994. Adsorbable organic halide removal mechanism in a pulp and paper mill aerated lagoon treatment system. Wat. Sci. Technol., 29(5/6):195–208.
Zender, J. A., T. R. Stuthridge, A. G. Langdon, A. L. Wilkins, K. L. Mackie & P. N. McFarlane, 1994. Removal and transformation of resin acids during secondary treatment at a New Zealand bleached kraft pulp and paper mill. Wat. Sci. Technol., 29(5/6):105–121.

Polychlorinated diphenyl ethers and chlorophenolic compounds in Salmon (*Salmo salar*) from the Arctic Teno River compared to the Baltic Sea

P. J. Vuorinen,[1] J. Koistinen,[2] J. Paasivirta,[2] M. Vuorinen,[1] & J. Hoikka[1]

[1] *Finnish Game and Fisheries Research Institute,*
P.O. Box 202, FIN-00151 Helsinki, Finland
[2] *Department of Chemistry, University of Jyväskylä,*
P.O.Box 35, FIN-40351 Jyväskylä, Finland

Keywords: PCDE, Chlorophenols, Fish, Arctic, Salmon

1. Introduction

Polychlorinated diphenyl ethers (PCDEs) and chlorophenolic compounds (CPs) can be generated in the incineration of wastes (Paasivirta *et al.*, 1986). PCDEs are impurities in technical chlorophenol products as well (Firestone *et al.*, 1972). PCDEs, along with CPs, can be transported long distances, as has been reported for many organochlorine compounds (reviewed by Wania & Mackay, 1993). In the Teno River area, there are no major, local sources of pollution.

Thus far, knowledge of PCDEs in environmental samples is very limited. They have only been reported in fish from the Great Lakes (Kuehl *et al.*, 1984; Huestis & Sergeant, 1992), in some marine organisms (Lake *et al.*, 1981) and in salmon and the white-tailed eagle in Finland (Paasivirta *et al.*, 1986).

Concentrations of organochlorine compounds (DDTs, PCBs, chlordanes, etc.) in two composite samples from Teno River salmon were given in an earlier paper (Vuorinen *et al.*, 1989). The results of the organochlorine compound analysis were also presented earlier, in Vuorinen *et al.*, (1992), using the same fifteen salmon as were used for this paper, which presents the results of the PCDE and CP analyses. The aim of the present study was to find the PCDE and CP concentrations in salmon caught during spawning migration in the Arctic-situated Teno River and compare the concentrations with those of Baltic salmon.

2. Material and methods

2.1 Sampling

In July and August of 1990, 15 female Atlantic Salmon (*Salmo salar* L.) from the Arctic Teno River (Fig. 1) were caught and sampled. For comparison, 14 female Atlantic salmon of the Baltic Sea were sampled at the Simojoki River (Fig. 1) at spawning time in October 1990. Mean weights, lengths and ranges are given in Table 1. Epaxial muscle samples (30 to 50 g) were taken from under the dorsal fin, excluding the red muscle tissue. Samples were wrapped in aluminium foil and stored frozen, at –60 °C, until analyzed.

Fig. 1. The sampling sites.

Table 1. Weight and length of salmon from the Teno and Simojoki Rivers. Means, ranges and number of sampled fish are given.

	Teno salmon	Simojoki salmon
Mean weight, kg	5.86	9.42
Range of weight, kg	1.7 – 10.6	6.7 – 12.6
Mean length, cm	85.5	96.2
Range of length, cm	58 – 106	90 – 103
N	15	14

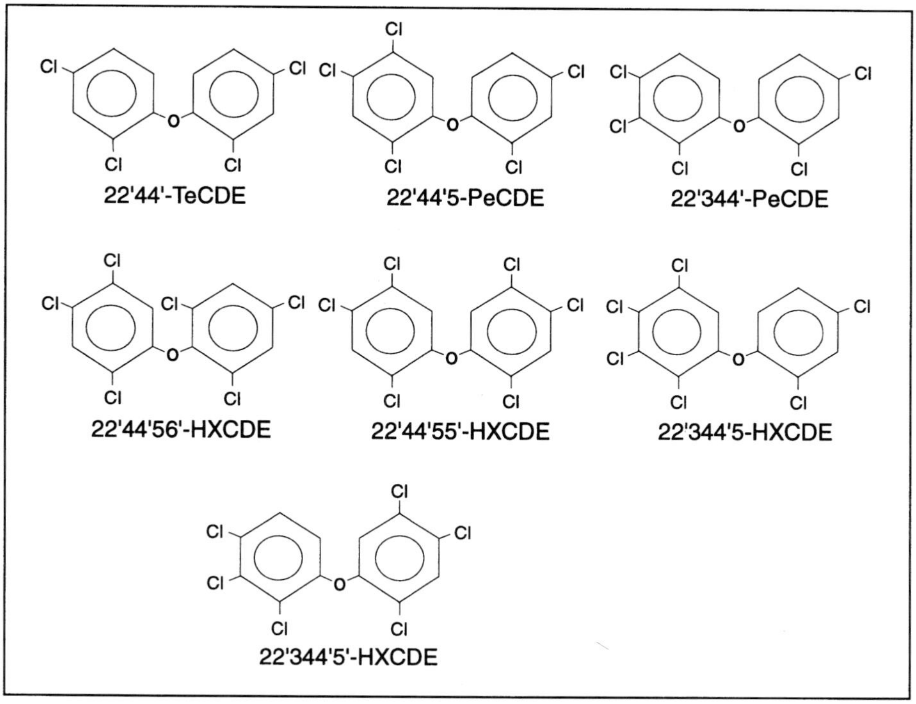

Fig. 2. The structures of the polychlorinated diphenylethers (PCDEs) that were detected in the muscle of the Teno salmon.

2.2 Analysis

For the PCDE analysis, 20 to 40 g of freeze-dried muscle tissue was Soxhlet-extracted for six hours with a solvent mixture of petroleum ether:acetone:hexane:diethyl ether (18:11:5:2; v/v/v/v). Fat content was determined gravimetrically and extract was cleaned up with a sulphuric acid treatment followed by florisil and carbon microcolumns. PCDEs were analyzed by gas chromatography/mass spectrometry using a Hewlett Packard (HP) 5890 gas chromatograph combined with a HP 5970 mass selective detector. The identification was based on authentic model PCDEs synthesised at the University of Jyväskylä. The total procedure is described in detail in Koistinen et al., (1993).

Chlorophenolics were extracted from an aliquot of muscle homogenised with sodium sulphate. The same solvent mixture as for PCDEs was used and the extraction was performed in a Soxhlet apparatus. The extracted fat was weighed and chloropenolics analyzed as depicted in Herve et al., (1988). Briefly, the hexane extract was shaken with a potassium carbonate solution. Polychlorinated anisoles (PCAs) and veratroles (PCVs) were collected by washing with hexane. Polychlorinated phenols (PCPs) remaining in the wash layer were acetylated. PCAs and PCVs were analyzed after being purified twice with a neutral alumina column.

Table 2. The mean (± SE) concentrations (ng g^{-1}; on the fresh weight basis) of polychlorinated diphenyl ethers (PCDEs) detected in salmon from the Teno River (Arctic) (N = 15) and as a comparison, those in salmon from the Simojoki River (Baltic) (N = 14). The statistical significance (p < 0.05) of the differences between the means and the Baltic/Arctic ratios of the means of the compounds are also given.

	Teno salmon	Simojoki salmon	P <	B/A
22'44'-TeCDE	0.039 ± 0.004	1.231 ± 0.139	0.05	31.6
22'44'5-PeCDE	0.047 ± 0.004	0.471 ± 0.050	0.05	10.1
22'344'-PeCDE	0.0007 ± 0.0007	0.096 ± 0.013	0.05	136.7
22'44'56'-HxCDE	0.009 ± 0.003	0.159 ± 0.015	0.05	17.1
22'44'55'-HxCDE	0.031 ± 0.019	0.207 ± 0.021	0.05	6.6
22'344'5-HxCDE	0.037 ± 0.004	0.111 ± 0.010	0.05	3.0
22'344'5'-HxCDE	0.007 ± 0.005	0.038 ± 0.005	0.05	5.7

2.3 Statistics

Of the parameters measured, the mean and the standard error of the mean (SE) are given. Statistical differences were analyzed by a one-way analysis of variance, followed by Scheffe's test of the significance (p < 0.05) of differences between means. Baltic/Arctic ratios were calculated from the mean values of the parameters. Pearson correlation coefficients were calculated for the weight, length and fat content of salmon muscle against PCDEs and CPs. All statistical calculations were made with the Statistical Analysis System (SAS, 1988).

3. Results

Of the 26 PCDE isomers measured, only seven (Fig. 2) were detected in Teno salmon muscle from the Arctic. Their mean concentrations on the fresh weight basis were significantly lower than the respective isomers of salmon living in the Baltic (Table 2), which also contained more PCDE isomers. The results of PCDEs in the individual salmon samples on the fresh weight basis have been presented elsewhere (Koistinen et al., 1993). However, mean concentrations of certain PCDE isomers did not differ between the Arctic and Baltic salmon on the fat weight basis (Table 3). None of the PCDE isomers correlated significantly (p > 0.05) with the weight or length of salmon (Table 4).

The mean concentrations of chlorophenolic compounds are given in Table 5 and their structures in Figure 3. The concentrations of 2,4-DCP, 2,3,4,6-TeCP and TeCG were significantly higher in the Teno salmon than in the Baltic salmon. However, the concentrations of 2,4,6-trichloroanisole, tetrachloroanisole and pentachloroanisole did not differ much between the Teno and Baltic salmons, and the tetrachloroveratrole concentration was significantly lower in the Teno salmon. None of the chlorophenolics correlated positively significantly (p > 0.05) with the weight, length or muscle fat content of salmon (Table 6).

Table 3. The mean (± SE) concentrations (ng g^{-1}; on the lipid weight basis) of polychlorinated diphenyl ethers (PCDEs) detected in salmon from the Teno River (Arctic) (N = 15) and as a comparison, those in salmon from the Simojoki River (Baltic) (N = 14). The statistical significance (p < 0.05) of the differences between the means and the Baltic/Arctic ratios of the means of the compounds are also given.

	Teno salmon	Simojoki salmon	P <	B/A
22'44'-TeCDE	0.923 ± 0.141	7.000 ± 0.647	0.05	7.6
22'44'5-PeCDE	1.031 ± 0.103	2.687 ± 0.244	0.05	2.6
22'344'-PeCDE	0.017 ± 0.017	0.544 ± 0.057	0.05	32.6
22'44'56'-HxCDE	0.173 ± 0.054	0.923 ± 0.083	0.05	5.3
22'44'55'-HxCDE	0.747 ± 0.483	1.186 ± 0.105	NS	1.6
22'344'5-HxCDE	0.811 ± 0.107	0.669 ± 0.086	NS	0.8
22'344'5'-HxCDE	0.167 ± 0.131	0.214 ± 0.026	NS	1.3

Table 4. The Pearson correlation coefficients of the polychlorinated diphenyl ethers detected in Teno salmon (N = 15) and weight, length and fat content. Those of salmon from the Simojoki River (N = 14) are also given. The statistical significance (p < 0.05) is given by an asterisk as a superscript.

	Teno River salmon			Simojoki River salmon		
	Weight	Length	Fat	Weight	Length	Fat
22'44'-TeCDE	0.334	0.320	0.102	0.339	0.258	0.517
22'44'5-PeCDE	0.326	0.187	0.605*	0.337	0.268	0.514
22'344'-PeCDE	-0.139	-0.110	-0.119	0.302	0.200	0.462
22'44'56'-HxCDE	0.229	0.101	0.507	0.407	0.323	0.426
22'44'55'-HxCDE	0.402	0.358	-0.129	0.257	0.125	0.493
22'344'5-HxCDE	0.286	0.261	0.381	0.507	0.419	0.157
22'344'5'-HxCDE	0.421	0.361	-0.154	0.290	0.229	0.492

4. Discussion

Chlorophenol compounds are formed, for example, in the chlorobleaching of pulp, chlorodisinfection of tap water and combustion of wastes. The survey of Finnish streams showed that the highest chlorophenol concentrations were not found near pulp mills with chlorine bleaching and therefore chlorophenol pollution was concluded to have originated mainly through the atmosphere (Vartiainen & Väänänen, 1993). The higher CP concentrations of Teno salmon compared to Baltic salmon were not due to analytical errors, but rather due to the special characteristics of the Arctic environment.

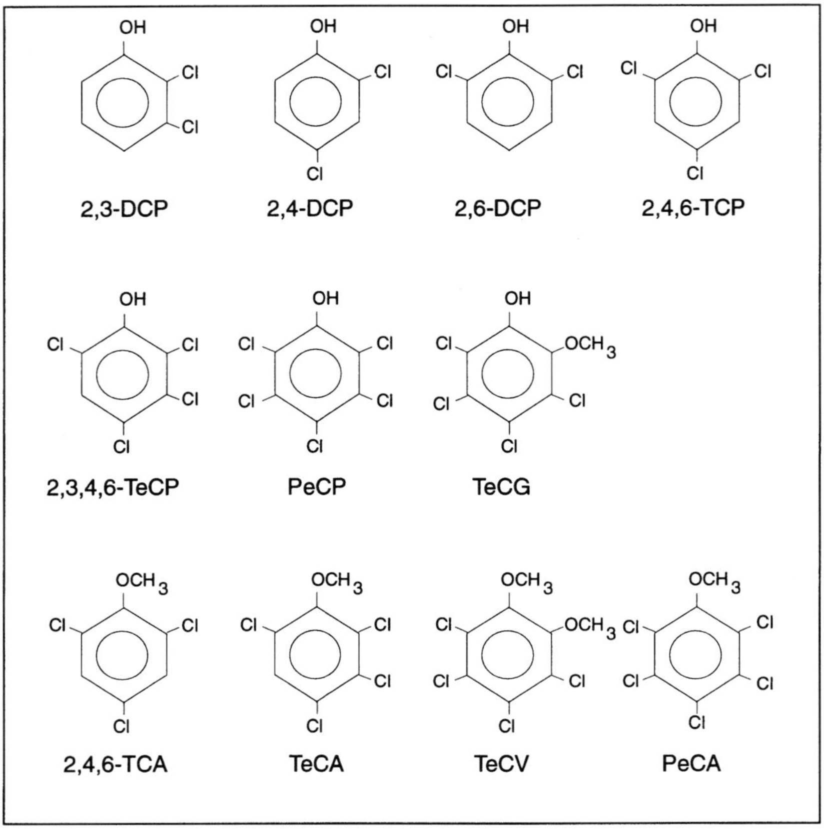

Fig. 3. The structures of the polychlorinated phenolics (CPs) that were detected in the muscle of the Teno salmon.

The decomposition of chlorophenols in the water column and metabolism in fish, might be significantly lower than in Baltic river estuaries. The accumulation and persistence of these pollutants could thus be higher in Teno salmon than in Baltic. It was also observed that concentrations of chlorophenols in streams in the Tenojoki area in northern Finland were relatively high compared to the more polluted parts of the country (Vartiainen & Väänänen, 1993).

Concentrations of PCDEs in Teno salmon were lower than in Baltic salmon because pollution in the Arctic area comes solely through long range transport, whereas the Baltic Sea has many local pollution sources.

The lack of a correlation between salmon size and fat content with concentrations of PCDEs, was unexpected because PCDEs are suspected to be recalcitrant against metabolism and very persistent in the environment (Norström *et al.*, 1976). On the other hand, nothing is known about the elimination of PCDEs in fish. Low concentrations of PCDEs in salmon muscle and great individual variations may, in part, explain the lack of a correlation. The fact that the concentrations of CPs did not correlate with salmon size or fat content is not surprising since CPs are relatively easily metabolised in fish (Oikari & Holmbom, 1986).

Table 5. The mean (± SE) concentrations (ng g^{-1}; on the fresh weight basis) of the polychlorinated phenols (PCP), guaiacols (PCG), anisoles (PCA) and veratroles (PCV) detected in salmon from the Teno River (Arctic) (N = 15) and as a comparison, those in salmon from the Simojoki River (Baltic) (N = 14). The statistical significance (p < 0.05) of the differences between the means and the Baltic/Arctic ratios of the means of the compounds are also given.

	Teno salmon	Simojoki salmon	P <	B/A
2,3-DCP	0.60 ± 0.48	5.79 ± 1.69	NS	9.65
2,4-DCP	15.60 ± 3.71	2.93 ± 1.87	0.05	0.19
2,6-DCP	36.13 ± 9.71	3.93 ± 1.60	NS	0.11
2,4,6-TCP	3.93 ± 1.27	3.64 ± 1.61	NS	0.93
2,3,4,6-TeCP	10.40 ± 2.16	1.71 ± 0.62	0.05	0.16
PeCP	3.93 ± 1.65	1.82 ± 0.62	NS	0.46
TeCG	0.80 ± 0.26	0.14 ± 0.10	0.05	0.18
2,4,6-TCA	2.12 ± 0.56	2.44 ± 0.79	NS	1.15
TeCA	2.18 ± 0.90	0.94 ± 0.23	NS	0.43
PeCA	0.53 ± 0.34	1.02 ± 0.40	NS	1.92
TeCV	0.26 ± 0.16	2.47 ± 0.84	0.05	9.50

5. Summary

In 1990, Atlantic salmon (*Salmo salar* L.; weight 1.7 to 10.6 kg; N = 15) from the Teno River were sampled and analysed for polychlorinated diphenyl ethers (PCDEs) as well as for chlorophenols, chloroanisoles and chloroveratroles. Concentrations were compared to those in Baltic salmon (weight 6.7 to 12.6 kg; N = 14).

Of the 26 PCDE isomers measured, only seven were detected in Teno salmon muscle (detection level 10 to 50 pg g^{-1}). The mean concentrations (in fat) of six of the PCDEs in Teno salmon were 3 to 78 percent of those of the respective isomers in Baltic salmon, in which more PCDE isomers were detected (detection level 10 to 200 pg g^{-1}). The mean chlorophenol concentrations were six times higher in Teno salmon than in Baltic salmon, while the concentrations of chloroanisoles were of the same order. Tetrachloroveratrole (TeCV) in Teno salmon muscle was ten percent of that in Baltic salmon.

The higher chlorophenol concentrations in the Teno salmon might reflect the very low decomposition rate of organochlorines in the cold Arctic climate and pure environment, with its low density of degrading micro-organisms.

Table 6. Pearson correlation coefficients of the chlorophenolic compounds detected in Teno salmon (N = 15) and weight, length and fat content. Those of salmon from the Simojoki River (N = 14) are also given. The statistical significance ($p < 0.05$) is given by an asterisk as a superscript.

	Teno River salmon			Simojoki River salmon		
	Weight	Length	Fat	Weight	Length	Fat
2,3-DCP	-0.232	-0.223	-0.032	0.015	-0.075	0.184
2,4-DCP	-0.600*	-0.634*	0.304	0.124	0.247	0.240
2,6-DCP	0.093	0.141	0.101			
2,4,6-TCP	-0.456	-0.375	-0.239	0.057	0.225	0.300
2,3,4,6-TeCP	-0.019	-0.031	-0.156	-0.199	-0.006	0.377
PeCP	0.056	0.067	-0.382	0.012	0.086	0.239
TeCG	-0.650*	-0.659*	-0.037	0.094	0.072	0.432
2,4,6-TCA	-0.302	-0.314	0.179	0.206	0.150	-0.275
TeCA	0.470	0.509	-0.129	-0.088	-0.002	-0.483
TeCV	0.170	0.131	0.257	0.160	0.058	-0.155
PeCA	0.082	0.093	-0.051	0.164	0.060	-0.601

Acknowledgements

Many thanks to the staff at the Teno field station of the Finnish Game and Fisheries Research Institute and the Oulu Regional Office of the Veterinary and Foodstuffs Institute, for sample collection. Kaija Hara and Tiina Rantio participated in the extraction of samples. Chlorophenols, chloroanisoles and chloroveratroles were determined by Leena Welling at The Institute for Environmental Research of the University of Jyväskylä. Ms. Kathleen Tipton checked the English language. The Ministry of Agriculture and Forestry and the Ministry of the Environment funded the study.

References

Firestone, D., J. Ress, N. L. Brown, R. P. Barron & J. N. Damico, 1972. Determination of polychlorodibenzo-p-dioxins and related compounds in commercial chlorophenols. J. Ass. Off. Anal. Chem. 55: 85–92.
Herve, S., P. Heinonen, R. Paukku, M. Knuutila, J. Koistinen & J. Paasivirta, 1988. Mussel incubation method for monitoring organochlorine pollutants in watercourses. Four-year application in Finland. Chemosphere 17: 1945–1961.
Huestis, S. Y. & D. B. Sergeant, 1992. Removal of chlorinated diphenyl ether interferences for analysis of PCDDs and PCDFs in fish. Chemosphere 24: 537–553.
Koistinen, J., P. J. Vuorinen & J. Paasivirta, 1993. Contents and origin of polychlorinated diphenyl ethers (PCDEs) in salmon from the Baltic Sea, Lake Saimaa and the Tenojoki River in Finland. Chemosphere 27: 2365–2380.
Kuehl, D. W., E. Durhan, B. Butterworth & D. Linn, 1984. Identification of polychlorinated planar chemicals in fishes from major watersheds near the Great Lakes. Environ. Internat. 10: 45–49.

Lake, J. L., P. F. Rogerson & C. B. Norwood, 1981. A polychlorinated dibenzofuran and related compounds in an estuarine ecosystem. Environ. Sci. Technol. 15: 549–553.

Norström, Å., K. Anderson & C. Rappe, 1976. Formation of chlorodibenzofurans by irradiation of chlorinated diphenyl ethers. Chemosphere 5: 21–24.

Oikari, A. & B. Holmbom, 1986. Assessment of water contamination by chlorophenolics and resin acids with the aid of fish bile metabolites. In: T. M. Poston & R. Purdy (eds.), *Aquatic Toxicology and Environmental Fate, Ninth Vol., ASTM STP 921*. pp. 252–267. Amer. Soc. Testing Mat., Philadelphia, Pennsylvania.

Paasivirta, J., J. Tarhanen & J. Soikkeli, 1986. Occurrence and fate of polychlorinated aromatic ethers (PCDE, PCA, PCV, PCPA and PCBA) in the environment. Chemosphere 15: 1429–1433.

SAS, 1988. SAS Institute Inc. SAS/STAT User's guide, release 6.03 edition. Cary, NC, USA, 1028 pp.

Vartiainen, T. & P. Väänänen, 1993. Chlorophenols in Finnish stream waters in August, 1990. In: J. Tuomisto & J. Ruuskanen (eds.), *Proceedings First Finnish Conference of Environmental Sciences, Kuopio, October 8–9, 1993. "Environmental Research in Finland Today"*, Finnish Society for Environmental Sciences. pp. 268–271. Kuopio University Publications C. Natural and Environmental Sciences 14.

Vuorinen, P. J., J. Paasivirta & J. Koistinen, 1989. Comparison of the levels of organochlorines in Arctic and Baltic wildlife samples. Pap. Pres. Global significance of the transport and accumulation of polychlorinated hydrocarbons in the Arctic, Oslo 18.–22.9.1989.

Vuorinen, P. J., J. Paasivirta, J. Koistinen & T. Rantio, 1992. Organochlorines in salmon (*Salmo salar* L.) from the Teno River. In: E. Tikkanen, M. Varmola & T. Katermaa (eds.), *Symposium on the state of the environment and environmental monitoring in northern Fennoscandia and the Kola Peninsula, October 6–8, 1992, Rovaniemi, Finland. Extended Abstracts*. pp. 186–188. Arctic Centre Publications.

Wania, F. & D. Mackay, 1993. Global fractionation and cold condensation of low volatility organochlorine compounds in polar regions. Ambio 22: 10–18.

Metals in cod (*Gadus morhua morhua* L.) from the Barents Sea

V. M. Savinov

*Murmansk Marine Biological Institute,
Kola Scientific Centre Russia Academy of Sciences 17,
Vladimirskaya st., Murmansk, 183023, Russia*

Keywords: trace elements, biomonitoring

1. Introduction

Determination of the distribution and accumulation of toxic elements within the various tissues of fish is an important aspect of environmental pollution control. Knowledge of these tissue's accumulation is necessary to identify specific tissue in different species that may be particularly affected by increasing environmental concentrations of toxic elements and to monitor metal accumulation in portions of fish consumed by humans.

Although at suitable concentrations some metals and metalloides, such as Co, Cu, Fe, Se and Zn are essential for metabolic activity, some metals (*eg* Pb) may also be toxic at higher concentrations and may inhibit enzymes by forming mercaptides with the sulphydryl groups which are responsible for catalytic activity. Consequently, most metals, whether essential or not, are potentially toxic to living organisms.

Monitoring programmes have been undertaken in several North European coastal areas, but not in the Barents Sea. The Helsinki Commission has been active in the Baltic Sea. The heavy metals content of fish and invertebrates has been monitored yearly since 1979 in the Finnish Sea area (Pertillä *et al.*, 1982; Tervo, 1987 and others). The International Council for the Exploration of the Sea (ICES, 1977) has carried out studies in the North Atlantic, North and Baltic Seas. In England and Wales, a coastal waters monitoring programme started in 1968 (Murray, 1979; Portmann, 1979; Murray, 1981). But little is known about concentrations of toxic metals in Barents Sea fish.

Cod is one of the most abundant species of fish in the Barents Sea region. Knowledge of heavy metal contamination in cod is a high priority. Investigations in the Baltic Sea have shown that cod exhibit area differences, which, with the exception of lead, follow the areal differences of metal concentrations in sea water (Pertillä *et al.*, 1982). These areal differences suggest that, in spite of extensive migration of cod, this fish is a better indicator species for aquatic pollution control than others. In Norway, cod also was proposed for a monitoring program (Julshamn *et al.*, 1978; 1985). This paper concerns the content of trace elements in different organs and tissues of cod from different regions of the Barents Sea.

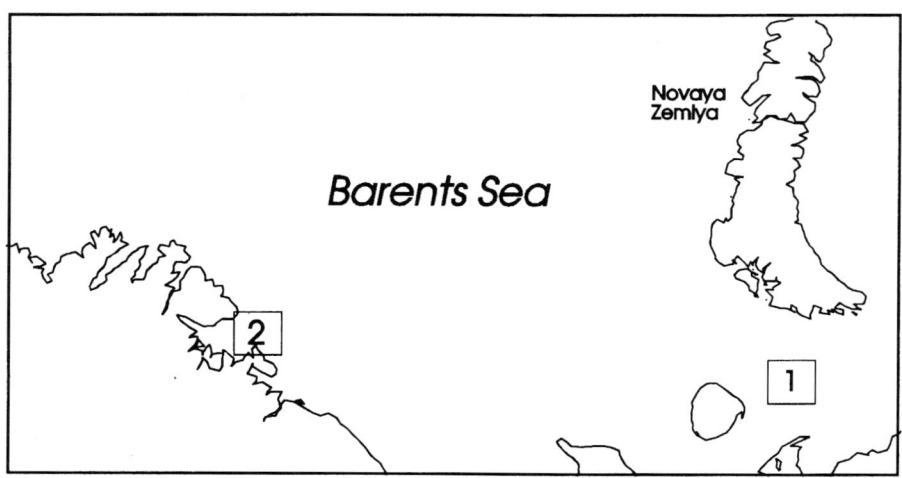

Fig. 1. Map of the Barents Sea showing location of sampling:
1 – Pechora Sea; 2 – Bolshaya Volokovaya Inlet

2. Materials and methods

Atlantic cod (*Gadus morhua morhua* L.) were collected in 1992 in eastern (Pechora Sea) and western (Bolshaya Volokovaya Inlet) areas of the Barents Sea. Sample areas are shown in Figure 1. Fish samples were collected from commercial bottom trawl catches. Prior to the selection of samples for analysis, the standard morpho-metric measurements were done. The age of the specimens was determined from the otoliths. The age ranged from four to six years; length from 32 to 90 cm.

Samples of cod organs and tissues were frozen and stored at –20°C until analysis. Sample were dried at a temperature of 105°C.

Tissues were digested with nitric acid. 'Suprapur' – quality reagents and 'Milli-Q' – water were used throughout. Cu, Ni, Cr, Cd and Pb were determined using a Perkin-Elmer 460 AAS equipped with an HGA – 400 graphite oven and deuterium background correction. Zn was measured by a Perkin-Elmer 360 AAS in propane/butane/air flame. As and Se were determined by AAS-30 with hydrid system MHS-10.

All concentrations are expressed as ppm wet wt.

Statistical analysis was carried out with SYSTAT 5.01, SYSTAT Inc. Differences between means ($P<0.05$) were determined by *t*-test (SYSTAT 5.01, SYSTAT Inc.)

3. Result and discussion

A summary of the trace element concentrations in muscle, liver and gill of cod caught from two regions in the Barents Sea are given in Table 1. A comparison of reported trace element concentrations in cod from different regions is presented in Table 2.

3.1 *Arsenic*

Arsenic in gill varies from 0.25 to 2.11 ppm. Maximum concentration was recorded in gill of cod from the western part of the Sea. It should be noted that the statistical

Table 1. Trace elements in organs and tissues of cod from the Barents Sea in 1992, ppm wet weight

Trace elements in organs and tissues of cod from the Barents Sea in 1992, ppm wet weight

Elements		Region					
		Pechora Sea			Bolshaya Volokovaya Inlet		
		Organ, number of samples			Organ, number of samples		
		gills n = 6	liver n = 6	muscles n = 6	gills n = 10	liver n = 10	muscles n = 10
As	mean	0.691	2.662	0.802	1.225	1.853	0.689
	S.D.	0.670	1.743	0.632	0.399	0.745	0.266
Cd	mean	0.023	0.207	0.018	0.060	0.233	0.022
	S.D.	0.009	0.098	0.010	0.021	0.086	0.008
Cr	mean	0.035	0.033	0.011	0.448	0.018	0.020
	S.D.	0.048	0.040	0.008	0.340	0.017	0.010
Cu	mean	0.440	5.988	0.257	0.379	4.565	0.140
	S.D.	0.217	1.945	0.071	0.116	2.204	0.033
N	mean	0.066	0.038	0.010	0.031	0.026	0.010
	S.D.	0.054	0.046	0.005	0.012	0.013	0.003
Pb	mean	0.009	0.014	0.004	0.010	0.013	0.010
	S.D.	0.015	0.010	0.004	0.010	0.007	0.004
Se	mean	0.312	0.128	0.026	0.848	0.184	0.106
	S.D.	0.208	0.096	0.012	0.149	0.188	0.052
Zn	mean	14.389	24.967	3.457	22.330	23.370	4.235
	S.D.	2.285	6.591	0.662	4.601	6.779	0.446

analysis has demonstrated significant differences between mean concentration in gill of male (1.533 ppm) and female (1.007 ppm) from this area. Significant differences between mean arsenic concentrations in gill of cod from the two regions was also found.

Highest values of arsenic concentrations in liver (4.73 ppm) and muscle (1.61 ppm) were found in fish caught in the eastern part of the Sea. However, arsenic may be present in cod at much higher concentrations than those detected in cod from the Barents Sea. Thus, cod muscle from the Norwegian coastal waters contains on average 5 ppm (Knutzen, 1987); from Oslofjord 4.2 ppm (Staveland et al., 1993) and from Scottish waters it varies from 0.7 to 4.6 ppm (Falconer et al., 1983). It is therefore assumed that the present arsenic concentrations represent normal physiological levels.

Table 2. Trace element concentrations in cod from different regions (ppm wet wt)

Region, years	Tissue	Cd	Zn	Cu	Ni	Pb	As	Cr	Se	Source
Coastal waters of England, 1970-1973	muscle	<0.2	3.9	<0.5		<0.1				Portmann, 1979
North Sea 1971-1972	liver	0.06	17.7	2.9	2.5					Wright, 1976
	muscle	0.11	6.1	1.0	2.3					
	gill		24.2	0.5	3.3					
North Sea, 1974	muscle	<0.2	3.7	<0.2-1.0		<0.3		<0.2-0.9		Murray, 1979
Barents Sea, 1975	muscle	<0.05	3.8	0.6		<0.5				ICES, 1977
Norwegian Sea, 1975	muscle	<0.05	4.0	0.8		<0.5				
Waters off W.Spitsbergen, 1975	muscle	<0.05	3.9	0.8		<0.5				
Barents Sea, 1975	muscle	<0.2	5.4	0.4		<0.2-0.4				Murray, 1981
North Sea, 1975	muscle	<0.2	3.6	0.38		<0.2-0.5		<0.2-0.5		
Castal waters of W.England, 1975	muscle	<0.2	3.2	0.5		<0.2-0.4		0.36		
Distant water W.Greenland, 1975	muscle	<0.2	16	6.0		<0.8		<0.2		
	liver	<0.2	5.0	6.0		<0.2		<0.2		

Table 2. Continued

Region, years	Tissue	Cd	Zn	Cu	Ni	Pb	As	Cr	Se	Source
Baltic Sea 1979-1981	liver	0.015-0.05	8.0-13.0	4.0-9.5		0.03-0.14				Perttilä et al., 1982
Baltic Sea 1982-1986	liver	0.012-0.051	10.0-20.3	2.9-11.8		0.01-0.06				Tervo, 1987
Scottish waters	muscle						0.7-4.6			Falconer et al., 1983
St.Lawrence	liver	0.05-0.26								ICES, 1984
Coastal waters of Norway	muscle	0.02	5	0.5	0.2	0.01	5	0.1	0.5	Knutzen, 1987
	liver	0.5	20	10	0.5	0.1	10	0.5	3.0	
Oslofjord, 1988	muscle						4.2			Staveland et al., 1993
North Sea	liver	0.02-0.66	6.6-43.0	1.7-15.0		0.02-0.08				NSTF, 1993

3.2 Cadmium

Cadmium concentrations in gill of cod ranged from 0.012 to 0.103 ppm. Significantly higher cadmium levels were found in gill of cod from the eastern area as compared to fish from the western area (P<0.001). The highest cadmium concentrations in liver (0.400 ppm) and muscle (0.036 ppm) were recorded in fish from the western part. No statistical differences in mean cadmium levels in liver or muscle were found between the areas. In muscle of cod from the western area, statistical differences between mean cadmium concentrations for male and female was observed.

Compared with other investigations, the hepatic cadmium level in cod from the Barents Sea was higher than that reported for the Baltic Sea (Pertillä et al., 1982; Tervo, 1987) and similar with data for the Canadian Arctic (ICES, 1984), the North Sea (NSTF, 1993) and slightly lower than those for Norwegian coastal waters (Knutzen, 1987).

3.3 Chromium

Tissue levels of chromium in cod from both areas were very low. Chromium concentrations varied from 0.004 to 0.036 ppm in muscle, from 0.005 to 0.110 ppm and from 0.002 to 0.950 in liver and gill, respectively. Significantly higher chromium levels in gill were found for cod from the western area (P<0.01). In this area, statistical differences between mean chromium concentrations in liver of male (0.006 ppm) and female (0.027 ppm) was also recorded. Compared with most other investigations, the concentrations of chromium in cod from the Barents Sea were low (Table 2).

3.4 Copper

Copper concentrations in liver of cod ranged from 2.03 to 9.76 ppm and were similar to those reported for the Baltic (Pertillä et al., 1982; Tervo, 1987) and North Seas (NSTF, 1993). However, the mean tissue concentrations in cod from the Barents Sea were lower than those previously reported for the coastal waters of Norway (Knutzen, 1987). The mean copper levels in muscle of cod from the eastern area were significantly higher (P<0.01) than the levels from the western area. It should be noted that copper levels in muscle of cod from both areas were lower than corresponding levels found in cod from other regions (Table 2).

3.5 Zinc

Zinc concentrations in gill, muscle and liver of cod ranged from 12.1 to 30.8, 2.7 to 4.8 and from 16.8 to 36.6 ppm, respectively, with the higher values in fish from the western area. The mean concentrations of zinc in gill and muscle of cod from the western area were significantly higher as compared with the eastern area. Reported zinc levels were generally in agreement with the concentrations found in this study (Table 2.)

3.6 Nickel and selenium

Few studies have dealt with nickel and selenium concentrations in cod. The present data indicates significantly lower concentrations in tissues of cod from the Barents Sea than from the North Sea and the Norwegian coastal waters (Wright, 1976; Knutzen,

1987). The muscle and gill of cod from the western area had significantly higher mean selenium concentrations than cod from the eastern area (P<0.005).

3.7 Lead

Lead concentrations in liver of cod from the Barents Sea varied from 0.003 to 0.033 ppm and were lower than corresponding levels in cod from the Baltic, North and Norwegian Seas (Table 2). In the case of lead measured in cod muscle, significant areal differences were detected.

Observed areal differences in trace metal concentrations in cod within the Barents Sea probably reflect different contamination levels in various parts of the sea.

In general, the concentrations of heavy metals in cod from the Barents Sea are low and reflect the normal physiological levels in fish.

Further studies of the contamination levels of heavy metals in marine organisms, comparisons between geographical regions, and temporal changes are needed in the Barents Sea regions. This is especially significant in connection with future oil and gas exploration on the shelf.

Knowledge of the concentrations of metals in biota of the Barents Sea is necessary to understand the natural biogeochemical flux of elements through the marine ecosystem and to predict the consequences of anthropogenic changes in the global flux.

4. Summary

This study presents results on levels of arsenic, cadmium, chromium, copper, nickel, lead, selenium and zinc in gill, muscle and liver of cod (*Gadus morhua morhua*) which were caught in 1992 on western (Bolshaya Volokovaya Inlet) and on eastern (Pechora Sea) parts of the Barents Sea.

Significant differences between mean concentrations of trace elements in cod from the western and the eastern parts of the sea were found for As, Cd, Cr, Se and Zn contents in gill and for Cu, Pb, Se and Zn contents in muscle of fish. Mean concentrations of all these elements with exception of Cu were higher in fish from the western area, probably reflecting different contamination levels in various regions of the basin. No statistical differences in mean levels of all measured elements in liver of cod were recorded between the areas.

Generally, trace element concentrations in cod from the Barents Sea were within the wide range for cod reported in the international literature.

Acknowledgments

I appreciate the assistance of analytical chemists at the Institute of Problems of Industrial Ecology Kola Scientific Centre Russian Academy of Sciences for their analysis of heavy metals. Valuable suggestions and comments by two anonymous referees are also gratefully acknowledged.

References

Falconer,C.R., R. J. Shepherd, J. M. Pirie,G. Topping, 1983. Arsenic levels in fish and shellfish from the North Sea. J. Exp. Mar. Biol. Ecol. 71: 193–203.

ICES, 1977. A baseline study of the level of contaminating substances in living resources of the orth Atlantic. Coop. Res. Rept. Int. Counc. Explor. Sea, 69, 83 pp.

ICES, 1984. The ICES Coordinated Monitoring Programme for Contaminants in Fish and Shellfish, 1978 and 1979 and Six-year Review of ICES Coordinated Monitoring Programmes. Coop. Res. Rept. Int. Counc. Explor. Sea, 128, 112 pp.

Julshamn, K., J. Haugsnes & F. Uthe, 1978. The contents of 14 major and minor elements (minerals) in Norwegian fish species and fish byproducts determined by atomic absorption spectrometry. Fisk. Dir. Skr. Ser. Ernaering, 1: 117–135.

Julshamn, K, K. E. Slinning, H. Haaland, B. Boe & L. Foyn, 1985. *Analysis av sporelementer og klorerte hydrokarboner i fisk og blaskjell fra Hardangerfjorden, og tilstotende fjordomrader hosten 1983 og varen 1984.* Fiskeridirektoratet, rapporter og meldingen, 6, 55pp.

Knutzen, J., 1987. *"Bakgrunnsnivåer" av metaller i saltvannsfisk. NIVA-rapport 0-85167 (1.nr. 1956)*, 25pp.

Murray, A. J., 1979. Metals, organochloride pesticides and PCB residue levels in fish and shellfish landed in England and Wales during 1974. Aquat. Monit. Pest., 2: 1–43.

Murray, A. J., 1981. Metals, organochloride pesticides and PCB residue levels in fish and shellfish landed in England and Wales during 1975. Aquat. Monit. Pest., 5: 1–40.

North Sea Task Force, 1993. North Sea Quality Status Report 1993. Oslo and Paris Commision, London. Olsen & Olsen, Fredensborg, Denmark, 132pp.

Pertillä, M., V. Tervo & R. Pormanne, 1982. Heavy metals in Baltic Herring and Cod. Marine Pollut. Bull. 13: 391–393.

Portmann, J. E., 1979. Chemical monitoring of residue levels in fish and shellfish landed in England and Wales during 1970–1973. Aquat. Env. Monit. Pest., 1: 1–75.

Staveland, G., I. Marthinsen, G. Norheim & K. Julshamn, 1993. Levels of Environmental pollutants in flounder (*Platichthys flesus* L.) and cod (*Gadus morhua* L.) caught in the Waterway of Glomma, Norway. II. Mercury and arsenic. Arch. Environ. Contam. Toxicol., 24: 187–193.

Tervo, V., 1987. Concentrations of metals in fish and benthic invertebrates in the Gulf of Finland and in the Gulf of Bothnia during 1982–1986. ICES C.M. 1987/E: 20 pp.

Wright, D. A., 1976. Heavy metals in animals from the North East coast. Mar. Pollut. Bull. 7: 36–38 p.

Biomobility of organic halogen compounds from contaminated soil – earthworms as a tool

M. Laine,[1] J. Jokela & M. Salkinoja-Salonen

Department of Applied Chemistry and Microbiology, P.O. Box 27, SF-00014 University of Helsinki, Finland

Keywords: bioaccumulation, sawmill soil, molecular weight distribution

1. Introduction

The fate of and hazard caused by organic halogen compounds in contaminated soil are affected by (bio)transformation reactions of the chemicals in soil. The compounds may volatilize, become mineralized or transformed by microbes. For example, chlorophenols used for wood preserving may be hydroxylated, methoxylated or polymerized by soil microbes at sawmill sites. Earthworms represent higher organisms in soil. They participate in degradation and transport of compounds in soil in addition to soil preparation. Hence, compounds taken up by earthworms may migrate further in the food chain and therefore may have wide impact on the environment.

In Finland, many old sawmill sites are polluted by the wood preserving chemicals. There is a need to make an effort to clean these sites. For biological decontamination of soil one must also take into consideration the intermediates and metabolites already formed by microbial biotransformation of the pollutants during the long residence time in the environment.

Decontamination of sawmill soil with fungus may lead to polymerization instead of mineralization of chlorinated compounds (Lamar & Dietrich, 1990; Bollag, 1992). Xenobiotics incorporated into soil are considered to be resistant against microbial release (Dec *et al.*, 1990) or biodegradation (Robinson & Novak, 1994) and their bioavailability to be small (Bollag, 1991). However, synthetic chlorolignin model compounds have been shown to be bioavailable to rats (Sandermann *et al.*, 1990). Thus, ecotoxicological significance of bound pesticidal residues needs further attention.

Bioaccumulation and toxicity of chlorophenols and related low molecular weight chlorinated compounds, to earthworms has been studied (Neuhauser *et al.*, 1985; Hague & Ebing, 1988; Van Gestel & Ma, 1988; Van Gestel & Van Dis, 1988; Haimi *et al.*, 1992, 1993). We studied the bioavailability of solvent extractable chlorinated compounds from contaminated soils using earthworms as model organisms. The earthworms were incubated in field soils, contaminated for decades by the chlorophenolic wood preservative (Ky-5) spills. Ky-5 consisted mainly of 2,3,4,6-tetrachlorophenol, 2,4,6-trichlorophenol, pentachlorophenol, and trace amounts of dimeric chlorophenol derivatives like chlorinated phenoxyphenols and dioxins as impurities (Kitunen *et al.*,

[1] Corresponding author, Present address: Finnish Environment Agency, Laboratory, Hakuninmaantie 4-6, FIN-00430 Helsinki, Finland

1985). In the following, the bioaccumulation of halogen containing contaminants from soil into the earthworms will be described.

2. Materials and methods

2.1 Characteristics of the contaminated soils

Soil dry weight was determined with an infrared drying balance and compared with the dry weight calculated after lyophilization. There was a good agreement between the two methods (difference ± 1 %). The soil organic matter was calculated from the ignition loss after 2.5 h at 850 °C. For the pH measurement 10 g of soil was mixed with 25 ml of 1 M KCl, shaken for 10 min (200 rpm), let to sediment for 1 hour and then the pH of the suspension was measured.

2.2 Incubation of the earthworms in the contaminated sawmill soils

Earthworms from kitchen compost *(Eisenia andrei)* were exposed to sawmill soils for 12 or 25 days at 22°C. Earthworms (ca. 30 worms per container) were kept in polyethene containers (0.9 l, two replicates) filled with contaminated soil (ca. 200 g). Fresh horse manure (100 g on day 0 and 50 g on day 12 per container) was added as feed and the humidity was maintained by irrigation (in Kaustinen soil, water content by weight was 50 per cent and in Toras soil 75 %, respectively). The dead earthworms were removed every third day. Containers with soil only (reference soil) were kept under the same conditions to observe any change in soil organic halogen content due to evaporation and/or microbial activities. The soils were not sterilized before the experiment. As a reference, the unexposed earthworms were analyzed.

2.3 Extraction of the organic halogen compounds from soil and earthworms

After incubation in the contaminated soil the earthworms were kept in a moistened filter paper in the dark at 12°C for one to two days to empty their gut content. The rinsed earthworms were killed with liquid nitrogen and homogenized with an Ultra-Turrax homogenizer in a 100 ml bottle. The homogenized earthworms were weighed (5 to 10 g) and freeze-dried. The freeze-dried earthworms were powdered and 20 ml tetrahydrofuran added. After three hours of shaking on a platform shaker (200 rpm), the extracts were allowed to settle. The supernatant was first filtered through a 5 µm filter (Millex-LS, Millipore) and then through a 0.45 µm filter (Nylon Acrodisc). While the samples were freeze-dried, the volatile compounds were lost.

The contaminated soils were freeze-dried and extracted as the earthworm homogenates to unify the sample preparation procedure. The soils which earthworms were introduced to, were sieved through a 0.5 cm sieve while the reference soils were mixed thoroughly, freeze-dried and analyzed as such.

2.4 Halogen measurements

The tetrahydrofuran extractable organic halogen content (EOX) of the extracts was analyzed with an Organic Halogen Analyzer (Euroglas, Delft, Holland) with EOX-equipment: 50 µl of the extract was injected into the oven where it was combusted at 850 °C under the flow of oxygen and argon and the halogen content was measured

Biomobility of organic halogen compounds 145

Table 1. Characteristics of the sawmill soils studied.

Soil Characteristics	Kaustinen	Toras
pH	3.9	6.3
Ignition loss (% (w/w))	7	55
Organic halogen content, µg Cl (g d. wt.)$^{-1}$	120 ± 20 (N=2)	134 ± 43 (N=2)

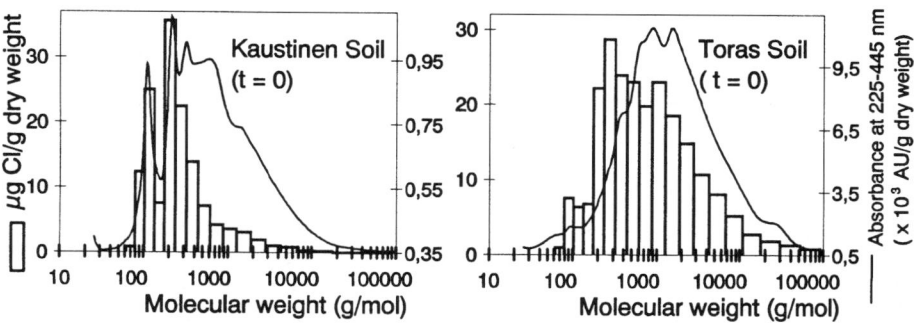

Fig. 1. Molecular weight distributions of halogenated (bars) and UV-absorbing (lines) compounds in sawmill soils Kaustinen (left panel) and Toras (right panel) (AU = Absorbance Unit).

microcoulometrically. Duplicate or triplicate injections were made from both replicate extracts. Tetrahydrofuran was used as a blank.

2.5 Molecular weight distribution analysis

The molecular weight distribution (MWD) of organic halogen compounds was analyzed by High Performance Size Exclusion Chromatography (HPSEC). The eluent used was tetrahydrofuran with the flow rate of 1 ml min^{-1}. Four Ultrastyragel (Waters) columns (10^4 Å, 10^3 Å, 500 Å and 100 Å as pore sizes) were used in series. Polystyrene compounds, lignin model compounds erol and bierol, vanillic alcohol and acetone were used as molecular weight standards (Jokela *et al.*, 1993). During the HPSEC run, the UV-vis-absorption in 225–445 nm and 264 nm was detected. After collecting fractions of 1 ml, the organic halogen content of each fraction was measured by evaporating tetrahydrofuran, combusting the residue in 1000 °C oven in a flow of oxygen and then measuring the halogen content microcoulometrically.

3. Results

3.1 Organic halogen compounds in contaminated sawmill soils

Soils from two sawmills were studied, one sandy (Kaustinen) and one with high organic matter (Toras). The characteristics of the two soils are shown in Table 1. MWDs of organic halogen compounds in Kaustinen and Toras soil extracts are illustrated in

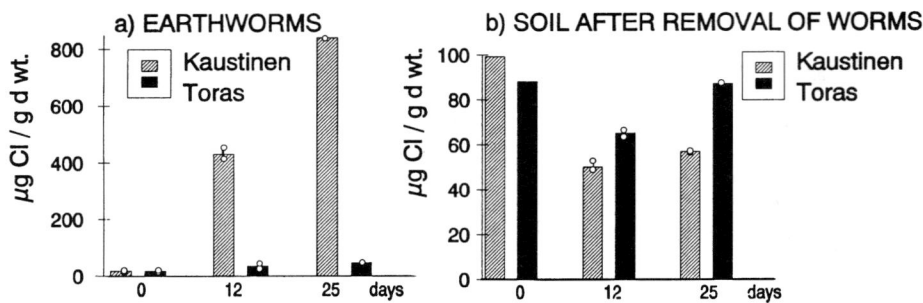

Fig. 2. Tetrahydrofuran extractable organic halogen concentrations a) in earthworms exposed to contaminated soils (Kaustinen and Toras) for 0, 12 and 25 d and b) in the Kaustinen and Toras soils 0, 12 and 25 d after the introduction of earthworms. Two replicate extractions were made except with earthworms (25 d) where only one extract was made. Minimum and maximum values of halogen contents are shown as dots.

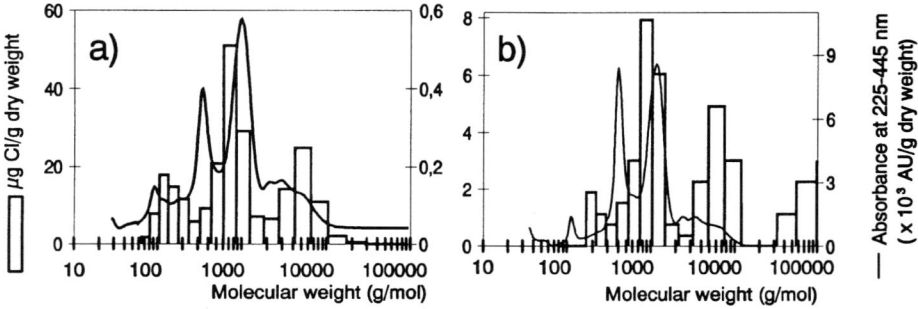

Fig. 3. Molecular weight distributions of halogenated (bars) and UV-absorbing (lines) compounds in earthworms after residing in contaminated sawmill soils for 12 d. a) Kaustinen, b) Toras.

Figure 1. The MWDs of halogen compounds were different in the soils. The organic halogen compounds found in Kaustinen soil were mainly in the size range of 100 to 1000 g mol^{-1}, although the wood preservative used by the mills was a technical chlorophenol mixture (Ky-5) of monomers and dimers with molecular weight below 500 g mol^{-1}. In the organic Toras sawmill soil, the halogen compounds appeared of even higher molecular size (100 to 10000 g mol^{-1}) than in the sandy Kaustinen soil. Hence, the chlorophenols of the technical wood preservative had undergone changes in the course of time, possibly by incorporation into soil humus.

3.2 *Organic halogen compounds in earthworms exposed to sawmill soils*

Are the high molecular weight organic halogen compounds present in contaminated sawmill soils biomobile? To answer this question, we exposed earthworms to two soils equally contaminated by organic halogen compounds (Table 1, Figure 2b). Figure 2a shows that earthworms effectively accumulated (840 µg Cl (g d. wt.)$^{-1}$) organic halogen compounds from the sandy Kaustinen soil. They also accumulated, to a smaller extent (35 µg Cl (g d. wt.)$^{-1}$), halogen compounds from the organic Toras soil. The bioconcentration factor (e. g. the ratio of concentration of halogen compounds per dry

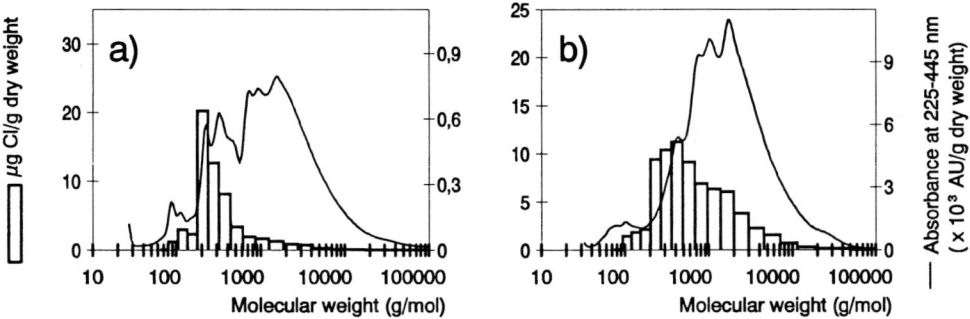

Fig. 4. Molecular weight distributions of halogenated (bars) and UV-absorbing (lines) compounds in the soils where the earthworms had been incubated for 12 d. a) Kaustinen, b) Toras.

weight earthworm to concentration per soil organic matter) was 30 in Toras earthworms and 600 in Kaustinen earthworms after 25 days of incubation. This indicates that bioaccumulation potential was much lower in organic than in sandy soil. The mortality of the earthworms was 14 percent in Kaustinen soil and 16 to 24 percent in Toras soil in 12 days. The mortality of the earthworms increased during the last 13 incubation days to 58 percent in both soils.

Next we characterized the earthworm accumulated halogen compounds by MWD (Figure 3). The MWDs of halogen compounds in all exposed earthworms had two major peaks with apparent molecular weights of 700 to 1000 g mol^{-1} and of 5000 to 7000 g mol^{-1}. When comparing the MWD of halogen compounds in the earthworms (Figure 3) with that in the soil (Figure 1) to which they were exposed, it was seen that the earthworms had material with higher molecular size than what was found in the soils. Figure 4 illustrates the MWDs of halogen compounds in soils where the earthworms had been incubated. The soil characteristics (pH, organic matter) did not affect the MWD pattern of organic halogen compounds accumulating in the earthworms.

4. Conclusions and discussion

The MWDs of halogen compounds differed in the two types of soil despite the presence of the same pollutant, the wood preservative Ky-5. In organic soil, the halogen compounds appeared in higher molecular size than in sandy soil. Phenoxyphenols and dioxins, (impurities in Ky-5), are even more resistant to biodegradation than chlorophenols (Valo & Salkinoja-Salonen, 1986). Even if the sawmill site had been abandoned for decades and chlorophenols may have partially degraded, the dimeric compounds do not explain the high molecular weight halogen compounds found in sawmill soils. Chlorophenols from the wood preservative may have bound to humus in the organic Toras soil to form the high molecular weight halogenated compounds monitored in this study. High organic matter increases absorption of, for example, pentachlorophenol to soil (Banerji et al., 1993; Warith et al., 1993). Xenobiotics are known to incorporate in humus (Bollag & Loll, 1983; Bollag, 1992) through abiotic or biotic reactions.

The earthworms accumulated tetrahydrofuran extractable halogen compounds up to 20 times more from sandy soil than from organic soil (Figure 2a). This indicates that soil organic matter diminished the bioaccumulation potential of halogenated compounds to the earthworms. Haimi *et al.* (1992), found bioaccumulation of chlorophenols by the earthworms to decrease with an increasing organic matter content of soil. Soil organic matter retards organic contaminants in soil, reducing their migration into ground water (Roberts *et al.*, 1985).

Toxicity of 2,4,6-trichlorophenol (artificial soil test) to *Eisenia fetida* was 58 mg (kg d. wt.)$^{-1}$ (Neuhauser *et al.*, 1985). According to Van Gestel and Ma (1988), toxicity of pentachlorophenol increased with decreasing organic matter in soil and with increasing soil pH. The mortality of the earthworms in the industrial field soil was rather high compared to that observed in an identical series of experiments at our laboratory where the earthworms were exposed to uncontaminated soil spiked with the wood preservative Ky-5 (Kauriala, M., unpublished data). For relevant results, the toxicity and bioaccumulation potential of pollutants in industrially contaminated soils must be studied using true field soils rather than soil reconstructed in the laboratory.

Halogenated compounds from earthworms incubated in contaminated soil appeared in large molecules. The MWDs of halogen compounds in all exposed earthworms followed the same pattern, having two major size classes: components with apparent molecular weights around 1000 g mol^{-1} and around 6000 g mol^{-1}, respectively. The nature of halogen binding to organic soil matter and to earthworm tissue components, as well as the individual chlorinated compounds (chlorophenols, chloroanisoles and phenoxyphenols) from the relevant SEC fractions is under study.

5. Summary

The bioavailability of organic halogen compounds in contaminated sawmill soils was studied using earthworms as model organisms. Earthworms were incubated in soils from old sawmills in order to find out whether and which of the halogen containing contaminants would migrate into the worms. Two different kinds of sawmill soils were used, one sandy soil and the other one with high organic matter (50 %). Earthworms from kitchen compost (*Eisenia andrei*) were incubated in contaminated soils spiked with horse manure for 12 or 25 days and the humidity was maintained by irrigation. After incubation, soils and earthworms were separated, freeze-dried, extracted with tetrahydrofuran and the extracts analyzed for the halogen compounds by high performance size exclusion chromatography. Despite the pollutant being of the same origin, (the wood preservative Ky-5), the molecular weight distributions of halogen compounds differed in the two soils: in organic soil, the halogen compounds appeared in higher molecular size than in sandy soil. The earthworms incubated in sandy soil accumulated twenty times higher concentrations (840 µg Cl/g d. wt.) of tetrahydrofuran extractable halogen compounds than the earthworms incubated in organic soil (35 µg Cl/g d. wt.), although the soils had equal content of organic halogen compounds. This indicates that organic soil matter diminishes the bioaccumulation potential of halogen compounds to the earthworms. The halogen compounds in the earthworms appeared in large molecules. The halogen compounds in all exposed earthworms appeared as two major size classes, one with an apparent molecular weight around 1000 g/mol and the other around 6000 g/mol.

Acknowledgements

We wish to thank Nordic Minister Council (Miljörisici, 69.13.02.00) and Maj and Tor Nessling Foundation for financial support of this study. Special thanks to Dr. Juhani Terhivuo from the Zoological Museum of University of Helsinki for identifying the earthworm species and to Dr. Jari Haimi from University of Jyväskylä for advice on how to handle the worms.

References

Banerji, S. K., S. M. Wei & R. K. Bajpai, 1993. Pentachlorophenol interactions with soil. Water, Air and Soil Pollution, 69:149–163.

Bollag, J.-M., 1991. Enzymatic binding of pesticide degradation products to soil organic matter and their possible release. In:Somasundaram, L. & J. R. Coats (eds.), ACS symposium series 459, *Pesticide transformation products, fate and significance in the environment*, Chapter 9, pp.122–132. American Chemical Society, Washington, DC.

Bollag, J.-M., 1992. Decontaminating soil with enzymes. Environ. Sci. Technol., 26(10):1876–1881.

Bollag, J.-M. & M. J. Loll, 1983. Incorporation of xenobiotics into soil humus. Experientia, 39:1221–1231.

Dec, J., K. L. Shuttleworth & J.-M. Bollag, 1990. Microbial release of 2,4-dichlorophenol bound to humic acid or incorporated during humification. J. Environ. Qual., 19:546–551.

Hague, A. & W. Eging, 1988. Uptake and accumulation of pentachlorophenol and sodium pentachlorophenate by earthworms from water and soil. The Science of the Total Environ., 68:113–125.

Haimi, J., J. Salminen, V. Huhta, J. Knuutinen & H. Palm, 1992. Bioaccumulation of organochlorine compounds in earthworms. Soil Biol. Biochem., 24(12):1699–1703.

Haimi, J., J. Salminen, V. Huhta, J. Knuutinen & H. Palm, 1993. Chloroanisoles in soil and earthworms. The Science of the Total Environment, Supplement 1993, Elsevier Science Publishers B. V., Amsterdam.

Jokela, J. K. & M. Salkinoja-Salonen, 1992. Molecular weight distributions of organic halogens in bleached kraft pulp mill effluents. Environ. Sci. Technol., 26(6):1190–1197.

Jokela, J., M. Laine & M. Salkinoja-Salonen, 1993. The effect of biological treatment on halogenated organics in bleached kraft pulp mill effluents by molecular weight distribution analysis. Environ. Sci. Technol., 27(3):547–557.

Kitunen, V., R. Valo & M. Salkinoja-Salonen, 1985. Analysis of chlorinated phenols, phenoxyphenols and dibenzofurans around wood preserving facilities. Intern. J. Environ. Anal. Chem., 20:13–20.

Lamar, R. T. & D. M. Dietrich, 1990. In situ depletion of pentachlorophenol from contaminated soil by *Phanerochaete* spp. Appl. Environ. Microbiol., 6 (10):3093–3100.

Neuhauser, E. F., R. C. Loehr, M. R. Malecki, D. L. Milligan & P. R. Durkin, 1985. The toxicity of selected organic chemicals to the earthworm *Eisenia fetida*. J. Environ. Qual,. 14(3):383–388.

Roberts, P. V., M. Reinhard, G. D. Hopkins & R. S. Summers, 1985. Advection-dispersion-sorption models for simulating the transport of organic contaminants. In: Ward, C. H, W. Giger & P. L. McCarthy (eds.), *Ground Water Quality*.pp. 427–429. John Wiley & Sons, Inc., USA.

Robinson, K. G. & J. T. Novak, 1994. Fate of 2,4,6-trichloro-(^{14}C)-phenol bound to dissolved humic acid. Wat. Res., 28(2):445–452.

Sandermann, H. Jr., M. Arjmaad, I. Gennity & R. Winkler, 1990. Animal bioavailability of defined xenobiotic lignin metabolites. J. Agric. Food Chem., 38:1877–1880.

Van Gestel, C. A. & W.-C. Ma, 1988. Toxicity and bioaccumulation of chlorophenols in earthworms in relation to bioavailability in soil. Ecotoxicol. and Environ. Safety, 15:289–297.

Van Gestel, C. A. & W. A. Van Dis, 1988. The influence of soil characteristics on the toxicity of four chemicals to the earthworm Eisenia andrei. Biol. Fertil. Soils, 6:262–265.

Valo, R. & M. Salkinoja-Salonen, 1986. Microbial transformation of polychlorinated phenoxy phenols. J. Gen. Appl. Microbiol., 30:505–517.

Warith, M. A., L. Fernandes, F. la Forge, 1993. Adsorption of pentachlorophenol on organic soil. Hazardous Waste & Hazardous Materials, 10(2):13–25.

Influence of soil and climatic factors on the kinetics of transformation of herbicides in soil

L. Torstensson & J. Stenström

Swedish University of Agricultural Sciences, Department of Microbiology
Box 7025, S-750 07 Uppsala, Sweden

Keywords: Decomposition, degradation, glyphosate, 2,4-D

1. Introduction

Herbicides have been used since the early part of this century. In spite of this fairly long history of chemical weed control, it is only recently that humans have become seriously concerned with the fate and behaviour of these chemicals in the environment. During this comparatively short period, a vast amount of knowledge on the environmental significance of chemicals has been generated. However, this knowledge mainly pertains to the appearance and effect of the substances under temperate and tropic conditions and very little under arctic-boreal conditions.

Herbicides on and in the soil are subject to photochemical, chemical and biological transformations (Torstensson, 1988). Most of these transformations are degradative and often remove the biological activity of the herbicide. The herbicide transformation rate is related to its distribution in the soil profile; to its chemical structure and formulation; and to physical, chemical, and biological properties of the soil.

Most herbicides are susceptible to photodecomposition under field conditions and a considerable amount of some herbicides may be transformed by this mechanism. Photochemical effects occur mainly before soil entry (that is, in the atmosphere or on plant surfaces), since diffusion and leaching in the soil normally ensure that herbicides are rapidly transported to sub-surface sites impenetrable by sunlight.

Non-biological reactions can contribute to essential stages in principally biological routes of degradation (Domsch, 1992). Probably, the two most important factors governing chemical transformation in soil are moisture content and pH, although other factors may also be involved, for instance, redox potential and free radical reactions. The biota in the soil may be partly responsible for the generation of these free radicals, *e.g.* via hydrogen peroxide produced by microbial extracellular oxidase enzymes.

The rates of transformation of herbicides in sterile and non-sterile soils show that most herbicides are transformed by microorganisms. The role of the soil microflora in such transformations and the influence of soil and climatic factors, are discussed in this paper.

2. Microbial transformation

2.1 *Microorganisms in soil*

The microbial population in the soil exists in a dynamic equilibrium governed by abiotic and biotic factors. The diversity of microbial species in the soil is great and many transformations of nutrients and energy are performed by microorganisms, in spite of the fact they occupy a volume of less than 0.1 percent of the soil.

Microorganisms can transform many chemicals: from simple polysaccharides, amino acids, proteins, lipids, etc., to more complex materials such as plant residues, waxes, and rubbers. Without the activity of microorganisms, such organic compounds would accumulate in the environment. Microorganisms are also able to transform chemical compounds synthesized by humans by a variety of mechanisms.

2.2 *Mechanisms of transformation*

Microbial metabolic activities need energy. An important criterion used in classifying microbial transformations of herbicides is whether or not the microorganisms derive energy from the process. This consideration is significant from a practical point of view, since it is important for prediction of the persistence of the herbicide. Metabolic decomposition of a herbicide thus means that the compound, by serving as an energy source, supports growth of the active organisms (Torstensson, 1987). Repeated application of a herbicide to the same field, often results in a faster decomposition because of the increased amount of degraders existing in the soil.

In cometabolic transformation the microorganisms are not able to use the herbicide as an energy source and therefore depend on accessible carbon sources for their general metabolic activity. Thus, cometabolic transformation is an incidental transformation of herbicides by microorganisms which grow on utilizable carbon and energy sources. It also describes transformation of non-utilizable substrates by resting cells. The rate of turnover by cometabolism is largely governed by the activity of the soil biomass, which in turn is often limited by the availability of carbon and energy sources (Torstensson, 1987).

2.3 *Kinetics*

To make adequate predictions of the fate and the persistence of herbicides in the soil, the kinetics of transformation have to be considered. Kinetic studies may also provide information on rate-limiting mechanisms.

2.3.1 *Metabolic decomposition*

The rate of metabolic decomposition depends on the initial microbial activity qN_0 *(i.e.* the product of the specific activity q and the initial amount of active microorganisms N_0) and the microbial growth rate μ (Stenström, 1989a). Such metabolic decomposition is then described by the equation

$$c = c_0 - \frac{qN_0}{\mu}(e^{\mu t} - 1) \tag{1}$$

Influence of soil and climatic factors

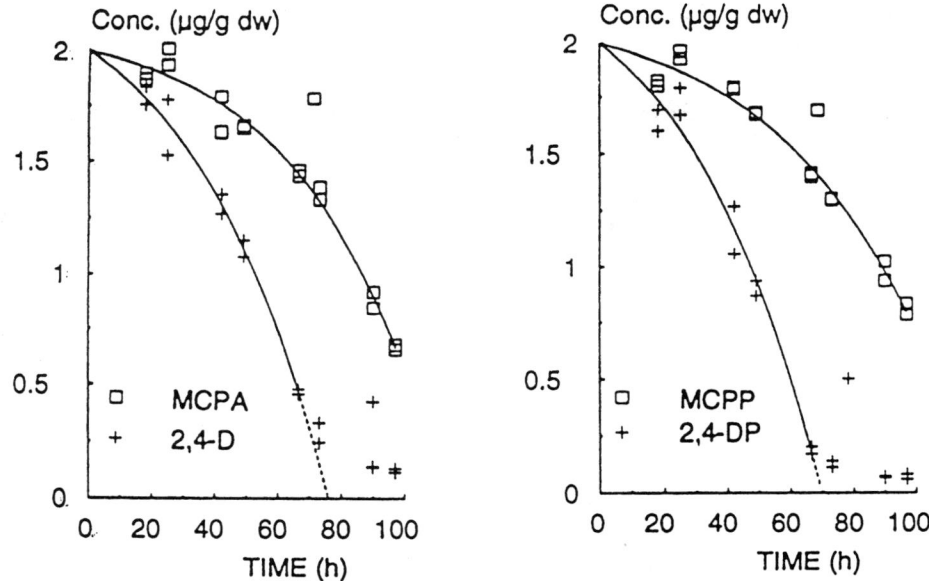

Fig. 1. Metabolic decomposition of the phenoxy acids MCPA, 2,4–D, MCPP and 2,4–DP in soil. Solid lines describe best fits of equation (1) and connect data that were included in the regression analysis (Stenström, 1992).

Table 1. Parameters ±SE of equation (1) obtained by non–linear regression of data on metabolic decomposition of phenoxy acids in soil (Stenström, 1992).

Phenoxy acid	qN_0 (pmole g^{-1}h^{-1})	μ (h^{-1})	Relative qN_0
2,4-D	44.75 ± 5.20	0.0226 ± 0.0032	2.7
2,4-DP	50.45 ± 7.17	0.0225 ± 0.0039	3.0
MCPA	16.66 ± 1.94	0.0245 ± 0.0020	1.0
MCPP	18.83 ± 2.15	0.0201 ± 0.0020	1.1

where c denotes the concentration of the herbicide at time t and c_0 its initial concentration.

Decomposition of the phenoxy acids (2,4-dichlorophenoxy)acetic acid (2,4-D), (4-chloro-2-methylphenoxy)acetic acid (MCPA), (±)-2-(2,4-dichlorophenoxy)propionic acid (2,4-DP) and (±)-2-(4-chloro-2-methylphenoxy)propionic acid (MCPP) was studied by Stenström (1992). The best fit of equation (1) to the data is shown in Fig. 1. According to the numerical values obtained for the parameters qN_0 and μ (Table 1), organisms that degrade the different phenoxy acids grow at approximately the same rate (mean value of μ: 0.0224 h^{-1}; generation time: 31 h) but the initial activity of organisms capable of degrading MCPA and MCPP, is about one third of those capable of degrading 2,4-D and 2,4-DP in the soil used.

Since the initial total activity depends on both q and N_0, variations in one or both of these parameters can give the difference obtained. Nevertheless, the results suggest that the main reason for different persistence of these herbicides in a certain soil, when they are decomposed metabolically, is not the microbial growth rate supported by the chemicals, which is approximately equal, but rather the initial microbial activity.

2.3.2 Cometabolic and chemical transformation

The equation for first-order kinetics is conventionally used to describe cometabolic and chemical transformation of organic compounds. However, in spite of often good fits to experimental data, application of this equation to processes in the soil has been questioned, since the first-order rate constant is often dependent on the initial concentration of the chemical (Hamaker, 1972; Stenström, 1989b) and since such kinetics sometimes fail to describe the transformation, especially when followed to low concentrations (Gustafson & Holden, 1990; Scow & Alexander, 1992).

An alternative to the equation for first-order kinetics for description of transformation data is the parabolic equation (Stenström, 1989b)

$$c = c_0 - k_1 \sqrt{t} \qquad (2)$$

which also often describes diffusion-limited processes (Aharoni & Sparks, 1991). The rate constant k_1 of equation (2) was found to be proportional to c_0 for degradation of low c_0 of linuron (3-(3,4-dichlorophenyl)-1-methoxy-1-methylurea) (Stenström, 1989b). Also, the equation for first-order kinetics and equation (2) fitted degradation data just as well, especially for the initial part of the degradation curve, and they often failed to describe the data in the later part of the curve. Differentiating equation (2) gives

$$-\frac{dc}{dt} = \frac{k_1}{2\sqrt{t}} \qquad (2')$$

Assuming that this rate law describes a process for delivery of the chemical to sites (*e.g.* microorganisms, the aqueous phase) where it can be transformed, but that the rate eventually also becomes first-order with respect to c, we obtain

$$-\frac{dc}{dt} = \frac{k_2 c}{2\sqrt{t}} \qquad (3')$$

and by integration

$$\ln c = \ln c_0 - k_2 \sqrt{t} \qquad (3)$$

Equation (3) is valuable for predictions of the persistence of chemicals in the soil to very low concentrations.

Equations (2') and (3') suggest that the following events occur when a chemical is

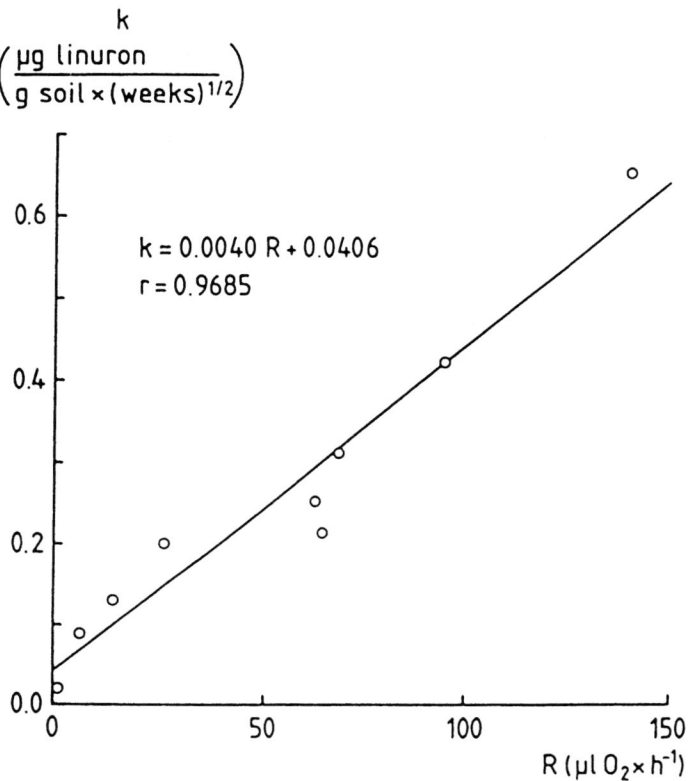

Fig. 2. Correlation between the rate of respiration (R) and the rate of decomposition (k) of linuron for some soils in a laboratory experiment (Torstensson & Stenström, 1986).

added to a soil (Stenström, 1992): The chemical may remain in solution, be adsorbed to mineral surfaces, be absorbed into organic material and migrate into aggregates, where it may be unavailable for degradation. Some of these processes are fast, but others take place on time scales comparable to degradation processes. The distribution of the chemical in and on respective phase will, of course, depend on the properties of the chemical and the soil, but it is probable that the 'load' (*i.e.* the absolute amounts distributed) and the rates of exchange between phases, are proportional to the initial concentration c_0. When high concentrations are available for degradation (*i.e.* in solution and at water/soil interfaces) the kinetics will be according to equation (1) or zero-order, depending on factors required for metabolic decomposition. Eventually, this easily degraded fraction is consumed by degradation or by continuing migration into catalytically inactive phases, and the degradation rate then becomes limited by the rate at which the chemical reaches active sites. Equation (2) is then valid. The degradation of the chemical at such sites prevents equilibrium from being reached between different phases, and thus constitutes a driving force for migration and thus also for degradation. The rate constant k_1 may then be correlated to the microbial activity, when cometabolic processes significantly contribute to this driving force. Eventually, the probability for a chemical to reach an active site is not just dependent on its migration rate, but also on its total remaining concentration, and equation (3) starts to be valid.

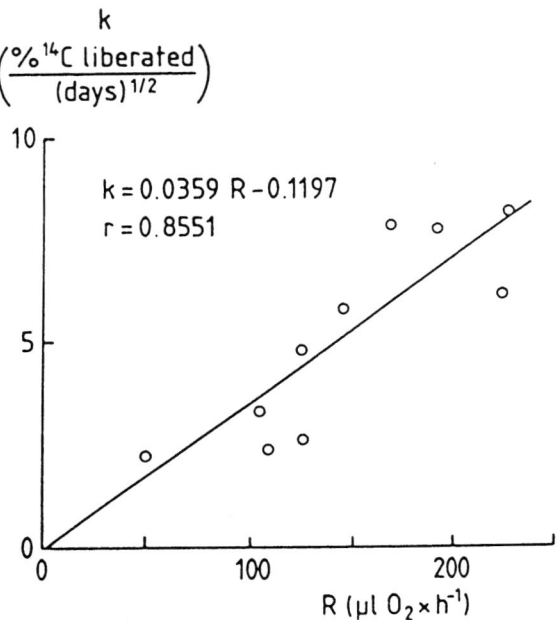

Fig. 3. Correlation between the rate of respiration (R) and the rate of decomposition (k) of ^{14}C–labelled glyphosate for some soils in a laboratory experiment (Torstensson & Stenström, 1986).

2.4 Rate of transformation

The rate of microbial transformation of herbicides in soil is a function of three variables: (a) the availability of the compounds to the microorganisms or enzyme systems which can transform them; (b) the quantity of these microorganisms or enzyme systems; and (c) the activity level of these organisms or enzyme systems (Freshe & Anderson, 1983). Edaphic factors, such as organic matter and clay content, moisture, temperature, pH, aeration, and nutrient status, are important as regulators. Inhibition of the microbial activity, by, for example, low or high temperature, drought, waterlogging, extremes of pH and xenobiotic substances, may result in the persistence in soil of potentially decomposable and mineralizable compounds.

Even with a great biomass of soil, there will be no degradation of a herbicide unless the biomass is active. Frehse & Anderson (1983) have shown that there was no correlation between the quantity of the microbial biomass and the rate of degradation of diallate (S-2,3-dichloroallyl di-isopropyl (thiocarbamate)) in soil. However, there was a direct correlation between the activity level of the biomass and the rate of degradation of the herbicide. Generally it can be said that environmental conditions which allow good growth and activity of microorganisms also favour the biodegradation of pesticides (Domsch, 1984).

The general biological activity level of the soil, measured for example as O_2 consumption or CO_2 evolution, is often a good indicator of the capacity of the soil to transform substances by cometabolism. However, this relationship has not always been found in earlier investigations, probably since the soils used in degradation experiments are normally dried, sieved and then remoistened again. These disturbances un-

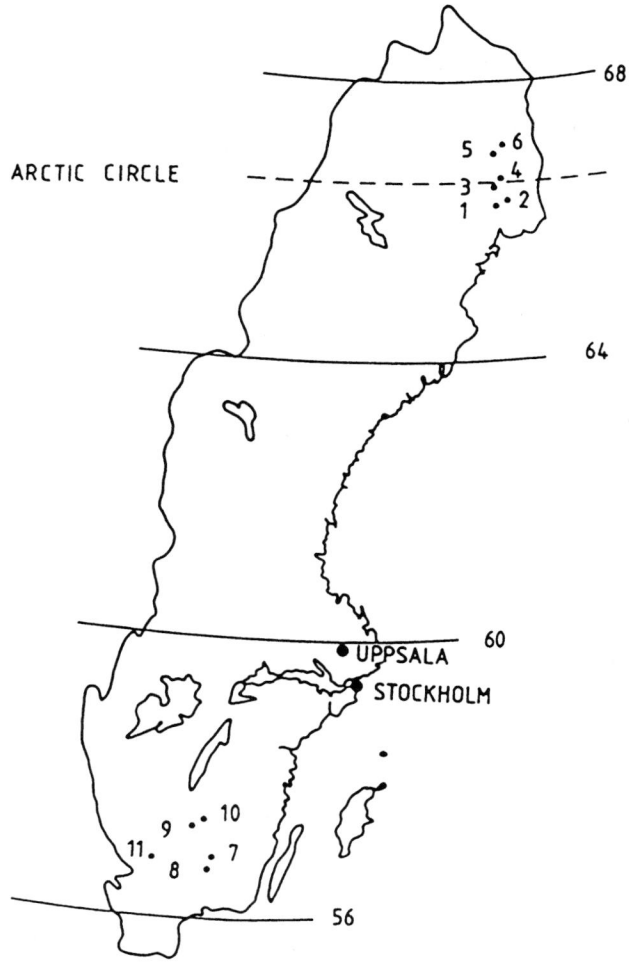

Fig. 4. Location of experimental sites in two climatically different areas of Sweden where the persistence of the herbicides glyphosate and 2,4–D was investigated (Torstensson *et al.*, 1989).

doubtedly affect the respiration, and high rates of respiration are often observed at the beginning of such experiments. However, after some time, the rate of respiration drops to a nearly constant level; the basic rate of respiration. Torstensson & Stenström (1986) made use of this experience and showed that a pre-incubation time of some days before the respiration rate was determined, gave a good correlation between the basic respiration rate and the transformation rate. This is expressed by the rate constant k_1 in equation (2), for the herbicides linuron and glyphosate (N-(phosphonomethyl)glycine) (Figs. 2 and 3).

3. Field experiments

The persistence in the field of the two herbicides glyphosate and 2,4-D was investigated after their application for brush control in conifer reforestation sites in two cli-

Table 2. Soil type and some characteristics of the uppermost 0– to 5-cm layer at the various experimental sites (Torstensson *et al.*, 1989).

Site	Soil type	Characteristics of the 0- to 5-cm layer		
		pH	Loss on ignition (%)	Basic respiration rate (μl O_2/h/sample)
1	Iron podsol	4.7	25	430
2	Iron podsol	5.4	57	494
3	Iron podsol	5.3	21	495
4	Iron podsol	5.0	60	409
5	Iron humus podsol	4.9	81	562
6	Iron podsol	4.9	40	527
7	Iron podsol	5.0	33	267
8	Brown soil	5.5	21	407
9	Iron humus podsol	4.5	40	382
10	Iron humus podsol	4.4	24	340
11	Iron podsol	5.5	11	330

matically different areas of Sweden (Fig. 4). Six sites (1 to 6) were located close to the Arctic Circle (about 66° N) and five sites (7 to 11) were located in southern Sweden (about 57° N). Some characteristics of the experimental sites are given in Table 2. Spraying was done in August according to the normal schedule used for coniferous reforestation. The persistence of the herbicides was assessed from the remaining concentration of the parent compounds only. Further details about the experiment are given by Torstensson *et al.*, (1989).

The mean amounts of glyphosate and 2,4-D found in the soil samples at various times after their application are shown in Figs. 5 and 6, respectively. The standard error of the mean was 4 to 16 percent at the different sites (n = 15).

Glyphosate reached low levels faster at the northern sites than at the southern sites during the weeks after the application. The mean amount of glyphosate per sample was 2.4 µg, 47 days after application at the northern sites. Not until 87 days after application, did the mean amount of residues reach the same low level at the southern sites.

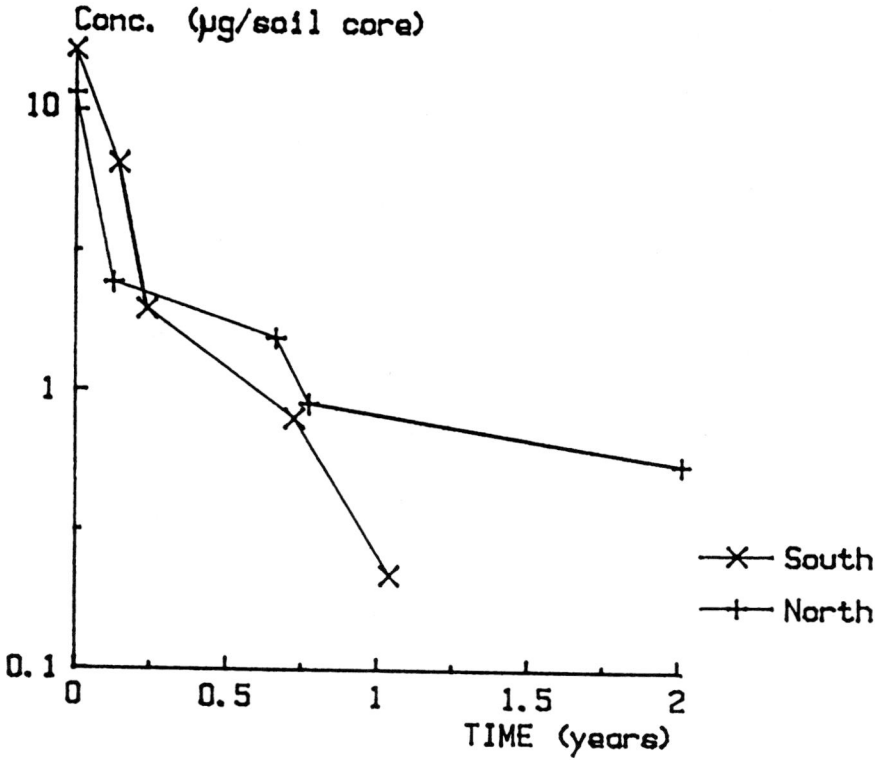

Fig. 5. Remaining concentrations of glyphosate in northern and southern parts of Sweden (mean values) (Torstensson et al., 1989).

The data on the decomposition of glyphosate from the time of application to the last sampling time in the first autumn were fitted to equation (2). However, since the rate constant k_1 is directly proportional to c_0 when c_0 is low (Stenström, 1989b), i.e. $k_1 = ac_0$ and $a = k_1/c_0$, and since the dose reaching the soil in the different field experiments varies due to the vegetation cover, the c_0-independent rate constant a, but not k_1, can be used to compare the decomposition of the herbicide in the different soils. A plot of a against the basic respiration rate gives a high correlation (Fig. 7). It is noteworthy that the northern soils have a higher basic respiration rate than the southern ones, and that this higher activity gives a higher capacity to degrade glyphosate.

During winter, with ground frost for about seven months at the northern sites and four months at the southern sites, the reduction in glyphosate level was small. Moreover, during the following summer, the disappearance was comparatively slow at all sites, and even two years after application small amounts (0.2 to 1.0 µg glyphosate per sample) were found at the northern sites. The slow disappearance of these minor residues of glyphosate is probably not due to insufficient capacity of the soil to degrade the herbicide, but rather to absorbed glyphosate being slowly released from binding sites in vegetation residues. The same is probably true for the small amounts of 2,4-D remaining two years after application; 0.1 to 0.7 µg per sample from the northern sites.

The concentration of 2,4-D decreased rapidly at all sites, and about seven weeks af-

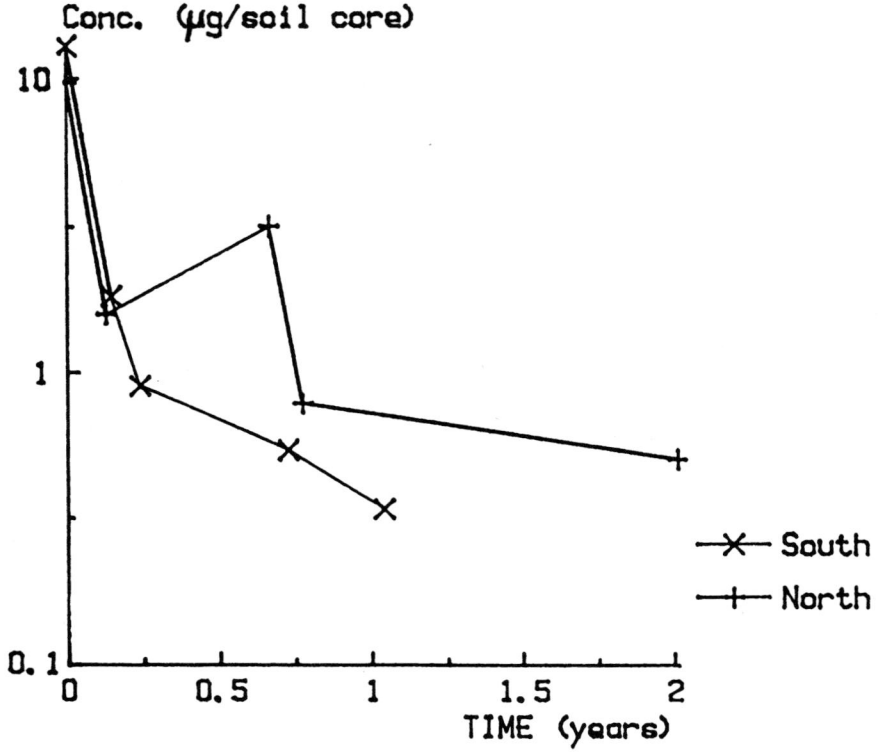

Fig. 6. Remaining concentrations of 2,4–D in northern and southern parts of Sweden (mean values) (Torstensson *et al.*, 1989).

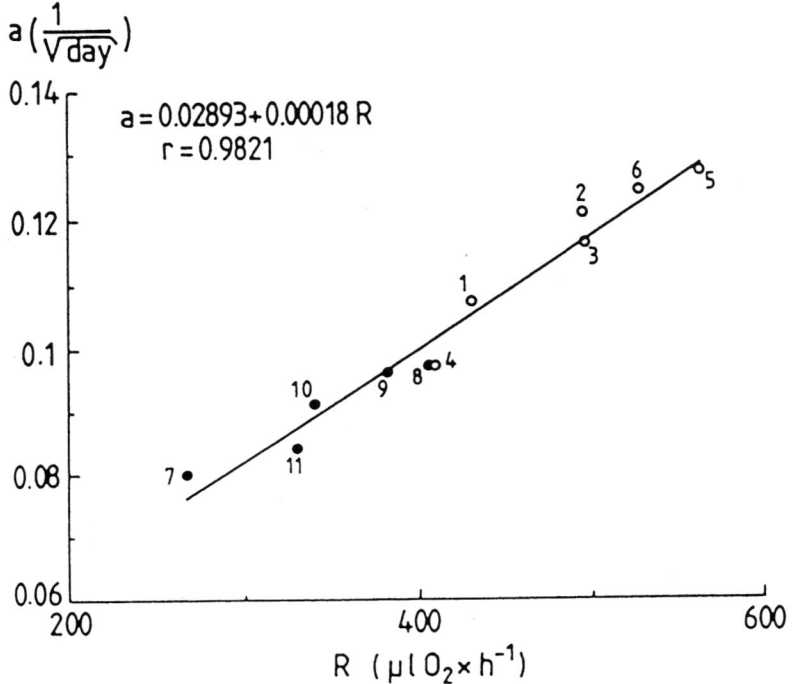

Table 3. Amounts of 2,4–D (mg/kg dry wt.) found in samples of birch residues at two experimental sites in northern Sweden (1 and 5) and two in southern Sweden (8 and 10) (Torstensson et al., 1989). ND = Not determined.

		Site			
Time	Vegetation residues	1	5	8	10
1 year	Leaves	20	11	<1	<1
	Branches	60	35	10	2
2 years	Leaves	<1	<1	ND	ND
	Branches	40	12	<1	<1
3 years	Branches	5	3	ND	ND

ter application only a few percent of the amounts originally applied were found. This suggests that 2,4-D is rapidly used as an energy source for metabolic decomposition by soil microorganisms. The important rate-determining edaphic and climatic factors – water, pH, and temperature – do not seem to have been limiting to any appreciable extent during the period.

Thus, it can be concluded that all soils investigated in this field experiment, both in the south and the north, have a good degradative capacity. The metabolically degraded herbicide 2,4-D is initially degraded at similar rates in the both areas. The cometabolically transformed herbicide glyphosate may initially be degraded as fast as or even faster in the northern area as compared to the southern area. However, the long period of frozen soil gives a longer total persistence of herbicides in the northern area compared to the southern area.

4. Summary

Herbicides on and in the soil are subject to photochemical, chemical, and biological transformations. The herbicide transformation rate is related to its distribution in the soil profile; to its chemical structure and formulation; and to the physical, chemical, and biological properties of the soil. Most herbicides are transformed by microorganisms in the soil.

The rates of transformation of herbicides are largely governed by the activity of the soil biomass. An important criterion used in classifying the enzymatic reactions involved in microbial transformations of herbicides is whether or not the microorganisms derive energy from the process, (i.e. metabolic and cometabolic transformation respectively). Metabolic transformation is the direct decomposition of herbicides when they serve as energy source to support growth. Cometabolic transformation is an incidental

←

Fig. 7. Correlation between the rate of respiration (R) and the capability to decompose glyphosate (a) in field experiments in northern (1–6) and southern (7–11) Sweden (Torstensson et al., 1989).

transformation of herbicides during the growth of microorganisms on a utilizable carbon and energy source. It also describes transformation of non-utilizable herbicides by resting cells. The energy derived from such a transformation does not support microbial growth.

The kinetics of transformation of herbicides in the soil suggest that the chemical often becomes available to the microorganisms by diffusion processes, *e.g.* in soil pores. The structure of the soil is thus an important factor for the transformation.

The persistence of one metabolically decomposed herbicide (2,4-D) and one cometabolically decomposed herbicide (glyphosate) was investigated after an application for brush control in conifer reforestation areas. Field experiments were carried out at different sites in the north and the south of Sweden. The 2,4-D disappeared rapidly from all sites. Initially, glyphosate disappeared faster in the northern soils than in the southern soils. However, its total persistence was longer in the northern area, probably due to the long period during which the soil remained frozen.

Low concentrations of both herbicides could be found in the soil even two years after application, probably as a result of adsorbed herbicides being slowly released from binding sites in vegetation residues.

References

Aharoni, C. & D. L. Sparks, 1991. Kinetics of soil chemical reactions – a theoretical treatment. In: D. L. Sparks & D. L. Suarez (eds.), *Rates of Soil Chemical Processes.* pp. 1–18. SSSA Spec. Publ. 27. SSSA, Madison, WI.
Domsch, K. H., 1984. Principles of pesticide-microbe interactions in soil. In: J. Szegi (ed.), *Soil Biology and Conservation of the Biosphere.* pp. 179–184. Akadémia Kiadó, Budapest.
Domsch, K. H., 1992. *Pestizide im Boden. Mikrobieller Abbau und Nebenwirkungen auf Mikroorganismen.* VCH, Weinheim. 575 pp.
Frehse H. & J. P. E. Anderson, 1983. Pesticide residues in soil – problems between concept and concern. In: R. Greenhalgh & N. Drescher (eds.), *IUPAC Pesticide Chemistry, Human Welfare and the Environment, Vol. 4, Pesticide Residues and Formulation Chemistry.* pp. 23–32. Pergamon Press, Oxford.
Gustafson, D. I. & L. R. Holden, 1990. Nonlinear pesticide dissipation in soil: A new model based on spatial variability. Environ. Sci. Technol., 24: 1032–1038.
Hamaker, J.W., 1972. Decomposition: Quantitative aspects. In: C. A. I. Goring & J. W. Hamaker (eds.), *Organic Chemicals in the Soil Environment.* pp. 253–340. Marcel Dekker Inc., New York.
Scow, K. M. & M. Alexander, 1992. Effect of diffusion on the kinetics of biodegradation: experimental results with synthetic aggregates. Soil Sci. Soc. Am. J. 56: 128–134.
Stenström, J., 1989a. Kinetics of decomposition of 2,4-dichlorophenoxyacetic acid by *Alcaligenes eutrophus* JMP134 and in soil. Toxicity Assessment, 4: 405–424.
Stenström, J., 1989b. Quantitative assessment of herbicide decomposition at different initial concentrations. Toxicity Assessment, 4: 53–70.
Stenström, J., 1992. Rate determining factors for the decomposition of pesticides in soil. In: J. P. E. Anderson, D. J. Arnold, F. Lewis & L. Torstensson (eds.), *Proc. Int. Symp. Environmental Aspects of Pesticide Microbiology, 17–21 Aug., Sigtuna, Sweden.* pp. 141–146. SUAS, Uppsala.
Torstensson, L., 1987. Microbial decomposition of herbicides in soil. In: D. H. Hutson & T. R. Roberts (eds.), *Progress in Pesticide Biochemistry and Toxicology.* pp. 249–270. Wiley, New York.
Torstensson, L., 1988. Microbial decomposition of herbicides in the soil. Outlook on Agriculture, 17: 120–124.
Torstensson, L. & J. Stenström, 1986. 'Basic' respiration rate as a tool for prediction of pesticide persistence in soil. Toxicity Assessment, 1: 57–72.
Torstensson, N. T. L., L. N. Lundgren & J. Stenström, 1989. Influence of climatic and edaphic factors on persistence of glyphosate and 2,4-D in forest soils. Ecotox. Environ. Safety, 18: 230–239.

Variation of cyanobacterial hepatotoxins in Finland

K. Sivonen,[1] M. Namikoshi,[2] R. Luukkainen,[1] M. Färdig,[1] L. Rouhiainen,[1]
W. R. Evans,[3] W. W. Carmichael,[3] K. L. Rinehart,[2] & S. I. Niemelä[1]

[1]*Department of Applied Chemistry and Microbiology, P. O. Box 27,
00014 University of Helsinki, Finland*
[2]*University of Illinois, Department of Chemistry, Urbana, Illinois 61801, U.S.A.*
[3]*Wright State University, Department of Biological Sciences, Dayton,
Ohio 45435, U.S.A.*

Keywords: *Anabaena, Microcystis, Oscillatoria, Nostoc*

1. Introduction

Toxic cyanobacterial (blue-green algal) blooms commonly occur in eutrophic lakes worldwide (Carmichael *et al.*, 1985; Carmichael, 1992, 1994). Toxic water blooms have caused several animal poisonings and are a threat to human health (Carmichael *et al.*, 1985; Gorham & Carmichael, 1988). Cyanobacteria produce neurotoxins and peptide hepatotoxins. In a survey conducted form 1985 to 1987 in Finland 45 per cent of 215 freshwater bloom samples were found to be toxic (Sivonen, 1990). Hepatotoxic blooms were more common than neurotoxic blooms as they are also elsewhere. Cyanobacterial hepatotoxins are either pentapeptide (nodularin) produced by *Nodularia spumigena* in brackish water or heptapeptides (microcystins) found in freshwaters. Microcystins (MCYST) have the general structure *cyclo*(-D-Ala^1-X^2-D-MeAsp^3-Z^4-Adda^5-D-Glu^6-Mdha^7-), where X and Z are variable L-amino acids, D-MeAsp is D-*erythro*-ß-methylaspartic acid, Mdha is *N*-methyldehydroalanine and Adda is (2*S*,3*S*,8*S*,9*S*)-3-amino-9-methoxy-2,6,8-trimethyl-10-phenyldeca-4,6-dienoic acid (Carmichael *et al.*, 1988). Toxins cause hepatocyte necrosis, which leads to pooling of blood in the liver. Death of the animals in acute poisoning is caused by hypovolemic shock (Beasley *et al.*, 1989). Recently microcystins and nodularin were shown to be inhibitors of protein phosphatases 1 and 2A (MacKintosh *et al.*, 1990; Matsushima *et al.*, 1990) and potential tumour promoters (Nishiwaki-Matsushima *et al.*, 1992). We have isolated several hepatotoxic strains of the genera *Anabaena, Microcystis, Oscillatoria* and *Nostoc*. The purpose of this study was to summarize the results of the isolation and characterization of hepatotoxins in the strains and selected bloom samples.

2. Materials and Methods

Cyanobacterial strains were isolated from different lakes in Finland. *Anabaena* strain 83/1 and *Microcystis aeruginosa* strain 972 originated from Norway and Russia, respectively. Toxin composition of eight *Anabaena* spp. strains (Namikoshi *et al.*, 1992a, 1992b; Sivonen *et al.*, 1992a, 1992d), five *Microcystis* spp. strains and two blooms

Table 1. Microcystins (MCYST) isolated and characterized from *Anabaena* spp. strains and their relative abundance (%).

Toxin	FABMS M+H m/z	\<br\>60[a]	\<br\>66[b]	*Anabaena* strain\<br\>83/1[c]	\<br\>90[d]	\<br\>141[d]	\<br\>202A1[d]	\<br\>202A2[d]	\<br\>186[e]
MCYST-LR	995	45		36	57	14			
[D-Asp³]MCYST-LR	981	10		18	15	28			
[Dha⁷]MCYST-LR	981						35	35	
[D-Asp³,Dha⁷]MCYST-LR	967						5	10	
[L-Ser⁷]MCYST-LR	999						4	10	
MCYST-RR	1038	35		12	28	20			
[D-Asp³]MCYST-RR	1024	10		24		38			
[Dha⁷]MCYST-RR	1024						45	45	
[D-Asp³,Dha⁷]MCYST-RR	1010						5		
[L-Ser⁷]MCYST-RR	1042						4		
[L-Ser⁷]MCYST-HtyR	1063		20						
[Dha⁷]MCYST-HtyR	1045		55						
[D-Asp³,Dha⁷]MCYST-HtyR	1031		15						
[Dha⁷]MCYST-HphR	1029		10						
[D-Glu(OCH₃)⁶]MCYST-LR	1009			5					
[D-Asp³,D-Glu(OCH₃)⁶] MCYST-LR	995			5					
[D-Asp³,L-Ser⁷]MCYST-XR	999					2			
unknown	936								x
unknown	976								x
unknown	998								x
unknown	1002								x
unknown	1016								x
unknown	1020								x
unknown	1024								x
unknown	984								x
unknown	1022								x
unknown	1054								x

Strain isolated from [a]L. Ylä-Keyritty, Finland; [b]L. Kiikkara, Finland; [c]L. Edlandsvatn, Norway; [d]L. Vesijärvi, Finland; [e]L. Vaaranlampi, Finland.
x = relative abundance not determined.
Abbreviations: Hty = homotyrosine; X = unknown amino acid (probably leusin homologue; Namikoshi et al., 1992b)

(Kiviranta et al., 1992; Luukkainen et al., 1994; Sivonen et al., 1992c), 13 *Oscillatoria agardhii* strains (Luukkainen et al., 1993), and one *Nostoc* sp. strain (Sivonen et al., 1992b) was studied. The strains were grown in batch cultures, in the inorganic nutrient medium Z8 at 22 °C with continuous illumination of about 50 microeinsteins $m^{-2}s^{-1}$. *Anabaena* and *Nostoc* strains were grown in a nitrogen free medium. The cells were harvested after 10 to 12 days of incubation and lyophilized. Toxins were extracted by water, or water with organic solvents, and purified by high-performance liquid chromatography (HPLC) and thin-layer chromatography (TLC). Amino acid composition was determined by the Waters Pico Tag method and/or gas chromatography (GC) on a chiral capillary column. Structures were assigned by fast atom bombardment mass

Table 2. Microcystins (MCYST) isolated and characterized from *Microcystis* spp. strains and bloom samples and their relative abundance (%).

Toxin	FABMS M+H m/z	Strain 98[a]	136[b]	199[c]	205[d]	972[e]	Bloom 116[f]	205[d]
MCYST-LR	995	21					1	
[D-Asp³]MCYST-LR	981						1	
[Dha⁷]MCYST-LR	981	69	2.5	16		29		
[D-Asp³,Dha⁷]MCYST-LR	967			6		14		
MCYST-RR	1038						96	
[D-Asp³]MCYST-RR	1024			3				3
[Dha⁷]MCYST-RR	1024	8	95	66	88	29		96
[D-Asp³,Dha⁷]MCYST-RR	1010				11	14		
[L-Ser⁷]MCYST-RR	1042		1	1	1			1
MCYST-YR	1045						1	
[Dha⁷]MCYST-YR	1031			9		14		
[Dha⁷]MCYST-FR	1015	1						
unknown	1035	1						
unknown	1040			1				1
unknown	1049						1	
unknown	1058		2.5	1	1			1

Strain or bloom originates from [a]L. Pyhäjärvi, Finland; [b]L. Hirsijärvi, Finland; [c]L. Rusutjärvi, Finland; [d]L. Mallusjärvi, Finland; [e]L. Kroshnosero, Karelia, Russia; [f]L. Pitkäjärvi, Finland.

spectrometry (FABMS), collisionally induced tandem mass spectrometry (FABMS/MS) and ^{1}H NMR spectroscopy (Kiviranta *et al.*, 1992; Luukkainen *et al.*, 1993; Namikoshi *et al.*, 1992a, 1992b; Sivonen *et al.*, 1992a, 1992b, 1992c, 1992d).

3. Results

Toxins isolated and characterized from *Anabaena*, *Microcystis*, *Oscillatoria* and *Nostoc* are shown in Tables 1–3 and Figure 1. All strains produced more than one toxin simultaneously. From *Anabaena*, *Microcystis*, *Oscillatoria* and *Nostoc*, 27 (19 of which

Table 3. Microcystins (MCYST) isolated and characterized from Oscillatoria agardhii strains and their relative abundance (%).

Toxin	FABMS (M+H)	Oscillatoria strain											
		18[a]	49[b]	97[c]	195[d]	209[d]	212[e]	213[d]	214[f]	223[g]	226[d]	CYA 126[a]	CYA 127[h]	CYA 128[h]
[D-Asp³]MCYST-LR	981		1		11	2	8	4			4	19		
[Dha⁷]MCYST-LR	981	1		1		2	4	4		1	1			1
[D-Asp³,Dha⁷]MCYST-RR	1010		1					5						
[D-Asp³]MCYST-RR	1024		78	99	88	96	88	87	97		95	81	10	1
[Dha⁷]MCYST-RR	1024	99	19							99			90	98
[D-Asp³,Mser⁷]MCYST-RR	1042		1		1				1					
unknown	1044								1					
unknown	1058								1					

Strain isolated from [a]L. Långsjön, Finland; [b]L. Valkjärvi, Finland; [c]L. Maarianallas, Finland; [d]L. Haukkajärvi, Finland; [e]L. Köyliönjärvi, Finland; [f]L. Östra Kyrksundet, Finland; [g]L. Kotojärvi, Finland; [h]L. Vesijärvi, Finland.

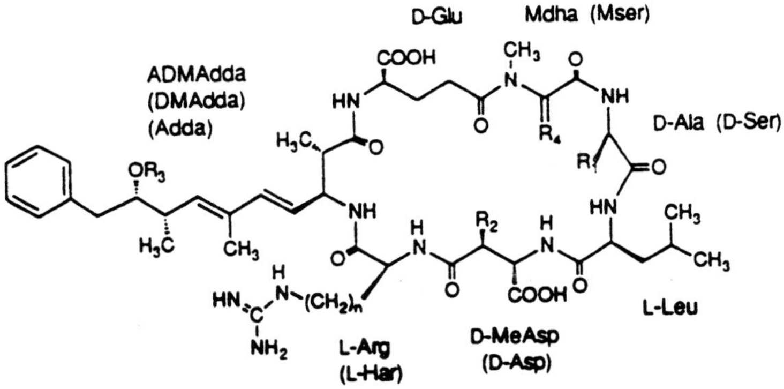

Variation of cyanobacterial hepatotoxins in Finland

	R_1	R_2	R_3	R_4	n
[D-Ser^1,ADMAdda^5]MCYST-LR	CH_2	$OHCH_3$	$COCH_3$	CH_2	3
[D-Asp^3,ADMAdda^5]MCYST-LHar	CH_3	H	$COCH_3$	CH_2	4
[ADMAdda^5,Mser^7]MCYST-LR	CH_3	CH_3	$COCH_3$	H,CH_2OH	3
[ADMAdda^5]MCYST-LR	CH_3	CH_3	$COCH_3$	CH_2	3
[ADMAdda^5]MCYST-LHar	CH_3	CH_3	$COCH_3$	CH_2	4
[D-Asp^3,ADMAdda^5]MCYST-LR	CH_3	H	$COCH_3$	CH_2	3
[DMAdda^5]MCYST-LR	CH_3	CH_3	H	CH_2	3

Fig. 1. Microcystin-LR (MCYST) homologs isolated and characterized from *Nostoc* sp. strain 152. Two new toxins, molecular weights of 985 and 1055, remain to be solved. Abbreviations: ADMAdda = 9-*O*-demethylAdda; DMAdda = 9-*O*-demethylAdda; Har = homoarginine.

were new), 16 (6 new), 8 (3 new) and 9 (9 new) toxins were found, respectively. The structures of 18 different new microcystins were determined and 19 remain to be solved. MCYST-RR and MCYST-LR and especially their demethylated (amino acid number 3 or 7 or both) variants were the most abundant and most frequently occurring toxins among *Anabaena*, *Microcystis* and *Oscillatoria* strains. All *Oscillatoria* strains produced only one major toxin (Table 3) while the strains of *Anabaena* (Table 1) usually produced two to four main toxins simultaneously. *Oscillatoria* strains produced only demethylmicrocystins. Microcystins that contain tyrosine were found only in *Microcystis* spp. samples (Table 2). New variants of microcystins were produced by a *Nostoc* with modified Adda, homoarginine or D-serine in place of D-alanine (Fig. 1). *Anabaena* 66 produced four microcystins, three containing homotyrosine (Hty) and one homophenylalanine (Hph) (Table 1). *Anabaena* strain 186 synthesized new compounds, the structures of which have not yet been determined.

4. Discussion

The structures of over 40 cyanobacterial hepatotoxins are known to date – about half of them were characterized in this study. In Finnish freshwaters, cyanobacteria produce a wide variety of microcystins. The variation of toxins as well as the number of species producing these toxins seem to be greater in lakes, than in the Baltic Sea or freshwaters in warmer climates. Only nodularin, a pentapeptide hepatotoxin, produced by *Nodularia spumigena* has been found in the Baltic Sea to date (Sivonen *et al.*, 1989). In Japan, 35 *Microcystis aeruginosa* and *M. viridis* strains were studied and only three microcystins (MCYST-LR, -RR and -YR) were found (Watanabe *et al.*, 1988). In the Finnish samples, demethyl microcystins seem to be more common than reported elsewhere. According to the literature, MCYST-LR is the most common microcystin worldwide (Carmichael, 1992). Structurally similar toxins are produced by different species but some strains produced only new varieties of microcystins. The qualitative and quantitative variation of microcystins was greatest among *Anabaena* and lowest among the isolates of *Oscillatoria*. Studies of biogenesis and genetics of cyanobacterial microcystins will be needed to reveal why so many varieties of these compounds are produced at a time and why certain compounds are found as the major toxins.

5. Summary

Microcystins (MCYST), cyclic heptapeptide hepatotoxins were isolated from several *Anabaena*, *Oscillatoria*, *Microcystis* and a *Nostoc* strains and two *Microcystis* bloom samples. Most of the samples originated from Finnish lakes. Toxins were isolated by high-performance liquid chromatography (HPLC) and thin-layer chromatography (TLC) and characterized by amino acid analysis, fast atom bombardment mass spectrometry (FABMS), collisionally induced tandem FABMS (FABMS/MS) and ^{1}H NMR spectroscopy. Two to ten microcystins were isolated from each strain and bloom sample. The most common toxins in these samples were MCYST-RR and MCYST-LR and their demethyl variants: [D-Asp^3]MCYST-LR, -RR, [Dha^7]MCYST-LR, -RR, [D-Asp^3,Dha^7]MCYST-LR and -RR. These same toxins were produced by strains which belonged to different genera such as *Anabaena*, *Microcystis* and *Oscillatoria*. All *Oscillatoria* strains and certain strains of *Anabaena* and *Microcystis* produced only demethyl microcystins. Two *Anabaena* strains and a *Nostoc* produced only new varieties of microcystins. The structural variation of microcystins was greatest among *Anabaena* and smallest among the isolates of *Oscillatoria*.

Acknowledgements

The research at the University of Helsinki was supported by grants from the Academy of Finland, the Maj and Tor Nessling Foundation, and the University of Helsinki. The research at Wright State University and the University of Illinois was supported by a grant from the National Institute of Allergy and Infectious Diseases (AI 04769) to K.L.R. and by a subcontract from this grant to W.W.C.. Mass spectrometry was provided in part by a grant from the National Institute of General Medical Sciences (GM 27029) to K.L.R.

References

Beasley, V. R., A. M. Dahlem, W. O. Cook, W. M. Valentine, R. A. Lowell, S. B. Hooser, K.-I. Harada, M. Suzuki & W. W. Carmichael, 1989. Diagnostic and clinically important aspects of cyanobacterial (blue-green algae) toxicoses. J. Vet. Diagn. Invest., 1: 359–365.
Carmichael, W. W., 1992. Cyanobacterial secondary metabolites – the cyanotoxins. J. Appl. Bact., 72: 445–459.
Carmichael, W. W., 1994. The toxins of cyanobacteria. Scientific American. Jan. –94: 64–72.

Carmichael, W. W., V. R. Beasley, D. L. Bunner, J. N. Eloff, I. Falconer, P. Gorham, K.-I. Harada, T. Krishnamurthy, M.-J. Yu, R. E. Moore, K. L. Rinehart, M. Runnegar, O. M. Skulberg & M. Watanabe, 1988. Naming of cyclic heptapeptide toxins of cyanobacteria (blue-green algae). Toxicon, 26: 971–973.

Carmichael, W. W., C. L. A. Jones, N. A. Mahmood & W. C. Theiss, 1985. Algal toxins and water-based diseases. CRC Crit. Rev. Environ. Control, 15: 275–313.

Gorham, P. R. & W. W. Carmichael, 1988. Hazards of freshwater blue-green algae (cyanobacteria). In: C. A. Lambi & J. R. Waaland (eds.), *Algae and Human Affairs*. Cambridge University Press, pp. 403–431.

Kiviranta, J., M. Namikoshi, K. Sivonen, W. R. Evans, W. W. Carmichael & K. L. Rinehart, 1992. Structure determination and toxicity of a new microcystin from *Microcystis aeruginosa* strain 205. Toxicon, 30: 1093–1098.

Luukkainen, R., M. Namikoshi, K. Sivonen, K. L. Rinehart & S. I. Niemelä, 1994. Isolation and identification of 12 microcystins from four strains and two bloom samples of *Microcystis* spp.: structure of a new hepatotoxin. Toxicon, 32: 133–139.

Luukkainen, R., K. Sivonen, M. Namikoshi, M. Färdig, K. L. Rinehart & S. I. Niemelä, 1993. Isolation and identification of eight microcystins from thirteen *Oscillatoria agardhii* strains and structure of a new microcystin. Appl. Environ. Microbiol., 59: 2204–2209.

MacKintosh, C., K. A. Beattie, S. Klumpp, P. Cohen & G. A. Codd, 1990. Cyanobacterial microcystin-LR is a potent and specific inhibitor of protein phosphatases 1 and 2A from both mammals and higher plants. FEBS Lett., 264: 187–192.

Matsushima, R., S. Yoshizawa, M. F. Watanabe, K.-I. Harada, M. Furusawa, W. W. Carmichael & H. Fujiki, 1990. *In vitro* and *in vivo* effects of protein phosphatase inhibitors, microcystin and nodularin, on mouse skin and fibroblasts. Biochem. Biophys. Res. Commun., 171: 867–874.

Namikoshi, M., K. Sivonen, W. R. Evans, W. W. Carmichael, L. Rouhiainen, R. Luukkainen & K. L. Rinehart, 1992a. Structures of three new homotyrosine-containing microcystins and a new homophenylalanine variant from *Anabaena* sp. strain 66. Chem. Res. Toxicol., 5: 661–666.

Namikoshi, M., K. Sivonen, W. R. Evans, W. W. Carmichael, F. Sun, L. Rouhiainen, R. Luukkainen & K. L. Rinehart, 1992b. Two new L-serine variants of microcystis-LR and -RR from *Anabaena* sp. strains 202 A1 and 202 A2. Toxicon, 30: 1457–1464.

Nishiwaki-Matsushima, R., T. Ohta, S. Nishiwaki, M. Suganuma, K. Kohyama, T. Ishikawa, W. W. Carmichael & H. Fujiki, 1992. Liver tumor promotion by the cyanobacterial cyclic peptide toxin microcystin-LR. J. Cancer Res. Clin. Oncol., 118: 420–424.

Sivonen, K., 1990. Toxic cyanobacteria in Finnish fresh waters and the Baltic Sea. Reports from Department of Microbiology, University of Helsinki 39. 87pp.

Sivonen, K., K. Kononen, W. W. Carmichael, A. M. Dahlem, K. L. Rinehart, J. Kiviranta & S. I. Niemelä, 1989. Occurrence of the hepatotoxic cyanobacterium *Nodularia spumigena* in the Baltic Sea and the structure of the toxin. Appl. Environ. Microbiol., 55: 1990–1995.

Sivonen, K., M. Namikoshi, W. R. Evans, W. W. Carmichael, F. Sun, L. Rouhiainen, R. Luukkainen & K. L. Rinehart, 1992a. Isolation and characterization of a variety of microcystins from seven strains of the cyanobacterial genus *Anabaena*. Appl. Environ. Microbiol., 58: 2495–2500.

Sivonen, K., M. Namikoshi, W. R. Evans, M. Färdig, W. W. Carmichael & K. L. Rinehart, 1992b. Three new microcystins, cyclic heptapeptide hepatotoxins from *Nostoc* sp. strain 152. Chem. Res. Toxicol., 5: 464–469.

Sivonen, K., M. Namikoshi, W. R. Evans, B. V. Gromov, W. W. Carmichael & K. L. Rinehart, 1992c. Isolation and structures of five microcystins from a Russian *Microcystis aeruginosa* strain CALU 972. Toxicon, 30: 1481–1485.

Sivonen, K., O. M. Skulberg, M. Namikoshi, W. R. Evans, W. W. Carmichael & K. L. Rinehart, 1992d. Two methyl ester derivatives of microcystins, cyclic heptapeptide hepatotoxins, isolated from *Anabaena flos-aquae* strain CYA 83/1. Toxicon, 30: 1465–1471.

Watanabe, M. F., S. Oishi, K.-I. Harada, K. Matsuura, H. Kawai & M. Suzuki, 1988. Toxins contained in *Microcystis* species of cyanobacteria (blue-green algae). Toxicon, 26: 1017–1025.

Human impacts on the Hudson Bay Region: Present and future environmental concerns

P. G. Sly

P. O. Box 2032, Piction, Ontario, Canada, K0K 2T0

Keywords: ecosystem, settlement, watershed, native, food-web

1. Introduction

In 1991, the Canadian Arctic Resources Committee (CARC), the Rawson Academy of Aquatic Science (RAAS), and the Environmental Committee of Sanikiluaq (an Inuit community on the Belcher Islands in south-east Hudson Bay), decided to jointly plan a research programme on existing and potential impacts in the bay of development occurring within the Hudson Bay watershed. Of particular concern are the likely cumulative impacts of development. The programme formally began in February 1992. The intent of this contribution is to summarize present scientific knowledge about the Hudson Bay region and to set the scene for further, more detailed research which will define probable long term impacts of development. The following text has been drawn from documents specifically prepared for the Hudson Bay programme (Sly, 1994), however, present discussion is somewhat limited by the length of this publication.

A detailed bibliography of published and unpublished material covering the Hudson Bay region has been prepared (RAAS, 1993), and a survey of traditional environmental knowledge (TEK) is being undertaken by residents of Hudson Bay basin communities (Okrainetz, 1992). The original science contribution is being used to assist in the collection and interpretation of TEK information.

Basin-wide overviews (Martini, 1986a; Stewart *et al.*, 1991) provide a broad base of knowledge about the region. Further, a number of recent meetings have reviewed the extent of knowledge about parts of the Hudson Bay basin and the effects and potential effects of cumulative impact on the region (Adam *et al.*, 1987; Allan *et al.*, 1992; Bunch & Reeves, 1992; Greig *et al.*, 1992; Lawrence *et al.*, 1992; Makivik Corp., 1991; Percy, 1990). Recent summaries (Shearer, 1991; Murray & Shearer, 1992) have described the current focus of research in the northern contaminant programme of the federal government. The views and ideas of many researchers are reflected in the present contribution which considers the impact of human activities and development on the Hudson Bay basin, as a whole.

While there may appear to be a significant amount known about the Hudson Bay region, it should be recognised that information is frequently available only from a few specific locations and the density of information is extremely low. Often the ability to link sets of information between sites and over time is very tenuous. The extent of knowledge, particularly about the ecology of the basin, is more apparent than real. Nutrient and toxic contaminant loadings in the region are generally known but the extent to which they have impacted the environment is poorly defined. In many ways, we don't know what we don't know.

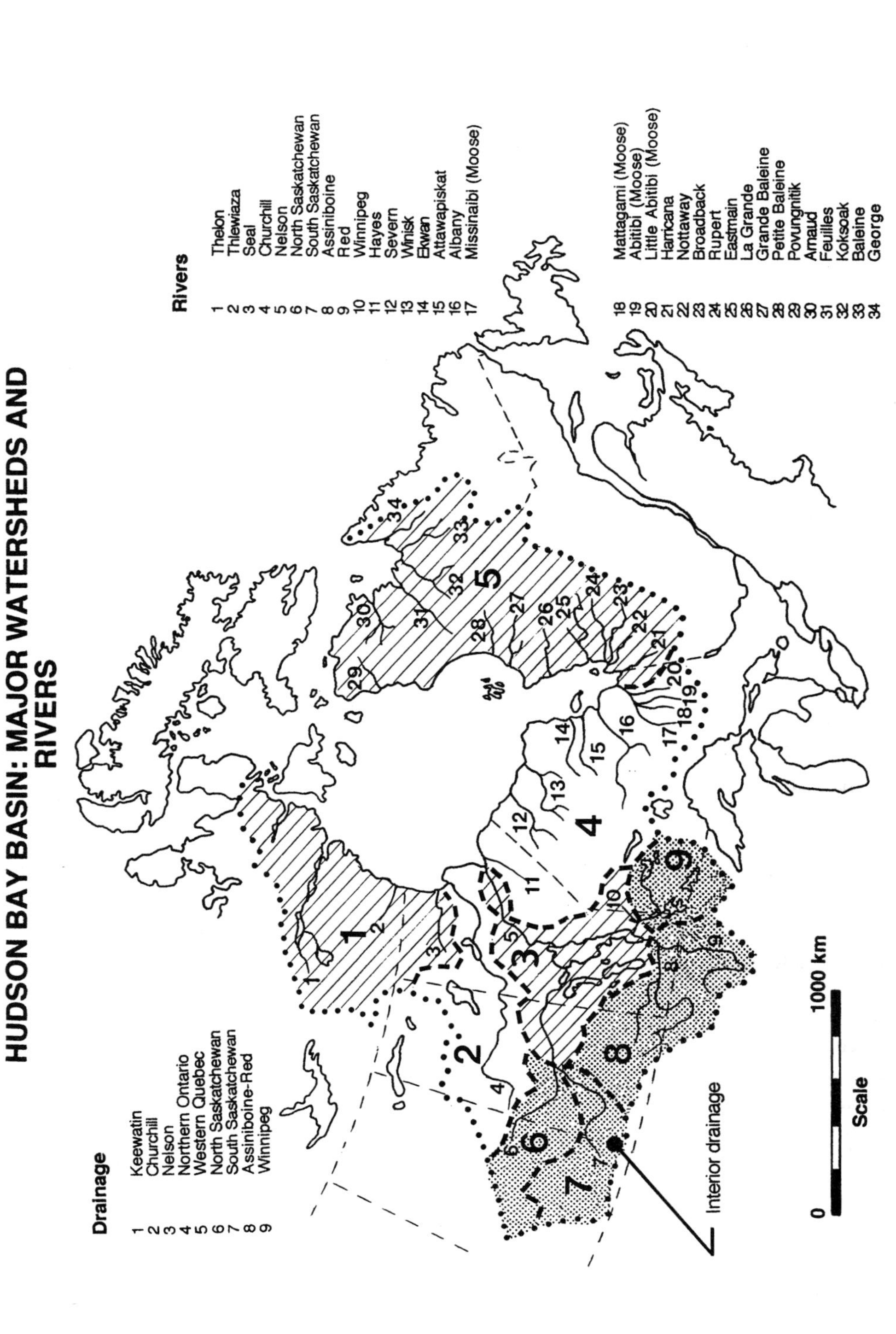

Fig. 1. Hudson Bay Basin: Major Watersheds and Rivers (Sources: Pearse et al., 1985; Statistics Canada, 1986).

2. The watershed

2.1 *General characteristics*

For the purpose of this contribution, the Hudson Bay basin is taken to include Hudson Bay, James Bay, Foxe Basin, and Hudson Strait. The surrounding area within the Hudson Bay watershed (Fig. 1) extends southward from the Arctic Circle for a distance of more than 2000 km. The extent, east to west, is more than 4000 km. Hudson Bay drainage covers about 39 percent the Canadian landmass and the Hudson Bay region includes both the basin and its drainage.

Throughout most of the watershed, relief is slight to moderate and it is only at the margins that elevations rise much above 700 m (Douglas, 1969). A complex of islands which increase in both elevation and relief from west to east, form the northern margin of the basin. The Frobisher Upland of southern Baffin Island, reaches a maximum elevation of about 1000 m, and the eastern Baffin Highlands exceed 2500 m (Fig. 2). Parts of the Labrador Highlands rise above 1000 m. The Severn and Abitibi Highlands, which form much of the southern limit of the watershed have maximum elevations between 300 and 400 m. To the west of the basin, land rises more or less continuously across the Manitoba, Saskatchewan and Alberta plains to reach an elevation of more than 800 m. In the southern part of the Rocky Mountains, maximum elevations of the western watershed exceed 3500 m. The extensive lowlands surrounding much of Hudson Bay reflect the process of isostatic recovery. Land is emerging from the sea in response to the release of geologic stress imposed by massive ice-loading during the glacial past (Douglas, 1969). The Hudson Bay Lowland is covered by extensive areas of wetland (Mortsch, 1990; Wickware & Rubec, 1989).

Five major drainage systems enter directly into Hudson Bay. These include the river drainage of west and northwest Quebec, northern Ontario, the Nelson River drainage, the Churchill River drainage, and the Keewatin drainage in the western part of the Northwest Territories (Fig. 1). In addition, four major drainage systems cover the southwest interior and feed into the Nelson River. These include the Winnipeg, Red and Assiniboine, South Saskatchewan, and North Saskatchewan drainage systems. The hydrology of the region is summarized in Table 1. Run-off reflects an abundance of supply in the western Quebec drainage, and the scarcity of water in the North and South Saskatchewan, the Nelson, and the Assiniboine-Red interior drainage systems.

Crystalline and highly altered rocks of the Canadian Shield underlie most of the Quebec and Keewatin drainage, and much of Ontario and Manitoba (Fig. 3). Platforms of old sedimentary rock formations and patchy glacial and post glacial deposits occur across parts of Quebec, much of northern Ontario and south-central Manitoba and Saskatchewan; they completely cover western Saskatchewan and Alberta (Douglas, 1969). The effects of bedrock geology are clearly seen in the hardness of surface water (Pearse et al., 1985). Minimum hardness occurs in the Keewatin drainage. A range of intermediate values characterize the variable nature of rock formations that are typical of the Quebec, Ontario, Winnipeg and Nelson and Churchill drainage systems, and values of maximum hardness occur in the South Saskatchewan drainage.

Natural vegetation is an effective indicator of climate (EWG, 1989), expressed as ecoclimatic provinces (Fig. 3). Ecozones are used to express a combination of landform and ecoclimate. Table 2 indicates the relative amount of each ecozone within the watershed. Much of the watershed is covered by boreal forest which is typical of moderate to cool summers and long, cold winters. Boreal Shield areas are predominant in Ontario (including the southern half of the Hudson Bay Lowland) and Quebec.

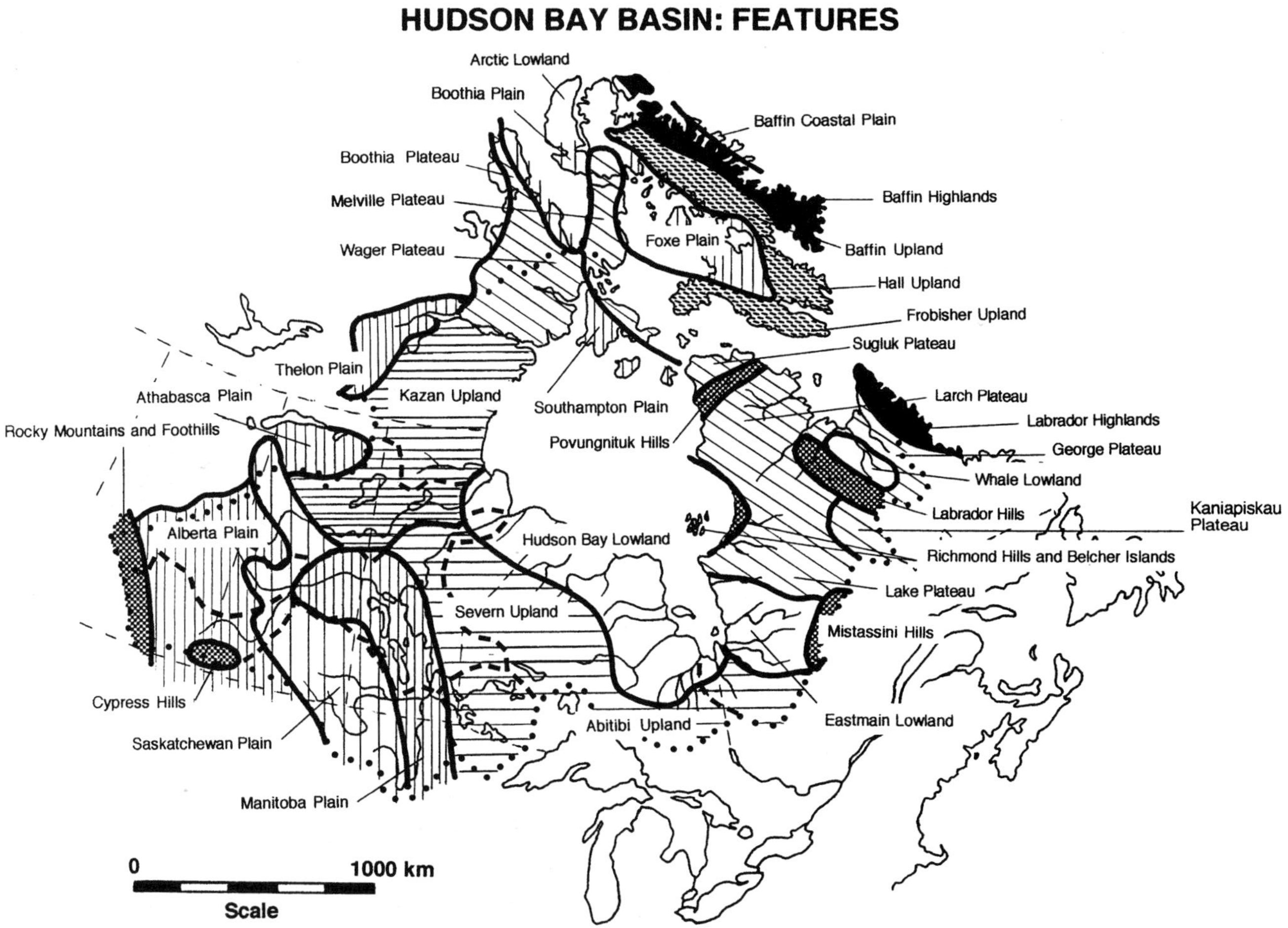

Fig. 2. Hudson Bay Basin: Features (Source: Douglas, 1969).

Table 1. Hydrology of the Hudson Bay Basin (Sources: Pearse et al., 1985; Prinsenberg, 1980; Statistics Canada, 1986).

Drainage System	Area (x 10^3 km^2)	Estimated reliable flow (x 10^6 m$^3 \cdot$ a^{-1})	Basin transfer, out (+) or in (-)	Present Flow estimate after transfers ** (x 10^6 m$^3 \cdot$ a^{-1})	Run-off (mm $\cdot$ a^{-1})	Hardness range (mg $\cdot$ L^{-1})
1) Keewatin	689	92,870		92,870	134.8	0 - 50
2) Churchill	298	33,905	-23,715	10,190	113.8	0 - 200
3) Nelson	363	11,222	+23,715	34,937 *	30.9	0 - 200
4) N. Ontario	694	125,394	-7,694	117,700	180.7	0 - 400
5) Quebec	950	404,300		404,300	425.8	0 - 200
6) N. Sask.	146	5,046		5,046	34.6	200 - 700
7) S. Sask.	170	4,636		4,636	27.3	200 - 1000
8) As. -Red	190	497		497	2.6	200 - 700
9) Winnipeg	107	9,328	+2,712	12,040	87.2	50 - 200

* Total discharge of the Nelson River into Hudson Bay (interior drainage from watersheds 6–9 and from the lower Saskatchewan-Nelson River system) is 55,087 × 10^6 m$^3 \cdot$a^{-1}.
** The total estimated discharge into Hudson Bay is 682,216 × 10^6 m$^3 \cdot$a^{-1}; this is equivalent to an average continuous flow rate of 21,634 m$^3 \cdot$s^{-1}.
The average ratio of low:high flow is 1:2 for rivers in the Quebec drainage, 1:2.5 for rivers in the Ontario drainage, and 1:3.0 for rivers in the Manitoba drainage.

Boreal Plain areas are predominant in Manitoba and Saskatchewan. Grassland occurs in the rain shadow of the Rocky Mountains and covers most of the Prairie. It is characterized by low precipitation and, often, by extremes of summer heat and winter cold. The Montane Cordillera includes both the foothills and the mountains of western Alberta, at the margin of the watershed. Taiga extends as a broad band from the Northwest Territories across the southern Hudson Bay and into Quebec. It includes the northern half of the Hudson Bay Lowland and the Eastmain Lowland (jointly referred to as the Hudson Bay Plain, Fig. 2). Taiga is present north of the continuous forest, where tundra vegetation occurs together with sparse stunted trees. Much of the Taiga Shield occurs in Quebec. Tundra and barren lands are typical of Arctic ecoclimatic provinces in which summers are cool and winters are long and very cold.

2.2 Population and water use

The total population of the region is more than 4.3 million (Table 3). This is about 18 percent of the population of Canada and, based on per capita income (Statistic Canada, 1986), the region provides about 16 percent of the national income. Considerable differences exist across the region. Incomes in the Prairie zone are more than double those in the Hudson Bay Plain. Populations are greatest in precisely those areas which have the least available supply of water. More than 86 percent of the people live within the interior drainage of the North and South Saskatchewan, and the Assiniboine and Red rivers (Pearse et al., 1985). Consumptive use is especially high in the South Saskatchewan and Assiniboine-Red River drainage systems. The major population centres in each drainage system and trends in recent population are summarized in Table 4. Increases are most noticeable in the drainage of the Saskatchewan and Nelson rivers.

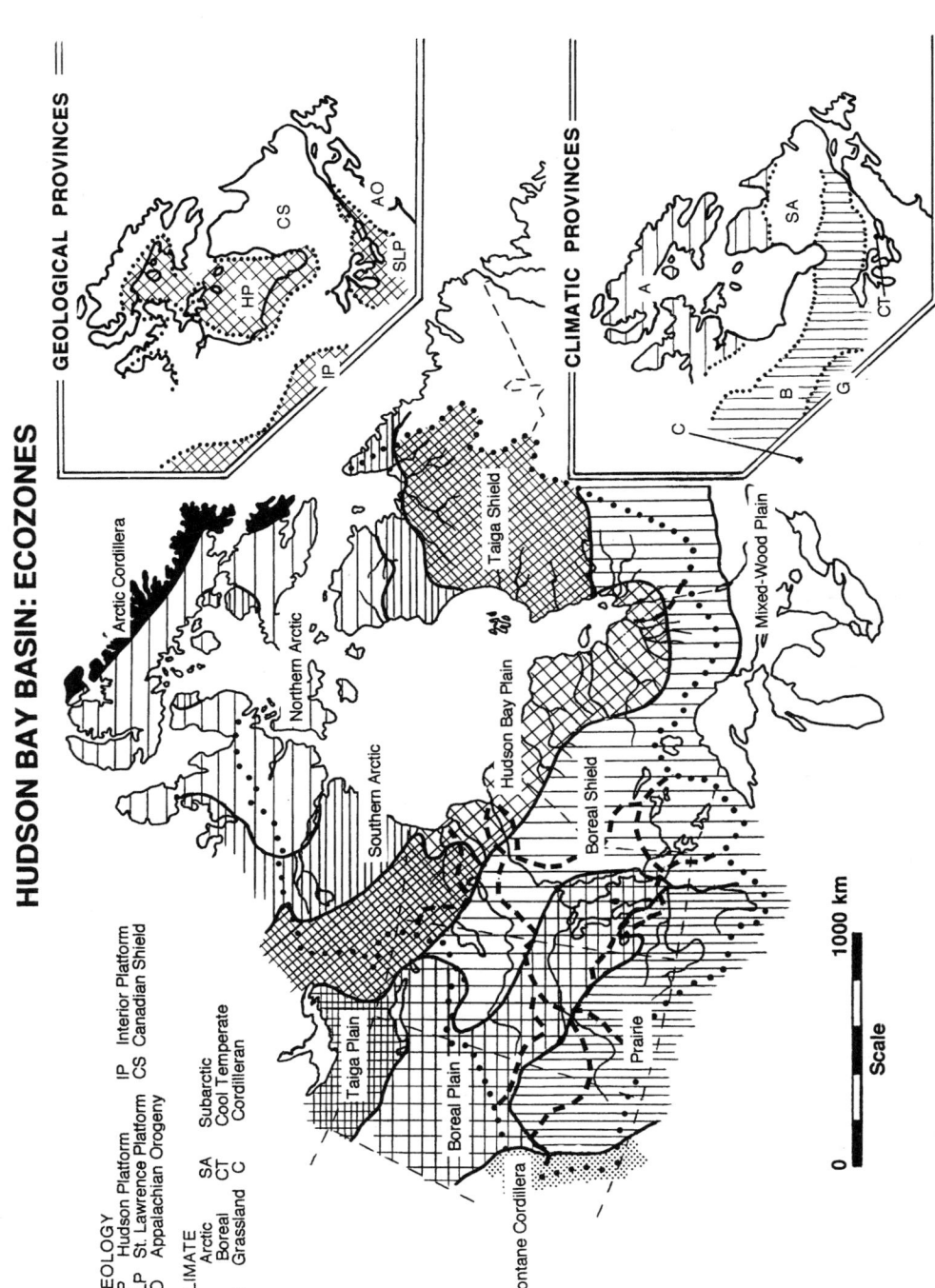

Table 2. Ecozones Within the Watershed (Sources: EWG, 1989; Statistics Canada, 1986).

Ecozone	Province (%)					
	N.W.T.	Alberta	Sask.	Manitoba	Ontario	Quebec
Boreal Shield			5	15	45	35
Boreal Plain		30	40	30		
Prairie		35	50	15		
Cordilleran		100				
Taiga Shield	15			25		60
Hudson Bay Plain			5	20	70	5
Southern Arctic	45					55
Northern Arctic	85					15

Values indicate approximate area (percentage) of watershed within each ecozone, see also Fig. 3.

Table 3. Hudson Bay Basin: Regional Water Use (Source: Pearse *et al.*, 1985).

	Population (x 1000)	Consumptive use as a % of total flow	All withdrawl as a % of total flow **	Munic -ipal	Rural resid.	Agricult -ural	Mining*	Manuf -acture	Thermo- electric*
				(Values as percent use in each drainage system)					
Drainage System									
1) Keewatin	5	n	n				100		
2) Churchill	68	n	n		100				
3) Nelson	224	<0.5	n	10.7	3.3	23.0	4.1	58.9	
4) N. Ontario	157	n	n	11.7			7.8	80.5	
5) Quebec	109	n	n				38.0	62.0	
6) N. Sask.	1,084	3	28	9.4	1.2	6.7	3.9	6.4	72.4
7) S. Sask.	1,282	36	56	10.2	0.8	76.1	0.3	4.7	7.9
8) As.-Red	1,300	42	204	15.3	2.1	18.6	1.4	2.4	60.2
9) Winnipeg	77	n	1	11.9	0.7	0.7	20.3	66.4	

*Includes washing and well injection, and related uses.
**High values, especially in the South Saskatchewan and Assiniboine-Red River drainages reflect recycling and re-use of supply.
n   Nil

2.2.1 Agriculture

Water withdrawals reflect the overriding importance of agriculture within the interior drainage, particularly the drainage of the South Saskatchewan and the Assiniboine and Red rivers. Farmland covers about 90 percent of the Assiniboine and Red rivers drainage, and about 63 percent of the Saskatchewan River drainage. It occupies about 17 percent of the Lake Winnipeg shore area, but less than 1 percent of the remaining area of the watershed. Table 5 provides a summary of farming practices. Between 1971 and 1981 there was an increase in farm size, a slight increase in cropland and major increases in fertilizer and biocide applications. Similar trends have continued through the past decade.

←

Fig. 3. Hudson Bay Basin: Ecozones (Sources: Statistics Canada, 1986; EWG, 1989; Wickware & Rubec, 1989).

Table 4. Major Population Centres and Recent Rates of Growth (Source: Statistics Canada, 1986).

Drainage System	Major Centres
Nelson:	Thompson.
N. Ontario:	Cochrane, Kapuskasing.
Quebec:	Radisson.
N. Sask:	Edmonton, Lloydminster, North Battleford, Prince Albert, The Pas*.
S. Sask:	Calgary, Lethbridge, Medicine Hat, Red Deer, Saskatoon, Swift Current.
As.-Red	Brandon, Moosejaw, Portage la Prairie, Regina, St. Boniface, Yorktown, Winnipeg.
Winnipeg	Ft. Frances, Kenora.

* The Pas is on the Saskatchewan River below the junction of the north and south branches.

Urban Population as % Total Population (A), and Total Population Change as % (B). All Values are Positive Unless Otherwise Noted.

	1971-76		1976-81	
	A	B	A	B
Nelson	66.2	2.1	61.4	-15.0
Winnipeg shore	67.3	3.0	68.8	1.7
Saskatchewan	72.6	10.1	76.7	18.4
Assiniboine	61.1	0.5	66.1	1.0
All others	56.2	3.8	51.1	4.9

Table 5. Farming Practices Within the Watershed (Source: Statistics Canada, 1986).

Ecozone	Wide Row	Close Row	Forage	Summer Fallow	Other Crops	Fertilizer	Herbicide	Pesticide
Boreal Shield	3 40	18 25	64	2	13	147	176	58
Boreal Plain	<0.5 330	54 19	22	16	8	138	142	152
Prairie	1 81	58 9	9	28	4	202	69	95
	(A) (B)	(A) (B)	(A)	(A)	(A)	(C)	(C)	(C)

(A) Area of farming practice relative to total farm area, expressed as % area.
(B) Percent increase in area of row farming (1971-1981).
(C) Percent increase in area of application (1970-1980).

2.2.2 Mining

A broad range of mining activities occur within the Hudson Bay watershed (Fig. 4) and most metal extraction is largely associated with surface or near surface exposures of Shield rocks (Fig. 3). Estimates suggest that about 98 percent of the total recoverable iron-ore remains in Ontario and Quebec. Between 89 percent (Saskatchewan) and 97 percent (Ontario) of total recoverable copper ore remains. Between 83 percent (Saskatchewan) and 96 percent (Ontario) of recoverable zinc, 98-99 percent of recoverable

→

Fig. 4. Hudson Bay Basin: Non-Renewable Resources (Sources: Douglas, 1969; Keith *et al.*, 1981; Johnson *et al.*, 1986; Statistics Canada, 1986; GSC, 1991; Canadian Mines Handbook, 1992).

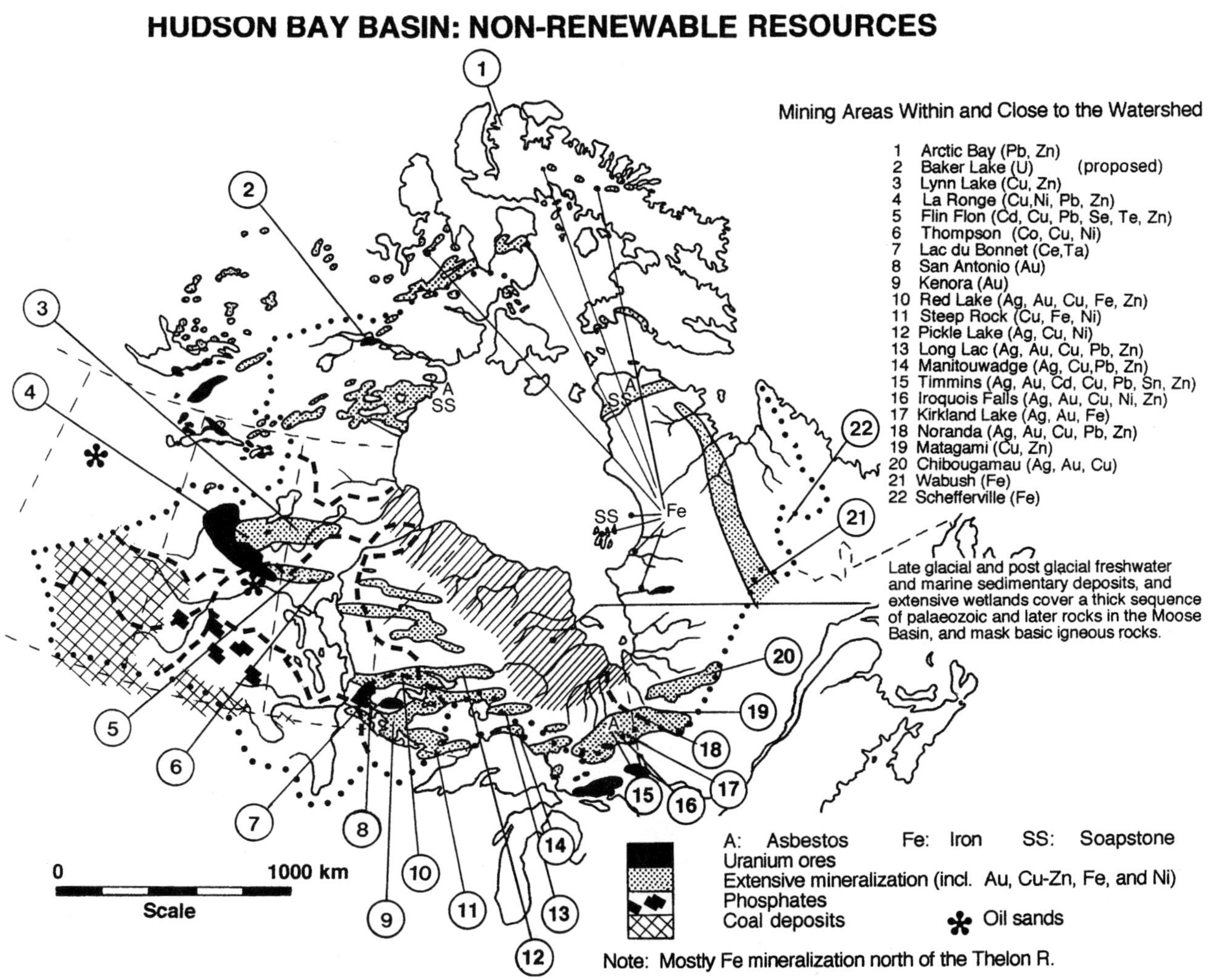
HUDSON BAY BASIN: NON-RENEWABLE RESOURCES
Mining Areas Within and Close to the Watershed
1 Arctic Bay (Pb, Zn)
2 Baker Lake (U) (proposed)
3 Lynn Lake (Cu, Zn)
4 La Ronge (Cu,Ni, Pb, Zn)
5 Flin Flon (Cd, Cu, Pb, Se, Te, Zn)
6 Thompson (Co, Cu, Ni)
7 Lac du Bonnet (Ce,Ta)
8 San Antonio (Au)
9 Kenora (Au)
10 Red Lake (Ag, Au, Cu, Fe, Zn)
11 Steep Rock (Cu, Fe, Ni)
12 Pickle Lake (Ag, Cu, Ni)
13 Long Lac (Ag, Au, Cu, Pb, Zn)
14 Manitouwadge (Ag, Cu,Pb, Zn)
15 Timmins (Ag, Au, Cd, Cu, Pb, Sn, Zn)
16 Iroquois Falls (Ag, Au, Cu, Ni, Zn)
17 Kirkland Lake (Ag, Au, Fe)
18 Noranda (Ag, Au, Cu, Pb, Zn)
19 Matagami (Cu, Zn)
20 Chibougamau (Ag, Au, Cu)
21 Wabush (Fe)
22 Schefferville (Fe)
Late glacial and post glacial freshwater and marine sedimentary deposits, and extensive wetlands cover a thick sequence of palaeozoic and later rocks in the Moose Basin, and mask basic igneous rocks.
A: Asbestos Fe: Iron SS: Soapstone
Uranium ores
Extensive mineralization (incl. Au, Cu-Zn, Fe, and Ni)
Phosphates
Coal deposits
Oil sands
Note: Mostly Fe mineralization north of the Thelon R.
0
1000 km
Scale
A
SS
Fe

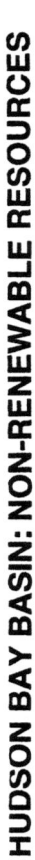

Table 6. Relative Areas of Ecoclimatic Provinces in the Hudson Bay Watershed: Values as Approximate Cover, to Nearest 25 percent (Sources: Forestry Canada, 1991; Statistics Canada, 1986).

Ecoclimatic Province	Drainage System								
	1 Keew.	2 Church.	3 Nelson	4 N. Ont.	5 W.Quebec	6 N. Sask.	7 S. Sask.	8 Ass.-Red	9 Win'peg
Cordilleran						< 25			
Prairie						> 50	100	100	
Cool Temperate									50
Boreal Forest		50	75	50/75	25/50	25			50
Taiga/H.B.P*.		50	25	25/50	50/75				
Arctic	100								

H.B.P*. Hudson Bay Plain

Forest cover: Within the Hudson Bay watershed, in each province, the Boreal Forest is the dominant timber resource. Large parts of the forest areas of Manitoba and Saskatchewan, and perhaps a quarter of Alberta's forest area lie within the Hudson Bay drainage. Nearly two-thirds of Ontario's Boreal Forest lies within the Hudson Bay drainage but only about one-third of the Boreal Forest in Quebec occurs within this drainage region. Cool Temperate Forest regions occur only in southern Ontario and Quebec where they represent a small part of the total forest cover.

Table 7. Effects of Human Activities and Natural Stress on Forest Areas (Source: Forestry Canada, 1991).

	Province				
	Alberta	Sask.	Manitoba	Ontario	Quebec
Forest area (x 10^6 ha)	37.7	23.7	34.9	80.7	94.0
Forest Land as a % of province	58.5	41.6	63.7	90.6	69.3
Productive Forest as % of Forest Land	62	38	40	53	90
Average annual harvest (1980-89) % of Forest Land	0.11	0.08	0.03	0.28	0.36
1990 Loss to insects and fire as % of Forest Land	4.0	2.0	0.1	20.5	1.1
All 1990 harvest and loss (disturbance) a % of Forest Land	4.21	2.08	0.13	20.78	1.46

lead and nickel, and a large proportion of total uranium reserves remain in place (Statistics Canada, 1986). About 60 percent of the known natural gas reserves, at least 98 percent of the oil sand reserves, and large quantities of coal remain. Annual coal production is about 0.5 percent of the total resource (Statistics Canada, 1986). Economic factors significantly influence the extent to which these resources are recoverable. Mining is a major user of water supplies in the western Quebec, northern Ontario, Nelson and Winnipeg drainage systems (Table 3).

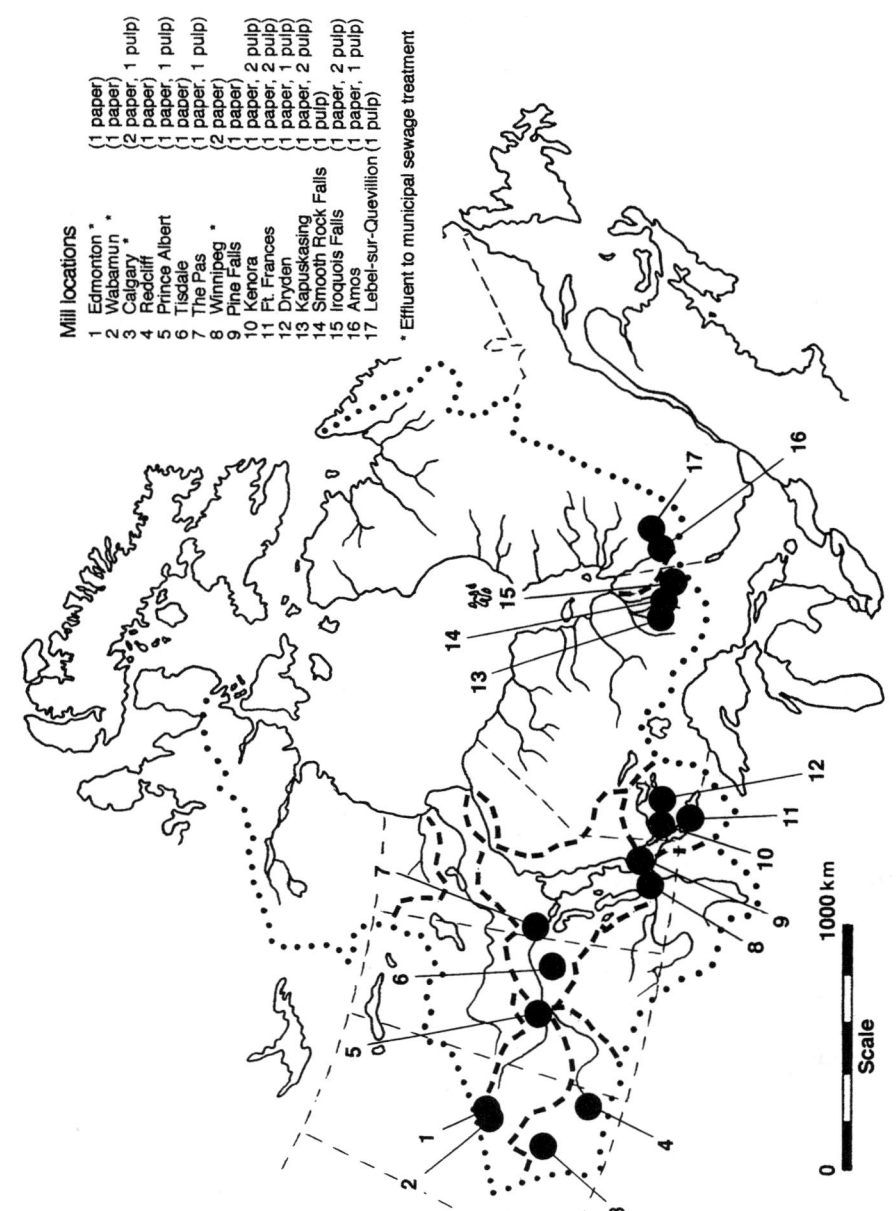

Fig. 5. Hudson Bay Basin: Pulp and Paper (Sources: CPPA, 1990; Sinclair, 1990; Anon., 1991; 1992).

2.2.3 *Thermal power*

Coal is used extensively to generate thermal power in the North Saskatchewan and Assiniboine and Red River drainage (Pearse *et al.*, 1985; Statistics Canada, 1986), and large power plants have been built close to sources of supply. Large volumes of water may be used to clean and process coal, and to operate the power plants (Table 3). Many medium and small size thermal power plants are oil-fired (widely distributed across the watershed), and diesel generators power many plants in more remote areas (including Inuit and Indian communities around the Hudson Bay basin).

2.2.4 *Forestry*

Forestry dominates the renewable resource industries of northern Ontario and western Quebec and accounts for about 36 percent and 53 percent, respectively, of total forest production within the watershed (Forestry Canada, 1991). Table 6 indicates the extent of forest cover within the basin and Table 7 documents the importance of different forms of disturbance which affect it. Virtually all forested areas are in the boreal ecozones (EWG, 1989) and, based on 1990 data (Forestry Canada, 1991), the annual harvest ranges from less than 0.1 percent to nearly 0.4 percent of Forest Land. In 1990, the smallest harvest was taken in Manitoba and the largest in Quebec. Productive Forest area is significantly less than Forest Land. Very large areas of forest can be affected by insect damage and fire. In Manitoba, for example, about 14 percent of the Productive Forest area was affected by fire between 1985 and 1990 (Natural Resources Manitoba, 1991). Clear cutting is the main harvesting technique, and this and forest damage may affect the quality of surface water discharge. Changes depend on duration and severity of disturbances and the size of area affected. Table 8 indicates, within the watershed as a whole, that much of the forest is now immature, the only large areas of mature forest remain in Quebec. Minimal site preparation for regeneration occurs in Ontario and Quebec; large areas in all provinces are dependent on natural regeneration. Hearnden *et al.* (1992) report that the application of artificial regeneration techniques has been increasing since the early 1970s.

2.2.5 *Manufacturing*

The use of water by manufacturing industries, particularly the pulp and paper industry (Pearse *et al.*, 1985; Statistics Canada, 1986) is heavily concentrated in the boreal ecozones of the western Quebec, northern Ontario, Winnipeg and Nelson drainage systems (Fig. 5). Because of its size and the nature of effluent materials, until recently, the pulp and paper industry has been particularly stressful to the environment (Sinclair, 1990; Statistics Canada, 1986). Table 9 indicates, most recently, that development in the watershed has tended to increase the number of industries which exert less stress. This condition has been brought about both by major reductions in contaminant release from mining, smelting, and pulp and paper manufacturing, and by the growth of other less stressful industries. Although more recent province-wide data have not been published (Statistics Canada, 1986), few industries should now appear as sources of high stress effects. Table 10 reports shifts of employment within the watershed and despite a lack of more current data, it probably gives a reasonable indication of continuing trends. In particular, declines have been experienced in agriculture and fishing and trapping. Increases have occurred in mining, manufacturing, construction, and service industries. Declines reflect changes in the resource base but other economic factors may be an over-riding influence.

Table 8. Forest Inventory and Site Preparation (Sources: Forestry Canada, 1991; Natural Resources Manitoba, 1991; Statistics Canada, 1986).

Ecozone Type	% Regeneration (<1m high)	Dominant Form Albta.	Sask.	Man.	Ont.	Que.
Boreal S/P	4-51	I		I	I	M
H.B.P.	2-29				O	M
Taiga Shield	0-19	I		I		M
Prairie	7-12	I/M	I	I		

H.B.P. Hudson Bay Plain;   S/P Shield and Plain
I Immature (> 1 m high but too small to harvest)
M Mature (ready to harvest)
O Overmature (stands in which mortality likely exceeds new wood growth).

Five Year Average Preparation (1976-80) as Percentage of Area Harvested, by Province

	Site Preparation	No Site Preparation	Planting and Seeding	Area Dependent on Natural Regeneration
Alberta	77.2	22.8	59.0	41.0
Sask.	34.2	65.8	36.3	63.7
Man.	21.9	78.1	8.8	91.2
Ont.	24.5	75.5	28.1	71.9
Quebec	3.8	96.2	10.2	89.8

Table 9. Trends in Industrial Stress and the Size of the Labour Force (Source: Statistics Canada, 1986).

	Stress			
Year	High (n)	Medium (n)	Low (n)	Total Labour Force
1973	69	277	2521	100,000-150,000
1976	67	1207	2371	111,500-112,000
1981	66	1484	3033	128,000-128,500

(n) refers to number of establishments in the entire area of the Hudson Bay watershed.

High Stress activities include: mining, oil/gas extraction, pulp and paper, iron and steel, smelting and refining, petroleum refining, industrial chemicals, and thermal power production.

Medium Stress include: food processing, pharmaceuticals, quarries, mills, cement, glass, fertilizer, textile manufacture, and metal fabrication.

Low stress include: finished clothing, printing, logging, wood processing, furniture making, electrical and machinery manufacture, construction, and packaging.

2.2.6 Impoundments and diversions

Within many parts of the Hudson Bay watershed, water is impounded to provide hydroelectric power, irrigation, domestic supply, and interbasin transfer (Fig. 6). Water shortages are a major concern throughout the North and South Saskatchewan, and Assiniboine and Red River drainage systems. Elsewhere, an abundance of water has made it possible to embark on massive reservoir construction in the western Quebec, and Nelson and Churchill drainage systems (Pearse *et al.*, 1985; Statistics Canada, 1986). Hydroelectric facilities installed and under construction, and planned additional

Table 10. Trends in Employment Throughout the Watershed (Source: Statistics Canada, 1986).

Year	Agriculture	Forestry	Fish./Trapping	Mines/Quarry/Oil	
1971-76	218,130	7,525	10,000	49,745	
1976-81	192,945	11,305	1,875	92,420	
	Manufactr.	Constr.	Service	Other	Ratio of Resource to Other Industries
1971-76	149,325	93,020	875,405	111,490	21.1:1
1976-81	214,445	179,390	1,420,545	62,290	15.2:1

Data expressed as numbers of individuals employed

capacity in the major watersheds of the Hudson Bay region are noted in Fig. 6. Large additional developments could take place in the west Quebec drainage and the Nelson River drainage but the scale of development in northern Ontario will be much smaller (Conservation et Protection, undated). Impoundments and diversions have affected the quality of surface water and discharge regimes (Brouard *et al.*, 1990; Guilbault *et al.*, 1979; Hecky & Guildford, 1984; Messier *et al.*, 1986; Rosenberg *et al.*, 1987). Freshwater discharge into Hudson Bay could be increased as a result of further diversions in northern Quebec. In recent years, attention has been drawn to the effects of reservoir formation including the production of methyl mercury (Jackson, 1987) and, most recently, methane (Rudd. In: House of Commons, 1991).

The combined effects of many human activities within the watershed have potential to affect the marine environment of the Hudson Bay basin and impacts are most likely to be felt in the nearshore region. Table 11 provides an indication of the rapid increase in the number of dams built in the watershed after 1940 and Table 12 provides a summary of the major characteristics of large reservoirs in the watershed. Large reservoirs are generally absent from northern Ontario drainage.

3. Settlement and land use

Today, the Hudson Bay coastal area is populated by Inuit, Indian (mostly Cree), and non-native (mostly European) peoples. The distribution of settlements is shown in Fig. 7. Except in the larger centres (Churchill, Moosonee, and Radisson), non-natives are a minority (Statistics Canada, 1986). Inuit populate most of the area north of Churchill on the west side of Hudson Bay. From Churchill, Indian communities extend south and east around the bay as far as Great Whale River (Grande Baleine) where Inuit (Kuujjuarapik) and Cree (Whapmagoostui) communities occur side-by-side. Northwards, again, all communities are Inuit.

Fig. 6. Hudson Bay Basin: Major Dams and Rivers (Sources: Pearce *et al.*, 1985; Statistics Canada, 1986; Burnham, 1990; Rougerie, 1990; Manitoba Hydro, 1992; Conservation & Protection, undated).

Hudson Bay Region: Present and future environmental concerns

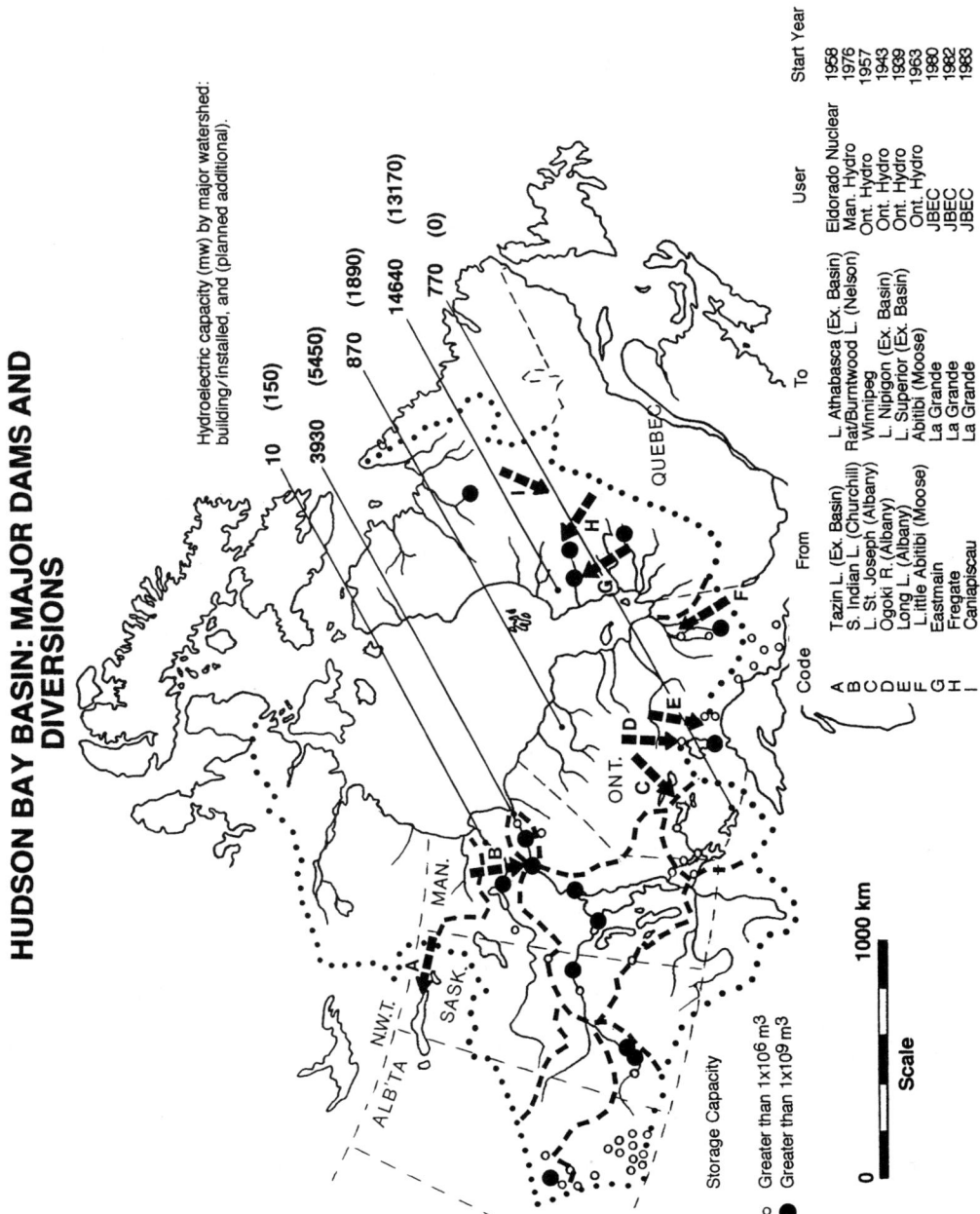

Table 11. Impoundment in the Watershed (Sources: Pearse *et al.*, 1985; Rougiere, 1990; Statistics Canada, 1986).

Drainage	Before 1940	1940-1960	1960-1982
Keewatin			
Churchill		4H	
Nelson	1H		13H, 1L, 1S
N. Ontario	7H		4H, 1L
W. Quebec			12H
N. Sask.			3H, 1S
S. Sask.	3L, 1S, 3I	11H, 1S, 23I	4L, 3S, 4I
As.-Red	1I	2L, 6S, 4I	4L, 2S, 5I
Winnipeg	6H	5H, 4L	

H Hydroelectric; L Level control; S Domestic/industrial supply; I Irrigation

Table 12. Major Reservoirs in the Watershed, 1984 (Sources: Pearse *et al.*, 1985; Rougiere, 1990; Statistics Canada, 1986).

Reservoir	Drainage	Total Area (km^2)	Year	Area Flooded (km^2)
Kelsey (Man.)	Nelson	706	1960	124
Stevens L. " "	Nelson	337	1970	236
S. Indian L. " "	Churchill	2391	1975	414
Notigi	Churchill	584	1975	431
W. Nelson Ch.	Nelson		1976	76
L. Diefenbaker	S. Sask.	430	1968	186
Cedar L. (Man.)	Sask./Nel.	3493	1965	1372
Tobin L. (Sask.)	Sask./Nel.	300	1963	246
Bighorn (Albta)	N. Sask.	62	1972	60
LG2 (Quebec)	W. Que.	2836	1979	
Opin.ca (Que.)	W. Que.	1036		738
LG3 Quebec)	W. Que.	1865	1982	
LG4 Quebec)	W. Que,		1984	
Caniapiscau (Que.)	W. Que.	4299		

3.1 Early cultures

Beginning about 10,000 years BP, Palaeo-Indian cultures moved north into southern Manitoba and Saskatchewan, following the retreat of glacial ice from the interior of the watershed (Symington, 1978). Closer to Hudson Bay, radio carbon dates suggest the presence of Archaic Shield Indian people in northwest Quebec (around Lake Caniapiscau) about 3500 years BP, and it is possible that parts of the bay were inhabited as early as 5000 years BP (Berkes & Freeman, 1986). The distribution of native people about the time of the 17th century European explorations of the region is shown in Fig. 8. On the west shore of Hudson Bay, Chipewyan Indians (Athapaskan) were found as far north as Churchill. Eastern Cree occupied almost the entire eastern shore of James Bay and Indian groups occupied more northern territory inland from the coast. Cree Indians occupied most of the southern part of the region (Dickason, 1992). Plains Indians occupied much of the interior watershed and the largest tract of Prairie was occupied by the Assiniboine.

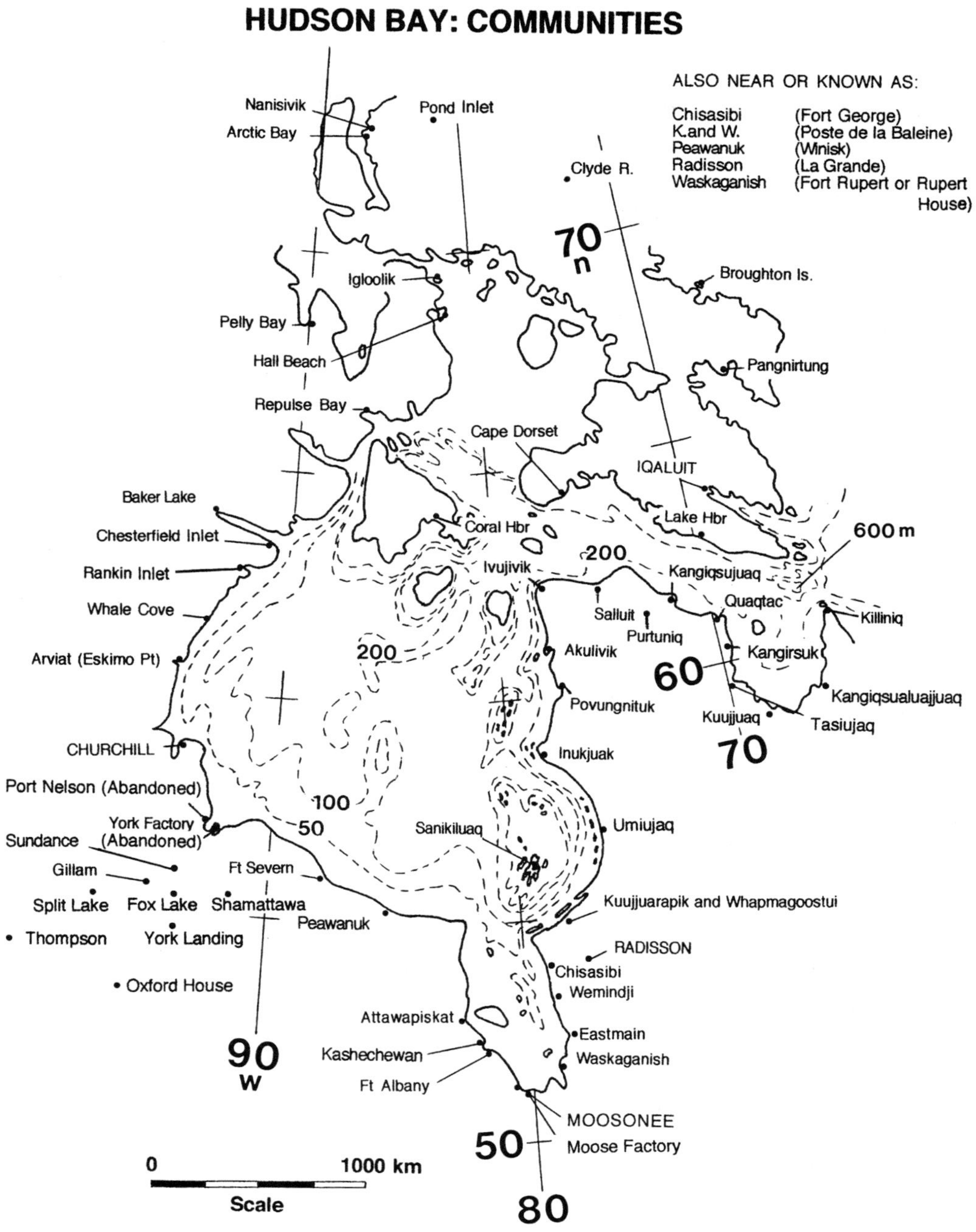

HUDSON BAY: COMMUNITIES

ALSO NEAR OR KNOWN AS:

Chisasibi	(Fort George)
K.and W. Peawanuk	(Poste de la Baleine) (Wnisk)
Radisson	(La Grande)
Waskaganish	(Fort Rupert or Rupert House)

Fig. 7. Hudson Bay: Communities.

Few historical ties appear to have existed between the Cree of the east and west shores of Hudson Bay. The dialect of the eastern Cree seems to be closest to the Montagnais of the north shore of the St. Lawrence River and central Quebec (Francis & Morantz, 1983). The James Bay Cree traded south through the Nottaway or Moose - Abitibi River systems into the Ottawa River. The Rupert and Mistassini and Sagunay

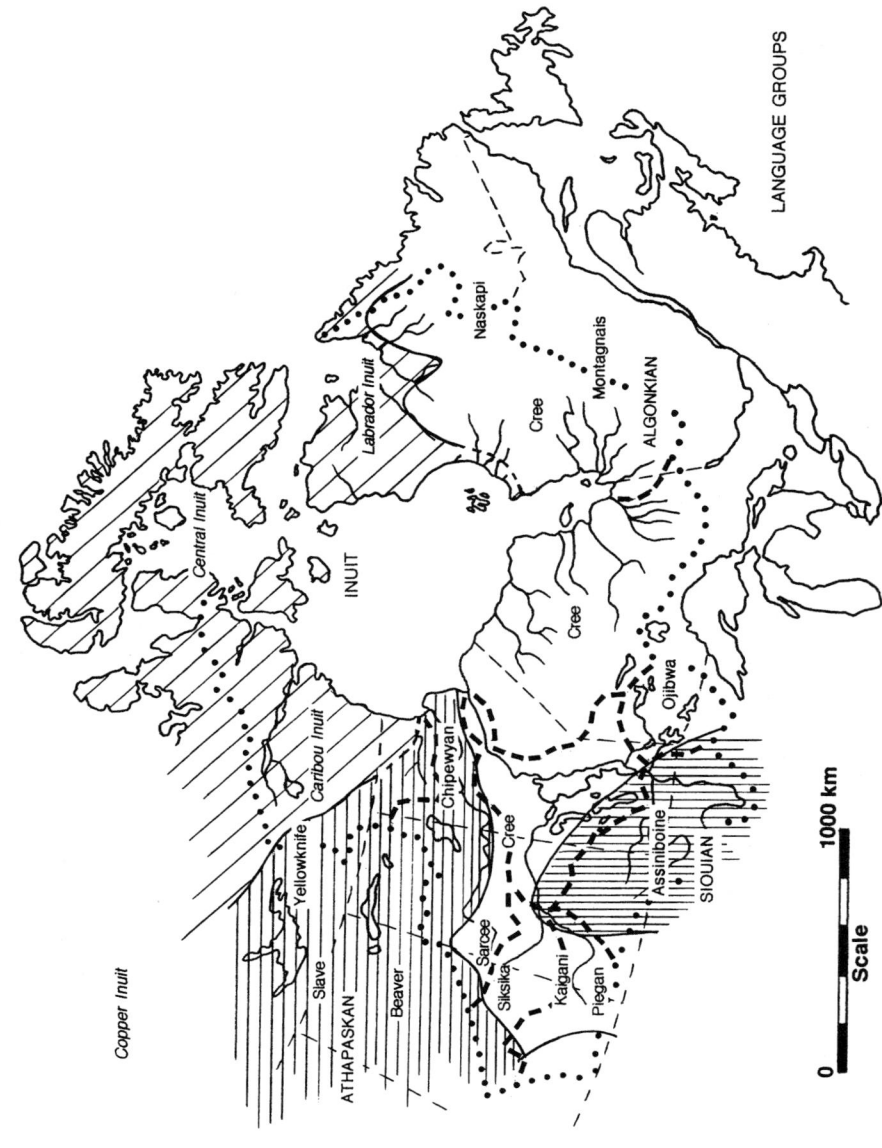

Fig. 8. Hudson Bay Basin: First Peoples (Sources: Mowat, 1970; Symington, 1978; Dickason, 1992).

River system provided an eastward trading route. Westward trade passed through the Mattagami River and the Lake Huron drainage or through the Albany River and into the Lake Superior drainage.

Several groups of Inuit inhabited the watershed north of the tree line but their southward migration was limited. They avoided the Indians (Dickason, 1992) with whom there were conflicts (NSL, 1970). Inuit are thought to have spread eastward along the coast from Alaska in a series of migrations which began more than 3000 years ago (Symington, 1978; Arnold 1982; McGhee, 1987). The Dorset culture is recognized by the presence of finely made artefacts. The succeeding Thule culture is thought to have brought with it the use of dogs for over-ice travel and techniques for whaling. The Thule culture which forms the basis for much of the recent Inuit culture is about 1000 years old in the western Arctic. It is a few hundred years less old in the eastern Arctic.

Coastal Inuit hunted whales, walrus, seals, and some polar bear, and they fished and harvested fish, shellfish and birds. Northern caribou were a major part of the diet of the Caribou Inuit and Chipewyan Indians (northwest Hudson Bay), and the Algonkian Indians (northern Quebec). Most of the Cree lived in or near woodland areas and were dependent on woodland caribou, moose, beaver, deer, fish, birds, and other small game. In the interior of the watershed, the Plains Indians were heavily dependent on bison.

Lifestyles and settlement patterns of all native people were adapted to match the distributions and behavioral patterns of their principal sources of food. There was a seasonal round of organized hunting and movement. Some Inuit communities became structured around centres of biological organization in the Arctic, particularly where there was proximity to open water during the winter and where marine circulation patterns or estuarine processes enhanced productivity. Other Inuit communities were dependent on more distant and dispersed resources; these people travelled great distances and lived in temporary camps and shelters. The woodland Indians were widely dispersed (Symington, 1978), reflecting the low densities of large game such as moose, woodland caribou, and deer (Moose were absent from the east shore of James Bay before this century, Francis & Marantz, 1983). More specifically, and based on the size of recorded trap lines and hunting areas, it seems that an area of about 20 km^2 is needed to support one hunter/ gatherer around southern James Bay. The wildlife resource base decreases rapidly, northward, and towards the southern part of Richmond Gulf (Lac Guillaume Delisle) a trap line of about 90 km^2 is required to provide equivalent support (Berkes & Freeman, 1986).

Estimates of the size of native populations at the time of early European contact are, at best, tenuous (F. Berkes, pers. comm.). The population of coastal Inuit may have been between 10,000 and 20,000 people. The population of inland Inuit (Caribou Inuit) was, likely, only a few thousand (NSL, 1970; Symington, 1978). There were probably less than 10,000 Indians living in the taiga but across the northern woodlands there may have been four or five times as many people. In the grasslands of the interior drainage, there may have been another 30,000 people.

3.2 Arrival of Europeans

Traces of an old settlement have been found in Ungava Bay. Initially, it was thought to be of Norse origin, dating from as early as 1100 A.D. This suggestion, however, remains controversial (Berkes & Freeman, 1986). An early record of Danish exploration in the Hudson Strait suggests that its existence may have been known as early as 1470 (Mowat, 1967). Certainly, Sebastian Cabot reported the existence of Hudson Strait in

1508 and Martin Frobisher is known for his ill-fated attempt to mine "fool's gold" from Frobisher Bay, Baffin Island (1576). Henry Hudson reported exploration of the bay in 1610.

By the mid-1600s French settlement was becoming firmly entrenched further to the south and around the St. Lawrence River and, within a relatively short time, several interests were seeking to expand the fur trade west and north. The French traders and explorers Groseillier and Radisson tried initially to obtain support for explorations in Hudson Bay and westwards from the French government, but were unsuccessful. They turned to English interests and following a successful passage to Hudson Bay and the sale of beaver pelts in London, King Charles II issued a charter for the Hudson Bay Company in 1670.

York Factory quickly became a key trading post and European settlement, nearby at Churchill, has been more or less continuous for two and a half centuries (Beals & Schenstone, 1968). On various occasions, between 1697 and the late 1700s, the French captured or sacked most of the Hudson Bay Company trading posts as the British and French sought to dominate the fur trade. Early trading posts were opened at the mouths of the Churchill, Severn, and Moose rivers, but for many years and around most of the bay trading was done from vessels. The company opened trading posts at the mouth of the Eastmain River in 1719 and Richmond Gulf in 1749. Some overland explorations took place northwest from Churchill during the late 1700's and seaward attempts to find a northwest passage continued. By the early 1800's, Hudson Bay Company ships were bringing supplies to several trading posts within Hudson Bay, and settlers were being brought through it to join the Fort Garry and Red River settlements in the interior drainage. The expansion of trade, whaling, and European settlement in the early 1800s marked the beginning of native exposure to growing cultural stress, as well as increasing environmental stress.

Church missions were established close to many trading centres around Hudson Bay. The Anglican Church developed a strong presence in Cree communities along the eastern shore, and the Catholic Church became dominant in western communities of the Cree (Beals & Schenstone, 1968). The Anglican Church also established an early presence in many of the Inuit communities. Rarely did Anglican missions penetrate far inland. Often, the Hudson Bay stores and the Anglican missions were located close together, as were the more inland centres of the rival Northwest Company and the Catholic Church.

The Hudson Bay Company and the Northwest Company (Montreal) amalgamated in 1821. In 1870, the Hudson Bay Company's rights to Rupert Land (all lands draining into Hudson Bay) were purchased by Canada and in 1880 Britain transferred all of its rights to the Arctic islands to Canada. Despite the potential political and economic interests associated with this vast territory, Canada did little to manage its resources and by the late 1800's, American whalers and traders largely controlled the region (Beals & Schenstone, 1968).

Whale oil was a vital commodity for Europeans and the early settlers in North America. In response to intense demand, whaling became a dominant activity in the eastern Arctic for more than a century. The Hudson Bay Inuit were extensively involved. Some whale oil was also prepared by Cree. Beluga were captured and rendered for oil at the mouth of the Eastmain River, and at mouth of both the Great Whale and Little Whale rivers, well into the 1800s (Francis & Morantz, 1983). Both as a result of cheaper substitutes and a lack of whales, the industry collapsed shortly after 1900 (Sergeant, 1986).

A broad range of contacts with traders, the military, and settlers exposed all native people to a number of diseases for which they had virtually no immunity. Tuberculosis, diphtheria, influenza, and measles were particularly virulent and continued to reduce native populations well into this century. The loss of the Sadlirmiut Inuit community on Southampton Island (1902-3) was almost certainly caused by disease spread from a nearby whaling station (Beals & Schenstone, 1968).

Disease was not the only major stress to effect native populations. The loss of their principal sources of food was equally if not more important. Marine mammals, musk-ox, and caribou were taken in large numbers to feed the crews of whaling ships. Trappers and traders made additional demands on caribou and other species. In some cases, depletion of stocks of large mammals continued into the 1960s. Estimates of the caribou population in northern Canada before the arrival of Europeans are uncertain but their numbers may have been in the millions. The extent to which their decrease reflected hunting practices is unknown and native hunters have challenged government estimates because of the inadequate survey techniques (Freeman, 1989). Large scale population fluctuations occur naturally and may have been confounded by changes in migration.

Uncontrolled human exploitation of resources and habitat disruption and destruction greatly exacerbated natural stress. Woodland caribou are distributed across northern Manitoba and Quebec, and small sedentary herds are present around Lake Bienville and Lake Caniapiscau (Quebec). Barrenland caribou occur in the Keewatin District of the Northwest Territories. Most herds migrate long distances between their winter and summer range and the populations are greatly influenced by the quality of grazing land and winter forage (Harrington, 1991). For example (Beals & Schenstone, 1968), ground burns to expose bedrock for mineral surveys have been exceedingly destructive of grazing and forage areas, and the movements of survey and exploration and development crews may have influenced patterns of animal behaviour over large areas (perhaps limiting successful feeding and reproduction). Further, female caribou seem to favour lowland areas near water for calving and early rearing (Harrington, 1991). The availability and stability of such habitat is therefore of additional concern, particularly where water level manipulation may alter the natural regime to which the caribou have adapted.

The sensitivity of the northern environment and associated ecosystems to human activities is becoming more widely recognized and reduced levels of stress have enabled substantial recoveries in many species. Early conservation initiatives, such as establishment of the Thelon Valley musk-ox sanctuary in 1927, were particularly important and may have saved this species from extinction. Nevertheless, although there has been no commercial whaling in the Arctic for most of this century and few bowhead whales (*Balena mysticetus*) have been taken during the past decade (H.E. Welch, Freshwater Institute, Winnipeg, Manitoba, pers. comm.), their population remains severely depressed (Sergeant, 1986).

3.3 Present native use of resources

The Cree and Inuit occupy traditional lands but, generally, land and resources are not individually owned (Berkes & Freeman, 1986). Rather, various members of each community hold responsibilities for leadership in hunting and harvesting (Scott, 1986). This includes both good hunting practices and the equitable distribution of resources. The use of snowmobiles and off-road vehicles has made it possible to retain access to many distant hunting and fishing sites, although most of the northern communities have become centralized.

Table 13. Native Harvest Practices in the Hudson Bay Area (Arragutainaq & Fleming, 1991; Beals & Schenstone, 1968; Berkes & Freeman, 1986).

Period:	Early Spring	Late Spring	Summer	Late Summer	Fall	Winter
Condition:	On-ice	Break-up	Open water	Freeze-up	On-ice	
INUIT						
	Ringed seal	Sea ducks (inshore)	Caribou	Narwal	Arctic hare	Fox and other small mammals
			Lake and river fish	Walrus	Ptarmigan	
	Canada geese (Hudson Bay and Strait)	Beluga (Hudson Strait)		All seals	Resident Eider and other ducks	Polar bear
		Ringed seal and Walrus (near ice edge)	Ducks and geese	Spawning fish		Ringed seal
	Snow geese (elsewhere)		Beluga	Berries	Cod	Shellfish
	Arctic char (in lakes)	Arctic Char (in lakes)	Shellfish			
	Shellfish	Fish netting at river mouths				
		Bird eggs				
		Shellfish				

In the Belcher Islands, Inuit also use local reindeer as a source of food throughout much of the year, though seldom during spring break-up or fall freeze-up.

CREE						
	Snow geese					
		Ducks and loons				
			Other geese		Geese	
	Lake fishing	--------------River and lake fishing--------			Lake fishing Beaver Caribou Moose Black bear Small mammal and other game	

The need to maintain intergenerational knowledge and leadership in hunting practices is of concern. Cree communities near the bay may include both "coasters" and "inlanders"and one individual is responsible for the use and management of various trap lines or hunting areas (Berkes & Freeman, 1986). The Inuit tend to have family hunting and fishing areas which may be harvested cooperatively. The entire environment is considered as their harvest area in which practices differ, depending upon the seasons and "target" species. This concept is summarized in Table 13. A summary of the northern Quebec Cree and Inuit subsistence fisheries around the bay is provided in Table 14. Although both groups harvest brook trout, lake trout, Arctic char, and lake whitefish, there are strong differences in harvesting patterns of these and other species.

Table 14. Cree and Inuit Subsistence Fisheries (Berkes et al., 1992b; Donaldson, 1984; Gamble, 1984; NHR, 1982a; 1982b; Roy, 1989).

Species	Cree	Inuit
	Approximate composition of harvest (%)	
Lake sturgeon	6	-
Atlantic salmon [1]	-	9
Brook trout	10	5
Lake trout	14	21
Arctic char [2]	<1	59
Whitefish (all) [3]	38	4
Northern pike	12	-
Suckers (longnose and white)	13	-
Burbot	2	-
Walleye	5	-
Arctic cod [4]	-	2
Sculpin	-	<1

1 Mostly taken in Ungava Bay; 2 Mostly taken in northern Hudson Bay; 3 Mostly taken in southern Hudson Bay; 4 Mosty taken from Ungava Peninsula.

Region	Data year	Harvest (kg)	Population	Average distribution ($kg \cdot cap^{-1} \cdot a^{-1}$)	Source
N. Quebec	1978	233,000	3,981	59	(NHR, 1982a)
Baffin	1982	370,000	6,889	54	(Donaldson, 1984)
Keewatin	1981-2	161,079	3,769	43	(Gamble, 1984)
E. James Bay *	1974-5	320,000	6,279	51	(NHR, 1982b)
W. James Bay and Hudson B.*	1990	191,246	6,470	30	(Berkes et al., 1992)

*   Cree communities

Both groups make extensive use of birds and the Inuit, in particular, make considerable use of marine mammals.

Subsistence fisheries are an important component of the native economy but considerable differences exist among communities, even in the same region. Reasons for this include preferences, alternative types of food, employment opportunities, and limitations on food consumption (for example, mercury content). Based on data from 1975 to 1982 (Berkes, 1990), per capita annual fish consumption (whole, wet weight) varied from a low of about 43 kg in the Keewatin District to a high of 59 kg in northern Quebec. In eastern James Bay, fish are also an important food source and represent about one-quarter of the subsistence harvest (per capita average about 51 kg). Prior to about 1970, fish were used as a principal source of food for dog teams but transport is now largely motorised.

In terms of traditional inland hunting and trapping, barren land caribou were taken mostly by the Inuit and Chipewayn Indians, and the woodland caribou were taken mostly by the Cree. Moose and black bear were also highly valued by the Cree. Other harvested species included marten, otter, lynx, snowshoe hare, muskrat, groundhog, porcupine, wolf and wolverine. In the northeast Manitoba, caribou, Arctic fox and beaver were key species; in northern Ontario, muskrat, beaver and marten were dominant; and in western Quebec muskrat, Arctic fox, marten, and seal were taken by the Cree (Ray 1990). Arctic fox was also important for the Inuit.

3.4 *Recent cultural influences*

The influence of North American society and culture affects Indian and Inuit communities throughout the Hudson Bay region. There is a south to north contact gradient and northern communities have least direct interaction with southern societies. Both Indian and Inuit communities have suffered from many forms of acculturation. To a significant degree, neither have been able to fully sustain their natural resource-based economies. After the collapse of the whaling industry, there was a shift in the economic focus of the region. There was a resurgence in the fur trade and both Inuit and Indian communities were able to benefit from this for a few decades. The Arctic fox was particularly important to the Inuit communities, as was the beaver to the Cree. Trading stores already served many northern communities and partly in response to the demand for fur, the Hudson Bay Company opened trading posts at several new locations, including Baffin Island and the Ungava Peninsula (1911), Southampton Island (1916), Chesterfield Inlet (1912), Repulse Bay (1921), Cape Smith and Povungnituk (1927), the Belcher Islands (1928) and northern Foxe Basin (1939).

Other forms of economic stimulus also drove development in the southern part of the basin. Rail connections were built to link Churchill with southern Manitoba in 1929 for the export of Prairie grains, and to Moosonee in 1931. Both ports also increased the availability of bulk shipments to northern communities. For Keewatin and north Hudson Bay communities, the annual sea lift of supplies is shipped from Churchill. Barge traffic supplies Fort Severn from Moosonee, and Quebec communities are supplied from Montreal and Quebec City. Early hydroelectric development also began in northern Ontario, during the 1930s (Fig. 6). Most recently, road connections have been established between Radisson and southern Quebec. Access roads are also extending linkages between hydroelectric developments on the lower Nelson River in Manitoba.

Under non-native control, harvests have been largely unrelated to natural productivity cycles in the Hudson Bay region. The First World War disrupted international trade and induced competitive fur sales on the North American market. The market collapsed in 1921 and wildlife stocks remained severely depleted because of over harvesting (Stephenson, 1991). During the Second World War, prices rose, but after an initial increase production again declined. Early legislation in the 1890s was passed in response to the decimation of stocks by white hunters who used poison-bait. Later legislation imposed bag limits and closed seasons, and reflected the growing damage to wildlife resources throughout the region (Ontario trapline legislation, 1935; Manitoba, 1940; Quebec, 1945; and the Northwest Territories, 1949). During the 1920s, many white trappers displaced native trappers from traditional lands and many Indians became too poor to trap and were dependent on subsistence hunting for survival (Ray, 1990).

Excessive hunting pressures depleted many key stocks throughout the Hudson Bay region and this, together with the impact of fur-ranching and a reduced demand for fur after the Second World War, caused many northern communities to become dependent on welfare (Beals & Schenstone, 1968). In the 1950s, the Canadian federal government began to centralize northern communities to provide better health, education, and nutritional care, in the belief that traditional hunting and gathering economies could no longer sustain Indian and Inuit populations. Welfare-dependent communities became common throughout much of the region and self-dependence was lost as a result of the extensive socio-economic changes that took place. By 1966 there were few native people living northwest of Churchill and those that were left at Baker Lake and Arviat (Eskimo Point) had no significant employment (Beals & Schenstone, 1968). Since the mid-1950s, the provision of support for health and nutrition, and education and skills training (Clancy, 1987) has improved in most northern communities and infant mortality has declined. The birth rate has increased and peaked in the early 1970s (an effect, also, of centralizing the settlements, Berkes & Freeman, 1986) and much of the population is now under 15 years old.

During the 1950s and 1960s, North American Military defence established several outposts across the north including the DEW Line (Beattie & Greeway, 1986) and some settlements, such as Hall Beach, owe their existence solely to defence interests. Inuit were also encouraged to resettle at more distant locations in the high Arctic to ensure a Canadian presence, and poorly planned relocations caused further and extreme hardships (Grant, 1991). When piston-driven aircraft began trans-Atlantic service there was a large influx of civilian and military personnel to the north and Iqaluit (Frobisher Bay), for example, became a major refuelling stop. With the advent of jet passenger service, however, refuelling became unnecessary and this severely affected the economic structure of the community. The problem was exacerbated because Inuit had been encouraged to move into Iqaluit from other communities and, in the process, gave up opportunities to maintain traditional economies elsewhere.

A specialized market for Cree and Inuit artwork supports a small number of skilled artisans (Beals & Schenstone, 1968), and mineral resource development may offer short to medium term employment. Hydroelectric development could provide medium to long term employment. Substantially greater involvement in renewable resource management (including parks) could provide a key activity for many people.

With exception of the relocation policy and some aspects of wildlife management, governments generally acted in what was thought to be the best interests of native people. But this was without clear understanding of the essential role of traditional hunting and gathering in the social and cultural fabric of northern society (Berkes & Freeman, 1986), or the need to prepare and enable people to cope with the enormous effects of change. Community adaptation to political and economic decisions made outside the area continues (Irwing, 1989) but this is not sufficient to establish a long term basis for societal renewal. The health and well-being of northern communities reflects both the adequacy and use of natural resources, and effective management and decision-making by indigenous peoples (Freeman, 1988).

Residents around the basin number nearly 35,000 people (Statistics Canada, 1986) and nearly three-quarters are Inuit. The proposal to form the Nunavut Territory in the eastern Arctic, in which Inuit will have a substantial political majority, is likely to have far-reaching influence on the rate at which new, sustainable, and long term objectives can be established. Throughout the Hudson Bay region, there is an overriding need to establish a vision of long term cultural and economic self-sufficiency. The vision should be largely the product of the Inuit (TFN, 1989) and Cree who should also be largely responsible for its development and management.

3.5 Recent economy

The 1975 James Bay and Northern Quebec Agreement (JBNQA) set aside land in three categories for use by Cree and Inuit, and provided additional cash settlements. About 5500 km^2 and 8300 km^2 of category I land were set aside for the exclusive use of Cree and Inuit, respectively. Much larger areas were set aside, additionally, where Cree and Inuit would have exclusive rights to traditional resources (category II). Cree and Inuit non-exclusive rights to traditional resources were also recognised over other areas (category III). By agreement, the Quebec government would be able to undertake or approve development in category II and III areas, subject to some forms of compensation (Scott, 1992).

A series of regional self-government structures were established for Cree and Inuit communities. In both, there has been extensive growth of employment in administration and social services, and collective enterprise. Privately owned entrepreneurial activities exist but are generally small. Socio-economic changes have occurred in both Cree and Inuit communities. The total population has increased considerably since 1975, the proportion of the population in the work force has increased, and most of the easily made jobs have been created (Scott, 1992).

Table 15 indicates changes in the Cree economy. Unemployment in the Cree population was reduced to 22 percent by 1989 but, between 1989 and 1994, 1800 new jobs will be required to avoid a return to increasing unemployment. Difficulties experienced in competing for business outside the Cree communities, however, may significantly limit this avenue of development and the effectiveness of employment initiatives. Personal income has increased under the JBNQA. Income security provides support for traditional hunting and trapping activities and this has reduced direct transfer payments (welfare, for example). Salaries (particularly for administrative and social services) now provide the main source of income. Through distribution of opportunities and resource sharing, Cree communities attempt to provide broad-based benefits and support for all members (Scott, 1992).

By comparison, Cree and Ojibway economies in northern Ontario are maintained mostly by a combination of transfer payments, and special governments grants and programs. Only a small part of the economic activity is derived from wages, and hunting and trapping. The traditional subsistence economy is based largely on harvesting for native consumption, with some external income from the sale of fur. Income from owner-businesses is limited because of the dominance of non-natives in this sector. The lack of opportunity for native employment is highlighted in a recent profile of Moose Factory where there is high under-employment and unemployment is about 40 percent (Stephenson, 1991). This is all the more crucial since there is a growing acceptance of the wage economy by younger natives and a preference for full or part-time employment, with hunting and fishing as recreational activities. The older generation is more likely to pursue hunting and trapping as a prime activity.

Cree and Ojibway communities are adapting to the wage economy. Although, in many ways, it appears to run counter to the native philosophy of collective good rather than individual gain (George & Preston, 1987). Native cooperatives and public sector employment (including administration, education, and wildlife and natural resource management) can provide a good basis for local economic development (George, 1989).

In Ontario, small scale, rather than large scale community-based development (as in Quebec) is occurring. Development activities include forest products, community stores, and tourism (George, 1989; Stephenson, 1991). Government still controls the li-

Table 15. Trends in Cree Income Under the JBNQA, 1971-1981 (Source: Scott, 1992).

	Percentage of Income	
	1971	1981
Hunting	61	43
Salaries	23	52*
Transfers	16	5
Increase in total income	1	4.3
Increase in total population	1	1.3
Increase in workforce	1	1.4

* Urbatique (1991), in a study of the Cree-Inuit centre of Whapmagoostui-Kuujjarapik, noted that about one-third of the total population was working. About 20% of the work force is involved in secondary employment (including building, maintainance and repair) and nearly 40% in tertiary employment (including administration, transportation, communications, finance and commerce).

cence and leasing of native lands and this has severely limited some forms of economic development. The recent formation of inter-band councils (Nishnawabe-Aski, and the Mushkegowuk Regional Council, for example) should provide a basis for more effective cooperation with both federal and provincial levels of government, and the administration of more accessible financial resources (George, 1989).

Relative to many Cree communities, the Inuit economy remains more dependent on traditional resources. Hunting and trapping provide a major (non-cash) contribution, and furs and art-work provide a limited but important cash return. Commercial activities are limited. Early attempts at cooperative enterprise were not welcomed by the Hudson Bay Company and government policy remained indifferent to opportunities for community-based development (Clancy, 1987). More enlightened policies have come into place over the past two decades. Government jobs provide full and part-time employment, and government assistance provides additional income. At Sanikiluaq in the Belcher Islands, for example, government jobs account for about 56 percent of the community income, business (mostly cooperative) 23 percent, social assistance 13 percent, transfer payments (including pensions and child care) nearly 7 percent, and fur sales are just over 1 percent. However, it is estimated that the cash equivalent value of traditional resources (particularly food) is worth more than 1.3 times the cash economy of the community (Quigley & McBride, 1987). Employment opportunities are a major concern for the Inuit and many Indian communities outside the JBNQA. In the Keewatin District, for example, unemployment is reported to average about 27 percent. In 1991, the Nunavut Planning Commission noted that about 250 new jobs will be required in this area, just to maintain the current level of employment between 1989 and 1995.

Within the Hudson Bay region, a number of studies have explored economic development and the role of the traditional renewable resource-based sector. In Sanikiluaq, economic development based on locally available eiderdown and recently stocked reindeer has been studied, as part of the overall economy which already relies heavily on fish, marine mammals and wildlife harvesting (Quigley & McBride, 1987). The most comprehensive study of sustainable community development within the region has been undertaken on the Ontario coast, covering 6,400 Cree people in eight communities. Undertaken jointly with the Mushkegowuk Tribal Council, this project has studied baseline socio-economic characteristics of Cree communities in a series of reports (for example Stephenson, 1991), historical factors (Schuurman *et al.*, 1992), obstacles to development (George & Preston, 1987; George, 1989), possibilities of joint resource management or co-management (Berkes *et al.*, 1991a, 1991b), traditional ecological knowledge of the Cree (Berkes *et al.*, 1992a), wildlife harvests and the economic significance of the traditional sector (Berkes *et al.*, 1992b), and land use and local government. These studies document a regional economy characterized by extensive land use and a bush harvest with a replacement value in 1990 of $8,400 per household, regional average (Berkes *et al.*, 1992b). This represents about one-quarter of the total economy, including wage and transfer payments. Viewed in a similar way (Quigley & McBride, 1987), the comparable land use and harvest of Inuit communities may exceed this value. The studies also show that realistic long term targets for sustainable development should involve a mixed economy that maintains a firm base on local renewable resources, with new resource-based activities, and locally administered services.

3.6 *Resource management*

Traditional Environmental Knowledge (TEK) is gathered by tapping the wealth of experience of Cree and Inuit; this knowledge is a proven and reliable means of resource management (Arrangutainaq & Fleming, 1991; Berkes *et al.*, 1992a). TEK is holistic and qualitative, it incorporates intuitive and ethical knowledge that is both locally specific and regionally extensive, and it takes into account variations over many years. TEK should be used to contribute to more effective co-management of wildlife and other natural resources.

There are many forms of co-management, from tokenism to real empowerment (Berkes *et al.*, 1991a, 1991b). At its best, co-management is linked to self-governance (Mulvihill & Jacobs, 1991) and recognises the need for sustainable development in northern communities.

Co-management is a major factor influencing the social and economic health of Cree and Inuit society. Under the JBNQA, land tenure is secure and community-based resource management is possible. Each community has its own hunter/trapper association and may have a small core of specialists to guide the resource management. This approach to resource management requires people on-site and the local employment that this provides benefits the community as a whole. Similarly, in Manitoba, the Keewatinowi Okimakanak Natural Resource Secretariat is a key source of scientific information and traditional knowledge used to support regional native resource management (Berkes *et al.*, 1991b).

Past management practices which generally excluded TEK and the community have often led to a high degree of dependence (Berkes *et al.*, 1991b). In Ontario, for example, the Ministry of Natural Resources retained control of trap lines and assigned them to individuals, and reassignment could occur without regard to traditional land tenure. Hunting areas were rarely rested. In contrast, it is generally believed that hunting has not increased on lands covered by the JBNQA, despite a significant growth in native population. Rather, a complex of social, cultural, and economic factors has influenced the use of local food supplies. As a result, dependence on wildlife resources has not increased and native harvesting seems not to have caused changes in most wildlife species.

Co-management will take different forms throughout the Hudson Bay region but to be effective it must recognize co-jurisdiction (Berkes *et al.*, 1991b; Mulvihill &

Jacobs, 1992) and the need to accommodate local and regional interests. For example, management of polar bears in Ontario should not be independent of their management in Manitoba or the Northwest Territories. This argument applies to many migratory species (birds, fish, mammals) their habitat, and food base.

3.7 *Parks and sanctuaries*

One of the more recent trends in northern land use has been the setting-aside of space for wilderness parks. Figure 9 shows the distribution of existing and proposed park sites and sanctuaries. In general, wilderness parks are established to regulate human access and to strictly limit natural resource development. Sanctuaries are usually more restrictive and are intended to eliminate or largely exclude human disturbance and hunting activities. Sanctuaries are designated for specific communities (game or large mammals, wildlife, and birds) in areas of key habitat.

At present, there are four wilderness parks around the basin, three in Ontario and one in the Northwest Territories. They cover an area of about 34,000 km^2. Additional areas are under consideration, including a number of waterway parks. Most parts of the watershed are provincially/territorially designated and through federal/provincial cooperative agreements national parks may be established (including Heritage Rivers). There are no marine parks in the basin but suggestions have been for marine parks off Rankine Inlet and the Churchill-Nelson area, and to include other areas of special interest (Stewart *et al.*, 1991). Additionally, there are many small, local, recreational, cultural, historic, and scenic-point parks throughout the region.

4. The bay and its ecosystem

4.1 *Geological and physical setting*

Two extensive reviews have been published in the past 25 years describing in considerable detail current knowledge about the Hudson Bay basin (Beals & Schenstone, 1968; Martini, 1986a). Exploration and investigation of Hudson Bay basin began during the early 17th Century but present understandings are largely based on knowledge of the region which has been gathered only during the past 50 years or so. Scientific information about many characteristics of the bay is often not available or inadequate but some comparison with other inland seas is possible. In particular, understandings drawn from the Baltic Sea provide useful insights into the effects of stress and cumulative impact (Elmgren, 1989; Rapport, 1989).

The total area of the Hudson Bay basin is about 1,650,000 km^2. Hudson Bay has an area of about 1,250,000 km^2 and James Bay has an area of about 50,000 km^2. Foxe Basin and Hudson Strait have areas of about 170,000 and 180,000 km^2, respectively (Fig. 10). The shape of the bay is strongly influenced by underlying geology, particularly structures within the Precambrian Shield. Two major sedimentary basins overlie the Shield and are filled with Palaeozoic and later sediments (Norris, 1986). The main sedimentary basin (Hudson Bay Basin) is centred nearly in mid-bay, and contains about 2000 m of sediment. A smaller sedimentary basin (Moose River Basin) is centred within the Hudson Bay Lowland (Fig. 2) and contains less than 1000 m of sediment. These sedimentary basins are separated by Precambrian rocks of the Cape Henrietta Maria Arch which extends northwards through the Hudson Bay Lowland and across the floor of Hudson Bay. To the east, numerous island groups between the

Figure 9: Symbols and explanations

Designated and Proposed Wilderness Parks

Québec
A   Monts Torngat et Rivière Koroc (4300)✷
B   Monts Pyramides (1600)
C   Confluence des Rivières de la Baleine et Wheeler (2000)
D   Baie aux Feuilles (5000)
E   Cratère du Nouveau-Québec (900)
F   Cap Wolstenholme (1300)
G   Monts de Povungnituk (3000)
H   Lac Guillaume-Delisle et Lac a l'Eau Claire (10 300)✷
I   Lac Burton-Rivière Roggan et la Pointe-Louis XIV (8400)
J   Péninsule Ministikawatin (700)

Ontario
K   Kesagami Wilderness Park (600)†
L   Polar Bear Wilderness Park (24 100)†
M   Opasquia Wilderness Park (4700)†

Manitoba
N   Churchill Wilderness Park✷
O   Nueltin Wilderness Park

Northwest Territories
Q   Wager Bay✷
R   North Baffin Island✷
S   Auyuittuq National Park (4500)✷†
T   Pitsutinu Tugavik. (*Tungavik*)
U   Sylvia Grinnel
V   Katannilik

Note: ( ) Approximate area of reserved space in km^2 (*Italics*) New name or spelling
      ✷ National Park status (all others as provincial or territorial designations)
      † Existing park (all others are either proposed or under consideration)

Designated and Proposed Waterways

Ontario
a   Missinabi River (middle and upper parts)
b   Little Current River (upper part)
c   Albany River (upper part)
d   Otoskwin-Attawapiskat River (middle and upper parts)
e   Winisk River (all)
f   Severn and Fawn rivers (upper parts)

Manitoba   g   Seal River (lower part)

Northwest Territories
h   Kazan River
i   Thelon River

Note:
- Sections of northern Quebec rivers are included within proposed wilderness and park areas
- Other rivers in northern Manitoba and the Northwest Territories may be proposed for Provincial/Territorial Waterway status and/or National Heritage status
- ( ) sections of rivers covered by designations, where known

Possible Marine Parks

Natural Areas of Canadian Significance, NACS   - k   Churchill-Nelson area   - l   Rankin Inlet - Marble Island

Note: - Additional small Natural Sites of Canadian Significance (NSCS) may be designated, as well as areas of unique marine conditions

Designated Sanctuaries

1   Dewey Soper (GS)
2   Bowman Bay (WS)
3   Cape Dorset.
4   North and South Twin Islands (GS)
5   Charlton Island (GS)
6   Boatswain Bay (B)
7   Hannah Bay (BS)
8   Akimiski Island (BS)
9   McConnell River (GS)
10   Thelon (GS)
11   Kaminuviak (*Qamanirjuaq*) Herd (Caribou S)
12   Harry Gibbons (BS)
13   East Bay (BS)

Note: GS - Game sanctuary   WS - Wildlife sanctuary   BS - Bird sanctuary

Hudson Bay Region: Present and future environmental concerns

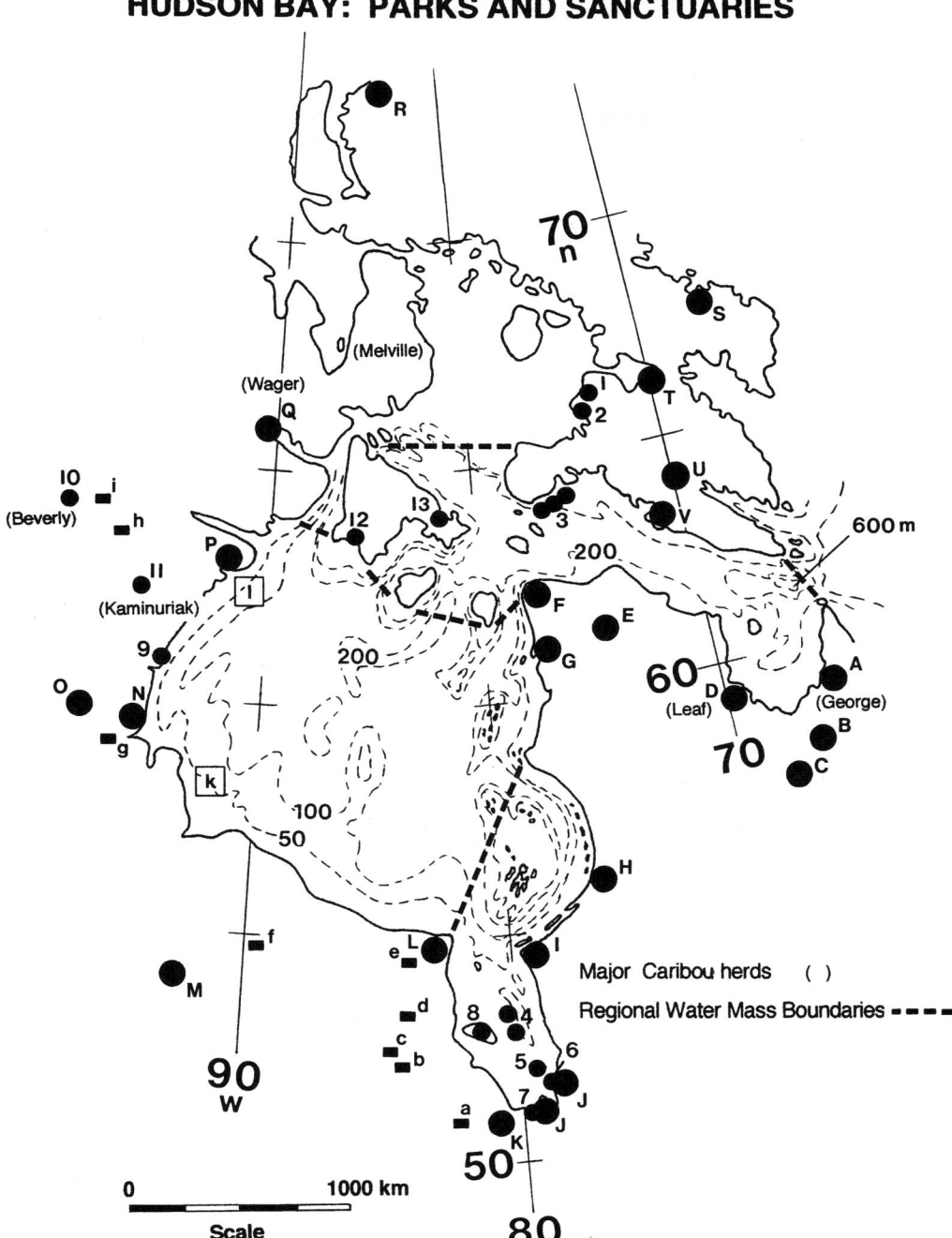

Fig. 9. Hudson Bay: Parks and Sanctuaries (Sources: Gouvernement du Québec, 1984; Government of Manitoba, 1986; Dunbar, 1988; Government of the Northwest Territories, 1989; Environment Canada, 1991; Stewart *et al.*, 1991; Government of Ontario, 1992a; 1992b; 1992c; Gouvernement du Québec, 1992; Thomas *et al.*, 1992; World Wildlife Fund Canada, 1992; Government of Ontario, undated).

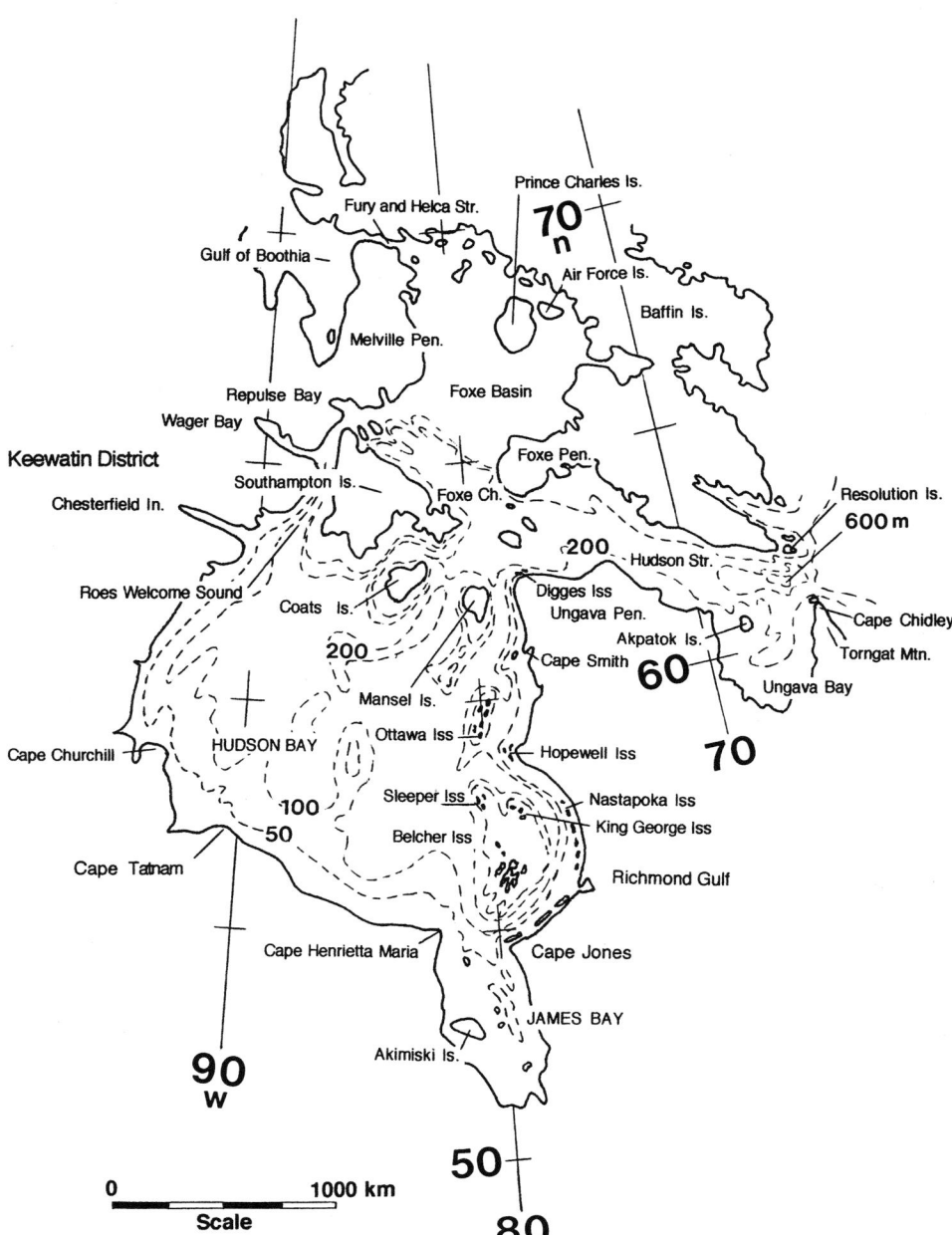

Fig. 10. Hudson Bay: Features.

Belchers Islands and Cape Smith reflect a belt largely composed of folded Precambrian sedimentary rocks. Precambrian rocks form most of eastern James Bay, and the east shore of Hudson Bay. Precambrian rocks are exposed around much of Foxe Basin and Hudson Strait where faulting and boundaries between major units of the bedrock largely determine features in the northern part of the basin.

Most of the shoreline of the Hudson Bay basin is of low to moderate relief (Martini, 1986b; Stewart et al., 1991). But, high and rugged shorelines occur around Fury and Hecla Strait, the northern opening of Foxe Basin, some promontories of Southampton Island, the south shore of Wager Bay, the Keewatin shore just south of Chesterfield inlet, parts of the eastern shore of Hudson Bay north of Cape Jones, Digges Island, and east along the Ungava shore and around much of Ungava Bay, including Akpatok Island. In many areas, particularly around Foxe Basin, the shoreline is backed by raised beaches (Martini, 1986b). Extensive beach ridge formations and mud flats continue south from Whale Cove to the south shore of James Bay. Much of the shoreline of Ontario and Manitoba is backed by tidal marshes and wetlands (Mortsch, 1990) and much of the shoreline of the Hudson Bay basin is depositional. Ice plugs often form near river mouths, they cause ice jams at spring break-up and flooding may be extensive in the lower reaches of the rivers (Martini, 1986b; Maxwell, 1986).

Discontinuous permafrost underlies much of the Taiga Shield and the Hudson Bay Plain (Fig. 3). Continuous permafrost begins north of the Nelson River mouth on the west shore of the bay and, on the east shore, north of Great Whale River (Fig. 1). Due to the influence of bay waters on soil moisture and temperature, permafrost is generally absent at the immediate shoreline around most of Hudson Bay and James Bay.

Hudson Bay (Fig. 10) is separated from Foxe Basin by a sill at a depth of about 185 m. South of this, the bay reaches a maximum depth of about 250 m and most of the nearshore east of Churchill and within James Bay is an area of shallow water (less than 50 m). The floor of Hudson Bay is crossed by several submarine valleys which may have formed, originally, as surface erosion features (Pelletier, 1986b). Foxe Basin is generally less than 100 m deep, except where the Foxe Channel deepens eastward into Hudson Strait. The strait also deepens eastward, and reaches a maximum depth of more than 600 m between Resolution Island and Cape Chidley. A sill, at about 300 m separates the strait from the open waters of the Labrador Sea. Large scale bathymetric features of the bay are usually associated with the presence of major rock units and often reflect deep-seated structural control. Late glacial and post glacial marine sediments provide a partial cover over much of the seafloor which has locally variable relief (Shilts, 1986). Sands and sandy gravels form the nearshore between the southern entrance to Roes Welcome Sound and Churchill, and extend offshore to depths of 100 m along the Ontario shore and into James Bay. Elsewhere, silts and sandy silts cover most of the Hudson Bay basin except in the deeper areas (greater than about 160 m) where clays and silty clays are predominant (Pelletier, 1986b).

4.2 Climate and oceanography

The Hudson Bay basin is distinguished from other enclosed seas at similar latitudes by its generally cold, harsh climate and extensive ice cover. These are a consequence of its mid-continental location, the inflow of Arctic Ocean waters, and limited exchange with the Atlantic Ocean (Grainger, 1963; Stewart et al., 1991). Whereas the North Sea, the Baltic Sea and the Bering Sea receive inputs of heat from both atmospheric and oceanic sources, there is little oceanic heat transfer into and out of Hudson Bay (Prinsenberg, 1986a). Received radiation is almost completely used to melt the ice and to warm the surface and near surface waters of the basin. Heavy winter ice combined with inputs of large volumes of freshwater, causes patterns of stratification and circulation within the water column of Hudson Bay that are more estuarine than oceanic.

The climate is dominated by Sub-Arctic and low Arctic continental conditions (Maxwell, 1986). Air flow across the basin is governed by the interaction of a generally persistent low pressure cell located near Baffin Island and Davis Strait, and high pressure areas that form over the Mackenzie and Keewatin districts of the Northwest Territories. These pressure systems strengthen in winter, steering cold, dry polar air south and east over Hudson Bay. During most of the year, when ice covers the basin, temperatures are low and winds are strong. Storm systems originating from the south carry moisture into the region during the spring and summer months, when the low pressure cell weakens and moves north. Hudson Bay exerts a significant influence on regional atmospheric conditions only during the ice free period. In summer, air over the bay and coastal areas is 10°C or less, while temperatures inland can reach 30°C or more. Stormy weather in the region is caused by re-establishment of a strong inflow of polar air during the autumn and the exchange of heat and moisture from the relatively warm and ice free waters of Hudson Bay.

The loss of continuous snow cover occurs in late April at Moosonee and at the end of June near Hall Beach (Maxwell, 1986). Continuous snow cover reforms in early November at Moosonee but in late September at Hall Beach. Foggy conditions are typical of many parts of Hudson Bay during the late summer - early winter period, and are most frequent in Hudson Strait. Mean daily air temperatures in January decrease northwest across the basin from -20°C in southeast James Bay to -33°C on the Melville Peninsula, and northwards in July from 15°C in southeast James Bay to 5°C in Foxe Basin (Maxwell, 1986).

At first, sea ice melts in place, and then through attrition and transport and exposure to slightly warmer freshwater discharge from the watershed. Early clearing takes place in James Bay and around the northwest and southeast margins of Hudson Bay. Clearing in the northeast is due to wind action, and in the southeast to the lower latitude and large spring run off. By early August most of the bay is clear (Markham, 1986). Ice begins to break up in Hudson Strait in late May but the strait may be congested by ice from Foxe Basin during June and July, and icebergs sometimes drift into Hudson Strait from the west Greenland icecap but bay ice occurs only as sheet-ice (Drinkwater, 1986; Prinsenberg, 1986c). Almost all of the bay is ice free by September. Sea ice normally enters Hudson Strait from Foxe Basin during late October and in situ freezing occurs along the shorelines in November, and in mid channel by late November. Ice forms, first, in Foxe Basin and extends south to cover most of the bay by late November or early December. Ice-ridging is most pronounced in Hudson Strait.

In parts of James Bay and Foxe Basin, autumn storms mix bottom sediment with the early forming sea ice. Ice movement continues throughout the winter and produces large areas of hummocky ice in Foxe Basin. Here, also, there are noticeable patches of clean white ice. Some of this is derived from Arctic ice which passes into the basin from Fury and Hecla Strait to mix with local sediment-laden ice (particularly from the eastern shore of Foxe Basin). Year to year variations in wind conditions significantly affect both the formation and breakup of ice in this area (Markham, 1986). Shore fast ice develops out to water depths of 20 m or more around much of the basin. Shore ice may extend from the northeast of James Bay to the Belcher Islands and, in some years, continuously to the Ottawa Islands and Cape Smith. Permanent areas of open water (polyanas) and shore leads are produced by the interactions of wind and tidal currents and persist along the west shore of Hudson Bay, from Churchill to north of Chesterfield Inlet. They also occur off the south and east shores of Southampton Island, the east shore of Coats Island, and to the east of the Belcher Islands (Markam, 1986). Polyanas also exist along the south shore of Baffin Island, west of Lake Harbour, and in Foxe Basin off the east shores of Prince of Wales Island, and between Hall Beach and Rowley Island (Prinsenberg, 1986c).

Hudson Bay and James Bay are largely enclosed bodies of water with limited tidal exchange. There is a net inflow of freshwater from James Bay to Hudson Bay, which is ultimately derived from surface run off. Peak freshwater discharge into both bays occurs during May and June. Inflow is augmented in late June and July by the input of a similar volume of freshwater derived from the melting of the ice cover in the bay (Prinsenberg, 1980, 1991). Nearly 40 percent of the freshwater discharge from the entire Hudson Bay watershed enters James Bay. After recent diversions for hydroelectric development (Messier *et al.*, 1986), about one third of this now comes from the La Grande River. Freshwater entering Hudson Bay gradually mixes with ocean water, and a relatively fresh surface current discharges from Hudson Bay along its eastern shore and into Hudson Strait.

The waters of Foxe Basin and Hudson Strait are vigorously mixed by tides. The minimum water depth in the Hudson Strait exceeds 200 m, and deep water mixing extends into the southwest part of Foxe Basin. Shallow northern and eastern parts of the basin remain largely beyond the influence of Atlantic water inflow. Arctic Ocean water flows into northern Foxe Basin, through the Fury and Hecla Strait, and onwards into the Northwestern part of Hudson Bay through Roes Welcome Sound.

All deep water in Hudson Bay is poor in nutrients and colder than -1°C. The water, originally, may have entered either as Arctic Ocean water or Atlantic Ocean water but has been "reworked". That is, it has gone through at least one freezing cycle and was involved in severe surface cooling and/or salt rejection below a growing ice cover. "Atlantic type" water is relatively warm (warmer than +1°C) and nutrient rich, but very little passes into Hudson Bay from Hudson Strait (Dunbar, 1958). Hudson Bay water below the seasonal pycnocline is termed "Arctic type" water. This, however, does not mean that it has come from the Arctic but, rather, that it has gone through a surface cooling process comparable to that of the Arctic. The distribution of the copepod *Calanus finmarchicus* in Hudson Strait, south Foxe Basin, and the extreme northeast of Hudson Bay closely parallels the limits of Atlantic type water (Grainer, 1963). The copepods *C. hyperboreus* and *C. glacialis* are both closely associated with Arctic type water (the former is found only at depths greater than 30-50m and the latter at water depths of 100m or less). Both of these species are widespread throughout the basin (Grainger, 1963).

The physical oceanography of Hudson Bay is characterized by persistent vertical stratification (Prinsenberg, 1986a) and a general counter-clockwise surface circulation (Prinsenberg, 1986b). The water column can be divided into a warmer, fresher surface layer and a colder, saline bottom layer. Temperature differences across the thermocline are greatest in summer. This thermal gradient deepens from a summer depth of about 20m to more than 90m in late winter. This is due to salt rejection from the growing winter ice cover.

A strong seasonal pycnocline occurs in Hudson Bay and this deepens from 20 to 50 m as summer progresses. It largely inhibits mixing between the two layers. Above the pycnocline, summer water temperatures reach 5-8°C and surface salinities vary from 10 to 30 °/oo, but the pycnocline is weak in shallow areas where tidal mixing is strong. Low salinities are particularly evident off river mouths, especially in James Bay. Vertical stability is strong in summer (Dunbar, 1958) but decreases in winter with reduced freshwater discharge and the effects of salt rejection from ice accretion. Salt rejection can effect the water column to a depth of more than 100 m in winter but the salt is returned to the surface during the spring and summer melt (Prinsenberg, 1991). Salt rejection is particularly significant in Foxe Basin (Prinsenberg, 1986c) and results in the presence of very cold (-1.7°C or colder) and saline bottom water. This underlies cold but less saline Arctic water (from Fury and Hecla Strait). The thermocline deepens from a shallow summer layer of 20 m to more than 90 m in late winter, due to vertical mixing generated by the salt rejected from the growing ice cover. The mean depth of the total freshwater content of Hudson Bay (inflow and ice-melt) is about 6.3 m. It has a south to north gradient, 8.1 m in James Bay, 6.9 m in the southern part of Hudson Bay, and 4.2 m near the outlet to Hudson Strait (Prinsenberg, 1986a).

The freshwater plumes entering the basin are highly stratified and can remain coherent up to 70 km from source, with widths of 20-30 km (Prinsenberg, 1991). The plume of the La Grande River is 2 to 5 m thick. It can limit vertical mixing and incorporation of bottom water nutrients into surface and near surface waters. Under-ice plumes may be twice their summer width and they are only partially coupled to the current flow of nearshore water which drags them slowly counter-clockwise along the coast. Plume area is related to discharge but there is only a weak relationship between orientation and atmospheric pressure gradients (Ingram & Larouche, 1987a).

The mean depth of the total freshwater content of Hudson Bay (annual inflow and ice-melt) is about 6.3m. It has a south to north gradient, 8.1m in James Bay, 6.9m in the southern part of Hudson Bay, and 4.2m near the outlet to Hudson Strait (Prinsenberg, 1986a). Freshwater in the surface layer enters as river discharge and leaves as a surface flow through Hudson Strait. Open water circulation is partly wind-driven. The combined effects of wind, tide and natural discharge produce the strongest circulation of James Bay waters during the summer (Prinsenberg, 1991). The discharge of freshwater from James Bay and the southern part of Hudson Bay reaches Hudson Strait by mid-September. The mean travel time of freshwater in the surface layer is between 10 and 20 months. In contrast, the residence time for water in the lower layer is 3 to 4 years near the edge of the bay and more than 6 years for bottom water in the centre of Hudson Bay. The counter-clockwise circulation in Hudson Bay reflects the Coriolis force; it is assisted by the prevailing atmospheric circulation and by density effects due to run off and ice melt. The influence of prevailing winds is greatly reduced when ice covers the surface of the bay (Prinsenberg & Ingram, 1991).

Tidal oscillations (Prinsenberg & Freeman, 1986) are reduced as much as 20 percent by ice cover. Ice cover also modifies the amplitude and phase of major tidal constituents (Lepage & Ingram, 1991). The main semi-diurnal tide enters Hudson Strait with an amplitude of about 4 m (greatest along the north shore). The amplitude is about 3 m at the western end of the strait, it decreases east to west within Foxe Basin (4.3 to 0.5 m) and anti-clockwise around most of Hudson Bay (1.3 m near Chesterfield Inlet to less than 20 cm along the north shore of Quebec). The inflow of Atlantic Ocean water heads west along the north shore of the Hudson Strait (Drinkwater, 1986), and warmer and less saline waters outflow along the south shore. Current velocities decrease from east to west within the strait, from about 2 m $\cdot$ s^{-1} to about 0.5 m $\cdot$ s^{-1}, and to about 0.25 m $\cdot$ s^{-1} in the eastern part of Foxe Basin. The strong tidal currents induce intense vertical mixing and provide a significant nutrient enrichment of surface and near surface waters. The mean flow of surface currents in the Hudson Bay gyre are about 5 cm $\cdot$ s^{-1}. A similar pattern of counterclockwise circulation occurs in James Bay, where a greater salinity gradient contributes to average surface currents of about 19 cm $\cdot$ s^{-1}. Tide-driven circulation dominates many parts of the basin (for example, Rupert Bay, Veilleux et al., 1992).

4.3 Nutrients and primary production

Arctic type waters have lower nutrient concentrations than Atlantic waters, but may be nutritionally rich relative to freshwater discharge. Because of its huge volume, Arctic water represents a significant reservoir of nutrients but biological productivity is low throughout most of the basin (Legendre & Simard, 1979). This reflects limited mixing between bottom and surface waters and a lack of nutrients in the discharge from surrounding drainage. During summer, the differences between nearshore and offshore water chemistry increase (Roff & Legendre, 1986). Offshore surface waters warm quite quickly but, in many areas, nearshore water temperatures remain depressed by coastal upwelling. In estuaries and near river mouths, salinities are kept low by freshwater discharge although, as a result of open water, summer plumes are less stable and the freshwater mixes more rapidly (Legendre & Simard, 1978, 1979; Roff & Legendre, 1986). Vertical nutrient fluxes are proportional to density stratification and tidal mixing is an important component of the nutrient supply to the photic zone (Prinsenberg, 1991). Upwelling brings a slight increase in the amount of nutrients available in surface waters, but surface and near surface concentrations of nitrate and phosphate remain low throughout the growing season (Because of ice conditions, there is little information about nutrient availability at the start of the growing season). Levels of nitrate become particularly low towards the end of summer (Legendre & Simard, 1979). The winter discharge of freshwater adds little nutrient to the nearshore. Concentrations of nitrate and nitrite, however, are two to three times higher in snow cover than in surface waters. Phosphates are also higher, and estimates suggest that about 10 percent of the nutrient flux to Hudson Bay may be derived from atmospheric sources (Roff & Legendre, 1986). This is comparable to the role of the atmospheric flux in Lake Superior (Upper Lakes Reference Group, 1977).

Nutrient levels increase just below the pycnocline, and summer concentrations of nitrate plus nitrite remain similar throughout much of Hudson Bay. They are considerably lower than concentrations in Atlantic Ocean waters which penetrate Hudson Strait. Low levels of nitrate plus nitrite probably reflect incomplete remineralization of inorganic nitrogen (caused by low temperatures and low inoculum of nitrifiers).

Low primary production in surface waters means that water clarity is generally high and that light penetration occurs to considerable depth. Indeed, there is a distinct chlorophyll a maximum just below the pycnocline, and to depths as much as 20 m below it (Anderson & Roff, 1980b). Concentrations of chlorophyll a, at this depth, may be more than 60 times greater than those in the surface water (Roff & Legendre, 1986). Fresh waters from the Canadian Shield are very low in phosphates (Legendre & Simard, 1978) and, in general, the highest levels of biomass are associated with patches of nearshore water enriched by upwelling. Relatively high levels of biomass also may be present in the inner parts of estuaries or further upstream, where particulate organic matter becomes available from specific terrestrial or lacustrine sources.

There is a diverse flora of ice algae in Hudson Bay and this source of primary production (Legendre *et al.*, 1981) is an extremely important component of the overall productivity of the bay, as in other areas of the Arctic (Welch & Kalff, 1975; Roff & Legendre, 1986). The annual cycle of primary production starts as ice begins to form. Ice algae stop growing as light levels decrease during the early part of winter. In the spring, ice algae resume growth. Ice algae are concentrated at the base of the floating ice where salt rejection and ice formation creates a thin layer of nutrient enrichment. In coastal areas, nutrients are also derived from the upwellings of deeper water. In this respect, tidal forcing and the presence of internal waves play an important role in maintaining the supply of nutrients (Gosselin *et al.*, 1985, 1986; Ingram *et al.*, 1989; Legendre *et al.*, 1982). The highest levels of chlorophyll a, in ice algae, occur in southeast Hudson Bay (up to 170 mg chl a $\cdot$ m^{-2} at locations off Chesterfield Inlet; Welch *et al.*, 1991). Ice algae comprise both pelagic and benthic forms, and benthic forms become dominant as spring advances and ice break-up begins. Welch & Bergmann (1989) noted that the development of ice algae could be synchronous over wide areas and that snow cover controlled light penetration through the ice. Maximum levels of 300 mg chl a $\cdot$ m^{-2} were reported near Resolute in 1986. The peak ice algae biomass under thin snow was estimated to equal 0.5t $\cdot$ ha^{-1}, dry weight.

As the ice melts in situ, airborne inoculation may produce early blooms of freshwater algae in shallow ponds which form on the surface of the sea ice. As the ice melts the ice algae community may also coexist for a short time with the start of the spring phytoplankton bloom. Deepening of the photic zone and pycnocline proceed slowly in spring during ice cover. As the ice melts, it reduces the salinity of the surface mixed layer and strengthens density stratification. This cuts off the upward supply of nutrients to surface waters and the spring bloom of surface phytoplankton quickly exhausts available nutrients (Roff & Legendre, 1986). Spring bloom values of chl a as high as 30 mg $\cdot$ m^{-3} have been reported from Chesterfield Inlet, 10 mg $\cdot$ m^{-3} in the Eastmain River, and 6.8 mg $\cdot$ m^{-3} in southeast James Bay. Bay-wide estimates suggest that offshore waters produce less than 30 g C m^{-2} $\cdot$ a^{-1} and nearshore waters about 70 g C m^{-2} $\cdot$ a^{-1}; productivity in the sub-pycnocline layer is at least 3 g C m^{-2} $\cdot$ a^{-1}.

The phytoplankton assemblages in bay waters have considerable diversity. They are made up of a mixture of Arctic, boreal, and temperate forms of diatoms and dinoflagellates. Zooplankton also comprise a mixture of marine and freshwater forms (Anderson & Roff, 1980a). Temperature has a marked effect on zooplankton growth and most species produce only one generation per year. Light is a major factor controlling productivity, and annual phytoplankton production is limited to an average of about 120 days. Depending on latitude, ice algae may add another 90 days to total primary production. In response to light, primary production is strongly pulsed and partial cropping by pelagic species allows more than 40% of the energy from phytoplankton to enter the food web through zoobenthos. The significance of the benthic food web is reflected in the considerable abundance of organisms on both rocky and sandy bottom substrates.

4.4 food web structure

Early surveys in Hudson Bay basin focused on nearshore fisheries (Comeau, 1915; Melville, 1915; Lower, 1915) and the 1930 Loubryne expedition (Hachey, 1931) surveyed the potential offshore fishery as well as undertaking oceanography. Oceanographic surveys by the *Calanus* (1949-1955) covered the northern part of the basin (Dunbar, 1958). Identifications and faunal lists covering biota from the basin have been compiled by several authors including Atkinson & Wacasey (1989a, 1989b), Bousefield (1955), Crossman & McAllister (1985), Dunbar (1954b), Dunbar & Hildebrand (1952), Hunter *et al.* (1984), Lubinsky (1980), McAllister (1990), and McAllister & Steigerwald (1986). These lists cover both benthic and pelagic invertebrates and fish.

Figure 11 provides a summary of some of the more important components of the Hudson Bay food web. The base of the food web is characterized by two distinct but linked components, one is associated with ice algae and the other with phytoplankton.

HUDSON BAY: FOOD WEB STRUCTURE *key species*

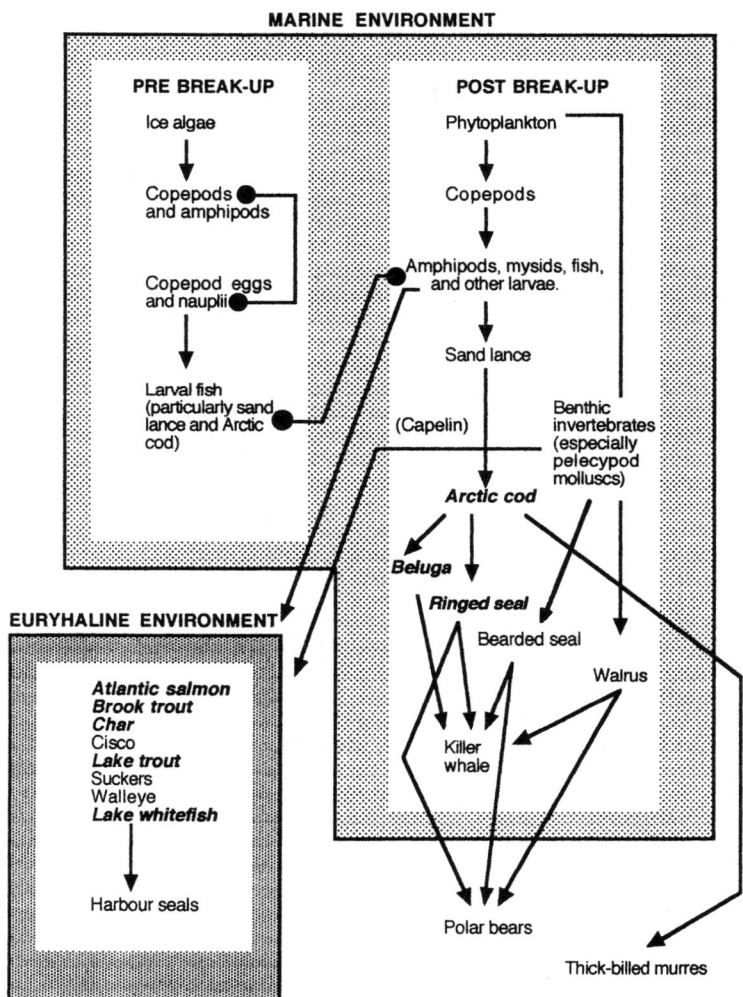

Note: Role of Capelin is uncertain

Fig. 11. Hudson Bay: Food-Web Structure (Sources: NSL, 1970; Morin & Dodson, 1986; Twitchell, 1991).

Ice algae provide a limited but crucial early pulse in the primary production cycle (Legendre *et al.*, 1981, 1982; Roff & Legendre, 1986). Amphipods feed on ice algae and form an important component in the under-ice food web. Runge & Ingram (1991) also noted the importance of ice algae to female copepods (*Calanus glacialis* and *Pseudocalanus minutus*). Large numbers of copepods feed on ice algae and the ingestion of algal material triggers an early start in their reproductive cycle. Copepod eggs and nauplii are a key component in the diet of many larval fish (Drolet *et al.*, 1991; Gilbert *et al.*, 1992) and, in particular, the larvae of sand lance (*Ammodytes hexapterus*) and Arctic cod or ogac (*Boreogadus saida*). The shift from endogenous to exogenous feeding is a critical period for fish. For sand lance and Arctic cod, the critical period appears to be during late May and early June, and their survival may be more dependent on availability of ice algae than the spring phytoplankton bloom. The sand lance includes both planktonic and benthic sources of food in its diet. In turn, it is an important source of food for Arctic cod which also feeds extensively on pelagic crustacea (Mikhail & Welch, 1989). The Arctic cod is probably one of most important species of fish in the basin (Morin & Dodson, 1986), particularly in the northern part (Welch, pers. comm.). It supports many top predators, including beluga whales (*Delphinapterus leucas*), ringed seals (*Phoca hispida*), bearded seals (*Erignathus barbatus*) and thick-billed murres (*Uria lomvia*). These in turn support killer whales (*Orcinus orca*), polar bears (*Ursus maritimus*), and humans. In southern Hudson Bay, the capelin (*Mallotus villosus*) may fulfil a similar role (Welch, pers. comm.). Copepods are a principal source of food for the bowhead whale and small numbers of bowheads are found in northern parts of the bay at times of copepod abundance (Grainger, 1959; Stewart *et al.*, 1991). In Hudson Bay, capelin are an important source of food for the Greenland cod (*Gadus ogac*), as noted by Mikhail & Welch (1989), but little is known about their life cycle or how they are linked to the food web of the basin.

Figure 11 is a greatly simplified representation of the food web and ignores many components which may have special but, as yet, little understood significance in different parts of the bay. Grainger (1982), for example, reported a high late summer abundance of ciliates near the Belcher Islands. They may be an important intermediary whose presence ensures effective coupling of the energy flow between nanoplankton and micro-phytoplankton (bacteria, organic detritus, small diatoms and small flagellates), and the larger omnivorous zooplankton.

The euryhaline environment is particularly important for a number of anadromous fish species, such as cisco (*Coregonus artedii*) and lake whitefish (*Coregonus clupeaformis*) which spawn and over winter in accessible freshwaters where they benefit from the thermal protection of the warmer water (Bodaly *et al.*, 1989; Roy, 1989). These same species also benefit from higher primary production in nearshore and estuarine areas. Upward mixing of more nutrient rich marine water may therefore enrich some otherwise nutrient-poor estuarine and nearshore areas (McAllister, 1991b; Prinsenberg, 1991). In terms of human food resources, the most important fish species present include Atlantic salmon (*Salmo salar*), brook trout (*Salvelinus fontinalis*), Arctic char (*Salvelinus alpinus*), lake trout (*Salvelinus namaycush*), and lake whitefish. Harbour seals (*Phoca vitulina*) are an important marine predator in the euryhaline environment.

4.5 Distributions of fish and their significance

The number of fish species present decreases from the truly marine to the freshwater environment in the bay: 51 species have been recorded from Hudson Strait, 34 from Hudson Bay, and 22 from James Bay (Morin & Dodson, 1986). The distribution of species also reflects the extent to which fish have been able to colonize the bay following deglaciation. Trawl surveys indicate that fish abundance and biomass decrease, east to west, through Hudson Strait and are many times greater in Davis Strait. Fish abundance and biomass almost certainly decrease further in Hudson Bay. Trawl surveys indicate that no fish were recovered from 65 percent of bay stations greater than 100 m, and from 22 percent of stations at depths less than 100 m.

In order of decreasing weight of catch, Greenland halibut (*Reinhardtius hippoglossoides*), Arctic cod, sea-snails, and cottids and agonids are most important in the eastern Hudson Strait. Greenland halibut, sea-snails and Arctic cod, and cottids and agonids are most important in Ungava Bay. Sea-snails, lumpfish (*Cyclopterus lumpus*), Arctic cod, cottids and agonids, and Greenland halibut are most important in western Hudson Strait.

Most fish species in Hudson Strait are benthic non-specialists whereas those in Hudson and James bays are opportunistic feeders, often with behavioral or physiological adaptations to cold temperatures. The anadromous cisco and the four-horn sculpin (*Myoxocephalus quadricornis*) are present in and around estuaries throughout most of the year. The longnose sucker (*Catostomus catostomus*) is also present year-round in very low salinity waters, while brook trout are present mostly during the autumn. Walleye (*Stizostedion vitreum*) are found only in the James Bay area and at the mouth of the Nelson River.

Arctic char are a key species in the rivers, lakes and streams of the northern parts of Hudson Bay. During the summer, char stocks tend to mingle at sea where a high proportion of their diet consists of amphipods and small fish (Adams *et al.*, 1989). They separate and return to their natal streams in the autumn. An invertebrate diet, particularly as an insectivore, makes char one of the few fish species capable of existing in northern streams where other forms of secondary production are severely limited by low temperatures and by hydraulic extremes (natural extremes of flow increase from west to east and northwards: for example, 2:1 on the lower Churchill River, 10:1 on the Nottaway River, and more than 300:1 on some on the tributaries to Ungava Bay). Low flow conditions during the fall migration are of particular concern where anadromous fish depend on shallow river channels to reach over-wintering lake habitat in northern parts of the Hudson Bay watershed (Power & Barton, 1987).

Suitable habitat for Atlantic salmon is limited and only two small populations of this species are present in the basin. A small stock of Atlantic salmon is present at the mouth of the Nastapoka River and a separate and larger stock supports a small but important Inuit fishery in Ungava Bay. Anadromous lake whitefish are present in most of the northern river mouths and estuaries (Baker, 1989; Bodaly *et al.*, 1989; Brousseau & Goodchild, 1989; Roy, 1989), and lake trout are also present. Capelin spawn nearshore along southern parts of the bay in June (and may be early in some estuaries, such as the Nelson River; D. Windsor, Manitoba Hydro, Winnipeg, Manitoba, pers. comm.). Modern Atlantic Ocean and Hudson Bay stocks of capelin do not appear to be linked by movements through Hudson Strait and capelin may be a relict Atlantic species in the bay (Dunbar, 1958). Fish species in the bay are still adjusting to postglacial environmental changes.

For many anadromous species, the rising spring discharge is used to assist passive migration from upstream sites to the estuarine environment where their larval forms depend upon the availability of early life stages of zooplankton for first feeding (Gilbert *et al.*, 1992). To what extent this strategy may also act to protect larval fish from predation by sight-feeders is unclear. Frazil ice and ice flows in the rivers at the time of break-up are often highly disruptive of the fluvial habitat. Gouging and plucking of bank and bed sediments and erosion under high flow events, not only strip the tributary channels of organic matter but also lead to high turbidities in receiving waters.

4.6 Marine mammals and polar bears

Seals are the most abundant marine mammal in Hudson Bay. Ringed seals are the dominant species, with a population of about half-a-million (Sergeant, 1986). These seals are most frequent near the coast on stable shore ice or stable sea ice (particularly around the Belcher Islands). Single pups are born in March or April, in lairs hidden under ice and snow on the stable ice. Later, the animals moult on the ice during the early summer. Their diet includes a wide variety of benthic and pelagic crustaceans and fish. In turn, they are a major source of food for polar bears.

Bearded seals are much larger than ringed seals. They are resident in fast and shifting pack ice, (often over shallow water). They are benthic feeders, depending on fish, crustaceans and molluscs, and they breed on the open ice in May. They are abundant in Hudson Strait, Foxe Basin and parts of eastern Hudson Bay.

Harbour seals are present on the west coast of Hudson Bay and inland where they may over-winter in accessible lakes. They are also found in the inland Seal Lakes (northeast of Richmond Gulf, Quebec) although, at this location, they are separated from the bay by a 50 m waterfall on the Nastapoka River. It is possible that this population is a relict from early postglacial colonization and its distinctiveness is presently under investigation. Only large lakes can support a resident seal population and it is estimated that the Seal Lakes might support about 200 animals (Sergeant, 1986). Harbour seals are also present in Ungava Bay where they may also migrate to inland lakes.

The harp seal (*Phoca greenlandica*) migrates into Hudson Strait and northern Hudson Bay during the summer. Hooded seals (*Cystophora cristata*) also migrate into Hudson Strait in summer to feed on deepwater fish and squid.

The Walrus (*Odobenus rosmarus*) is most common in Foxe Basin, in the area of the large polynya (south east of Igloolik), but it also occurs in small numbers around many of the islands in eastern Hudson Bay (north from the Belcher Islands). The walrus uses ice surfaces and flat rocks (close to deep water) to haul out. It feeds on bivalve molluscs and, occasionally, other seals.

The total population of beluga whales in Hudson Bay is about 25,000 (Stewart *et al.*, 1991). During the summer, most are present in west Hudson Bay (Lawrence *et al.*, 1992). Historically, beluga are thought to have occurred in large numbers along the east shore of the bay, as well, but the population has not recovered from past commercial harvesting. A total of nearly 20,000 whales occur off the Winisk, Severn, Nelson, Churchill, and Seal river estuaries where they show considerable fidelity to each estuary. Greatest numbers are present off the Nelson and Churchill rivers. Earlier estimates (Sergeant, 1973) suggested a population of only half this size. Beluga are widespread and occur elsewhere in the basin (Mansfield, 1990; Reeves & Mitchell, 1988).

The whales enter the estuaries in May, before break up is complete. In summer, the freshwater plumes provide much warmer water in the nearshore and this may stimulate the belugas' thyroid hormones which promote growth and moult (St. Aubin & Geraci, 1989; St. Aubin *et al.*, 1990). The estuaries appear to be used as nursery areas but most calving takes place outside the estuaries (Baker, 1989). Beluga whales calve in June and July, and remain along the southwest shore of Hudson Bay until late August when they begin to migrate northward. Most are thought to over-winter in Hudson Strait. Moving on and off shore with the tide, there is summer feeding on capelin, lake whitefish, cisco, pike, suckers, Arctic char, squid, polychaetes, and decapods. It seems unlikely that the estuaries are a principal source of food supplies since their productivity could not sustain such large numbers of whales (Baker, 1989).

The summer kill of bowhead whales by whalers in the late1800s was concentrated in the Northwestern part of Hudson Bay, west of Southampton Island. Though now in small numbers, bowhead are thought to migrate into the bay from Hudson Strait in April, and to extend their feeding along the edge of the ice. They utilize the rapidly expanding population of copepods which feed on the increasing biomass of phytoplankton. Feeding can continue around the bay and Foxe Basin until late summer. At Igloolik, in Foxe Basin, for example, adult copepods do not reach maximum densities until late August or early September (Grainger, 1959). Bowhead whales were never very numerous in the bay and the main population (numbered in thousands) makes use of the much greater productivity of Davis Strait.

Narwhals (*Monodon monoceros*) are present in small numbers. They appear to migrate into deeper parts of northern Hudson Bay during the summer, from both the western (Fury and Hecla Strait) and eastern (Hudson Strait) openings to the bay.

Polar bears are present throughout the coastal region of Hudson Bay in three to five distinct sub-populations (Stirling & Ramsay, 1986). James Bay has the most southerly known population. Polar bears show a strong tendency to use the same on-ice hunting areas, and the same on-land breeding and resting areas, year-after-year. As ice breaks up, the bears move ashore. While adult males remain close to the shore, females and cubs move further inland. Juveniles and sub-adults occur with both groups. Tagging indicates that a few bears make extremely long distance migrations (mostly juveniles or sub-adults), these occur throughout the eastern Arctic (including Labrador and Greenland). Arctic polar bears usually wean their cubs at about 2.5 years, but in Manitoba and Ontario cubs are weaned at 1.5 years. Early weaning allows breeding at two-year intervals for the southern population but at three years for the northern populations. Towards the southern limit of their distribution, polar bears may have larger litters.

Although several factors likely influence breeding of the bears and the weaning of cubs, nutritional controls are particularly important (Stirling & Ramsay, 1986). A successful spring hunt of seals is likely a prerequisite for healthy adults returning to land and their ability to nurture cubs during the following three to four months. In the spring, ringed seals are caught in their lairs on the fast ice. However, it is thought that bears have to reach a critical weight before they can break into the lairs. As a result of both faster growth and an earlier ice-melt, the southern bear cubs appear to reach critical weight much earlier than in more northerly populations. It is unclear to what extent differences in the size of seal populations or the seals themselves may influence the rate of growth of bear cubs. In the more northerly parts of Hudson Bay, bearded seals comprise a greater proportion of the bears' diet. While the bearded seals are larger than ringed seals, they are more difficult to catch and this may reduce their net nutritional value.

In Manitoba and Ontario, polar bears make use of earth dens, often dug against permafrost. It is thought that this practice assists thermal regulation but it may also lessen problems associated with insects. Bears feed very little while on land and conserve most of their energy by resting. It is well known that polar bears visit the garbage dump at Churchill but most of the animals which use this potential source of food are thought to be nutritionally stressed. Most of the polar bears that visit the dump are juveniles or females with cubs, and these animals have a high rate of metabolism. Bears return to hunt on the ice in November.

4.7 Birds of the offshore and coastal areas

Many different species of seabirds use the offshore resources of the basin, but few actively breed in the area and most are present for only a few months of the year. The lack of breeding fulmars (*Fulmarus glacialis*) and black-legged kittiwakes (*Rissa tridactyla*) could be due to inadequate quantities of suitable food, and this may have allowed development of large colonies of thick-billed murres which otherwise might not be competitive (Morrison & Gaston, 1986). Although abundant (population about two million birds), the murres could be sensitive to even modest variations in climatic conditions. They invest a high proportion of their energy budget in egg production and chick feeding (Gaston, 1985a). In total, there are five breeding colonies of murres on Akpatok, Digges, and Coats islands, where they nest on ledges on near-vertical cliffs. Because of the large size of some colonies, many birds use distant feeding areas which may be near their limit of practical excursion. The amount of non-coastal fish in diet is highest, the size of the fish caught is smallest, and chick growth is slowest at Digges Island (Gaston, 1985b). The murres feed mostly in central and eastern Hudson Strait and the extreme northeast of Hudson Bay. They feed underwater and at depths thought to be between 40 and 100 m, based on the species of fish caught. They consume Arctic cod, capelin, sand lance, other small fish, and amphipod and mysid crustacea. Most other seabirds in this area are surface feeders and may be more dependent on zooplankton.

Herring gulls (*Larus argentatus*) and glaucous gulls (*Larus hyperboreus*) are widely distributed throughout the bay and nest mainly in coastal areas. Breeding groups of Arctic terns (*Sterna paradisaea*) and black guillemots (*Cepphus grylle ultimus*) are also widely distributed. Large numbers of Sabine's gull (*Xema sabina*) occur in Foxe Basin and west Baffin Island where there are several breeding colonies. Many species of sea birds and waterfowl feed extensively on small fish and invertebrates in shallow coastal waters throughout the bay (Morrison & Gaston, 1986).

The most abundant breeding duck is the common eider (*Somateria mollissima*) which nests in colonies on low-lying offshore islands throughout the area (Abraham & Finney, 1986; Nakashima & Murray, 1988). Two distinct groups of this diving duck are recognised: northern eiders (*S.m. borealis*) breed in Hudson Strait and Foxe Basin, and migrate to Newfoundland and the Gulf of St. Lawrence in winter; non-migratory Hudson Bay eiders (*S.m. sedentaria*) nest and are most abundant on offshore islands of eastern Hudson Bay (including the Belcher Islands), and winter in the polynyas and open waters of Hudson Bay and James Bay (Abraham & Finney, 1986; Reed & Erskine, 1986). King eider ducks (*Somateria spectabilis*) are less abundant. They nest on low flat tundra in the northern part of the basin where some may overwinter in polynyas, most migrate to the Atlantic coast (Labrador south to Maine) or southwest Greenland (Abraham & Finney, 1986).

Common loons (*Gavia immer*) are present around southern Hudson Bay, James Bay, Ungava Bay, and south Baffin Island. Yellow-billed loons (*G. adamsii*) are present along parts of the Melville Peninsula, and Arctic and red-throated loons (*G. arctica, G. stellata*) are present throughout the Hudson Bay region. The loons nest on freshwater lakes, near the coast, and feed mostly in the marine nearshore. Common loons and red-throated loons require particular habitat (Savard, 1988), and a significant part of their breeding area may have been lost as a result of reservoir formation and water level manipulation in the James Bay region (Rimmer, 1987).

Three species of scoters, the surf scoter (*Melanitta perspicillata*), black scoter (*M. nigra*), and the white-winged scoter (*M. fusca*) congregate in very large numbers in the coastal waters of James Bay and southeastern Hudson Bay to moult in late July and August. All breed in low densities, mainly on inland high-boreal lakes in northern Quebec and mainland Northwest Territories (Bellrose, 1980; Savard & Lamothe, 1992). Red-breasted mergansers (*Mergus serrator*) are common summer residents throughout the area, and common mergansers (*M. merganser*) occur mainly in James Bay and southern Hudson Bay. Oldsquaw (*Clangula hyemalis*) breed around Hudson Bay but are rare in James Bay. The harlequin duck (*Histrionicus histrionicus*) can be found at the mouths of some fast flowing streams around the basin but it is thought to be very sensitive to habitat change (Rimmer, 1992). Other species of diving ducks may occur in appreciable numbers during migration or the moult period, especially in James Bay. They include greater and lesser scaup (*Aythya marila, A. affinis*) and common goldeneye (*Bucephala clangula*).

Waterfowl make extensive use of several different types of coastal habitat (Dignard et al., 1991). Sea ducks feed mainly on benthic invertebrates. Eiders and scoters, in particular, feed on bivalve molluscs (*Mytilis edulis* and *Macoma balthica*). Several species of surface feeding ducks make intensive use of the James Bay shoreline and parts of the southern Hudson Bay shoreline during both spring and fall migrations (Morrison & Gaston, 1986). Northern pintail (*Anas acuta*) is widespread and other species which are locally abundant include American black duck (*A. rubripes*), green-winged teal (*A. crecca*), mallard (*A. platyrhynchos*), and American widgeon (*A. americana*). The tidal flats and marshlands provide a major source of small invertebrate foods, including molluscs (particularly *Macoma balthica*), gastropods (particularly *Hydrobia minuta*), and dipteran larvae (infauna of short grass, *Pucinellia phryganodes*). Black ducks feed extensively on *Potamogeten filiformis* (Rimmer, 1987).

The Hudson Bay Basin is a major staging and breeding area for geese. Lesser snow geese (*Anser caerulsecens*) nest in several large colonies along the south and west shores of Hudson Bay, on Southampton Island and southwest Baffin Island. In total, these areas support more than a million breeding adults (Boyd et al., 1982; Reed et al., 1987). Most of the snow geese also make heavy use of coastal lowlands during migration, especially along southern Hudson Bay and western and southern James Bay (Thomas & Prevett, 1982). Probably the entire population of Atlantic Brant (*Branta bernicla hrota*) stage in James Bay during the spring and fall migrations, and most breed on Southampton Island and in Foxe Basin. Several different populations of Canada geese (*Branta canadensis*) are present. The subspecies *B.c. interior* breed in coastal lowland areas throughout James Bay and eastern Hudson Bay, and members of this group have different winter destinations: those from the east coasts go mostly to the U.S. Atlantic seaboard; those from southwest James Bay and Akimiski Island go to Tennessee; and those from southern Hudson Bay move south into the Mississippi valley. Canada geese in western Hudson Bay, Southampton Island and Foxe Basin are smaller in size (*B.c. parvipes* and *B.c. hutchinsii*), and most winter in Oklahoma and Texas. Some giant Canada geese (*B.c. maxima*) use Hudson and James Bay lowlands for moulting but, generally, they breed further south. Snow geese feed on sedges (*Cyperaceae*), arrowgrasses (*Juncaginaceae*), horesetails (*Equisetaceae*), and grasses (*Gramineae*). Brant geese feed mostly on eel grass (*Zostera marina*) and if there is not enough available, on short grass (*Puccinellia phryganodes*) (Rimmer, 1987). Geese also feed on *Carex subspathacea* and *Triglochin palustris* vegetation.

Many species of small shorebirds are present in large numbers in the tidal flats and marshes during summer months, and many parts of the Hudson Bay and James Bay shorelines are critical as stop-over points for migrants to Arctic breeding grounds. These transients are heavily dependent upon appropriate time and space linkages for successful passage (Morrison & Gaston, 1986). Indeed, throughout the food web, the significance of timing in the life-cycles of so many different species in the Hudson Bay region is widely shared and the significance of this can not be over stated.

HUDSON BAY BASIN : LOADINGS AND STRESS

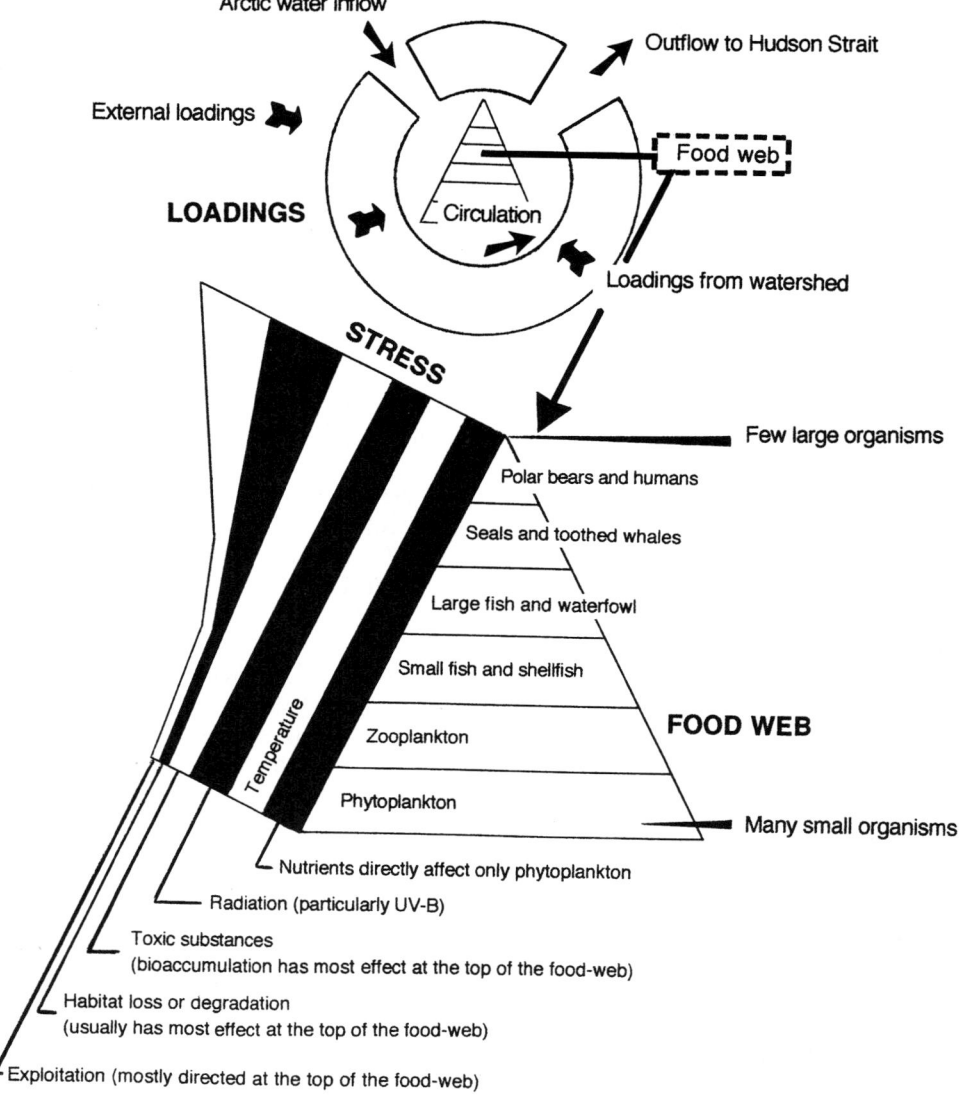

Fig. 12. Hudson Bay Basin: Loadings and Stress.

5. Concerns

5.1 *Identification of concerns*

There is no boundary which separates the social and economic forces behind the arti-

Hudson Bay Region: Present and future environmental concerns 217

Table 16. Hudson Bay Basin: Matrix of Activity/Stress on Bay and Coastal Areas.

Activity	Nutrient Balance	Temperature	Radiation (UVb)	Toxic Substance	Habitat Change	Exploitation
Forestry	L	L	-	L	L	L
Agriculture	L	-	-	L	-	-
Mining						
(local)	L	-	-	M-H	M-H	-
(distant)	-	-	-	L	L	-
Oil/gas						
(local)	L	-	-	L-H	L-H	-
(distant)	-	-	-	L	-	-
Pulp and Paper	L	-	-	L-M	-	-
Hydroelectric						
(local)	M	M-H	-	M-H	H	-
(distant)	-	L-M	-	L-M	-	-
Transmission	L	-	-	L-M	L-M	-
Air transpt.	-	-	-	L	L-M	M-H
Shipping	L	-	-	L-H	L-H	M-H
Roadway	L	-	-	L-M	M-H	M-H
Rail transpt.	L	-	-	L-M	M-H	M-H
Constr. sites	L-M	-	-	M-H	L-M	M-H
Tourism/Rec.	-	-	-	-	L	L-H
Settlement (inc. population growth)	L-H	L-M	-	L-H	M-H	M-H
LRTAP	L	-	H	L-H	-	-
Global warm. (distant)	L-M	L-H	M-H	-	L-H	L-H

Relative Stress: L Low; M Medium; H High

LRTAP Long Range Tansport of Atmospheric Pollutants

facts of development from the effects of development. Therefore, social, economic and environmental factors need to be addressed together. It is also recognized that social and economic factors are frequently the underlying cause of stress even though they are not the immediate source of it.

Throughout the Hudson Bay region, various forms of global and local stress interact to induce cumulative impacts on the whole ecosystem and many of the major forms of stress comprise more than one component. Figure 12 provides examples of several forms of stress which may affect different levels of production within the food web of the Hudson Bay region.

Table 16 provides a summary of the probable importance of different human activities relative to stress in the Hudson Bay basin and its coastal region. It is a guide to focus attention on what types of stress are likely to be associated with what activities. The presentation is qualitative and implies nothing about the size, technology, operating practices, or site-specific conditions required for specific comparison. Activities include those within and external to the Hudson Bay watershed.

5.1.1 Agriculture

Agricultural activities, as noted in Table 3, are largely concentrated in the interior wa-

Table 17. Tributary Water Quality: Lowest Monitoring Points (Source: Conservation and Protection, 1992).

Parameter	Severn	Winisk	Attawap.	Albany	Moose	La Grande	G. Baleine	Nastapoka
Specific Conductance (USIE cm^{-1})	135-574	106-256	101-297	99-616	92-612	3-23	3-31	17-26
Turbidity (JTU^{-1})	1-260	<1-11	<1-31	<1-6	<1-390	<1-2	<1-6	<1-2
DOC (mg L^{-1})	7.6-15	7.5-18	0.2-17	0.1-20	0.1-22	3.7-5.5	-	-
KN (mg L^{-1})	0.4-2.9	0.3-1.2	0.3-1.1	0.2-1.4	0.2-3.2	0.2	-	-
Nitrate+Nitrite (ug L^{-1})	20-770	5-2900	5-320	10-2510	16-6690	10-380	10-180	10-90
Total Alkalinity - unfiltred (mg L^{-1})	11-105	9.5-127	30-124	1.0-159	22-159	2.6-5.0	3.7-7.9	4.4-5.8
Acidity (pH)	3.0-8.5	4.6-8.4	7.3-8.5	6.8-8.5	6.6-8.3	6.1-7.5	6.3-7.0	6.6-7.2
PCB's (ng L^{-1})	0-9	9-11	9	9	9-15	9-7	3-47	9-40
Total copper (ug L^{-1})	1-57	1-70	1-33	1-9	1-37	1-23	1-3	1-15
Total mercury (ng L^{-1})	20	20	20-37	20	50	1-6	20-230	20-310
Total nickel (ug L^{-1})	1-24	1-2	<1-3	2-16	1-32	<1-2	1-2	1-3
Total lead (ug L^{-1})	<1-9	1-9	<1-3	<1-33	1-43	1-170	<1-19	<1-11
Total zinc (ug L^{-1})	1-62	1-80	1-48	1-104	1-84	<1-55	1-38	1-52

Underlined data indicate that upper values exceed typical (though not extreme) nearshore water quality of Lake Superior (for pH values, exceedence is in the lower values). Period of data recording: from start (varies between stations) to November 21st, 1992.

tershed. Agricultural practices may increase nutrient levels in tributary drainage (largely from fertilizer and animal wastes). Herbicide and pesticide residues (mostly applied to cereal production) may also enter ground and surface water drainage. Trace quantities of biocides have been reported from tributaries in some of the lower parts of the watershed (including both agricultural and forest applications) but at exceedingly low concentrations. Due both to the rising costs of chemicals and recognition of potentially adverse human and environmental health effects (Environment Canada, 1991), future applications may well decrease and favour less persistent formulations. Also, because of the presence of many impoundments in the interior watershed (Fig. 6, Table 11), direct loadings to coastal waters are likely to be negligible.

5.1.2 *Forestry*

The Boreal Forest extends northwards into the Eastmain Lowland (Fig. 2) but is absent in the Hudson Bay Lowland (Fig. 3). Most of the forest lands lie at some distance from the coastline. Forest harvesting (at present, largely by clear cutting) can have a large local and regional impact on both ground and surface water regimes, increasing both nutrient and soil loss from the watershed (Evans *et al.*, 1991). Because of distance from the bay and the presence of many impoundments on major tributaries, timber harvesting is likely to have little effect on the bay. Changes in habitat may alter the availability of fish and wildlife over large areas of the drainage basin, and the application of insecticides to control forest pests may result in run off and downstream transfer (Environment Canada, 1991). The extent of contaminant migration is uncertain, as is the extent to which such applications may effect aquatic invertebrates and other species dependent on them as a source of food. The availability of logging roads, particularly for use in recreation and hunting adds significantly to wildlife disturbance and exploitation.

5.1.3 *Mining*

Mining is widely practiced in the watershed but most centres are well inland from the coast (Fig. 4). Mining (copper, gold and zinc) is important in the Nottaway River drainage but its impact on estuarine and coastal waters is not established. There is no evidence of major or widespread mineral or metal loading to the bay (based on Environment Canada Water Quality records), but significant local contaminant effects are known to occur at several sites. Table 17 provides a summary of water quality characteristics of tributaries entering the bay, reported at their lowest sampling points. High values are noted but almost all of them are due to naturally occurring events (particularly, snow-melt). Records of water quality are generally too limited to provide adequate statistical summaries but the majority of sample data lie much closer to the lower than the upper reported values. The data indicate relatively high natural levels of specific conductance, nitrogen and total alkalinity in the northern Ontario drainage, and relatively high levels of copper also occur throughout the reported data. Extremes in pH may reflect the composition of precipitation and, or, discharge of naturally acidic water from the drainage area.

Mining activities at or close to the bay have the potential to induce high but localized levels of environmental stress (particularly from toxic effects and habitat change). Long term concerns are likely to be focused on the construction and management of tailings and waste pond facilities, and disposal of equipment. There are potentially recoverable deposits of iron, nickel, and other metals at many locations around the bay, but most are not economically viable at present. Coal deposits occur in the Moose drainage of northern Ontario but their development is unlikely to be economic for some time. Peat can also be used as a fuel for generating thermal power and considerable thicknesses occur throughout the watershed, particularly in northern Ontario. Both coal and peat removal would alter large areas of habitat and could affect surface and ground water hydrology over even larger areas. Other forms of mineral development could occur, particularly in Shield areas.

5.1.4 *Oil and gas*

Oil and gas exploration has taken place throughout much of the bay but without finding an economically viable resource. Oil and gas have been located in the Arctic Islands and it is possible that gas could be brought south by pipeline, and a route down the west side of Hudson Bay has been proposed. Although the potential for local stress on habitat and from toxic spills (exploration and development stages) is high, present and anticipated activities are unlikely to produce more than a small direct stress on the bay. Indirect stress through other associated economic developments could be of much greater significance.

5.1.5 *Pulp and paper manufacturing*

Until the early 1970s, the significance of chlor-alkali plants was not apparent as a contributor to mercury in fish and wildlife. Many tonnes of metallic mercury were released into rivers and lakes from some plants which produced chlorine for the bleaching of pulp and paper. Large releases occurred over many years into the English-Wabigoon River system (Winnipeg River drainage), and the Nottaway River. Although the process has been phased-out, the effects of residual mercury are likely to persist for many decades. The pulp and paper industry remains a major source of environmental contaminants, both as a result of effluent emissions which exert high biochemical oxygen demand on receiving waters and the discharge of naturally occurring forest toxins and toxic substances (produced by in-plant processes such as chlorine-bleaching). A wide range of organic compounds are released.

Even though contaminant transport from pulp and paper mills can occur over distances of 1000 km and more (Sly, 1992), it is unlikely that direct transport over such distances will occur within the watershed because multiple impoundment increases sedimentation and water residence time. There is no evidence to suggest that a significant problem exists within the bay at present (McCrea & Wickware, 1986) and with increasingly stringent effluent controls, current levels of stress should decrease. Bioaccumulation and bodily transport in migratory species is, however, possible.

5.1.6 *Diversions*

Within the Hudson Bay watershed, there are several tributary diversions that provide additional water supply for irrigation, power production, and water level control. Most of the diversions lie wholly within the watershed (Fig. 6) and concerns relating to them (as, for example, James Bay region, Chenier, 1991) are largely addressed as a component of hydroelectric development (following). It should be noted, however, that schemes exist for water diversion out of the Hudson Bay region, on a massive scale. The GRAND (Great Replenishment and Northern Development) Canal scheme is the largest such proposal, and seeks to divert the entire freshwater discharge entering James Bay south to the North American heartland (Milko, 1986). The GRAND Canal would divert more than 60 percent of the freshwater inflow from Hudson Bay. Major concerns exist with respect to thermal stratification and heat flux, nutrient cycling, the availability of key habitat for waterfowl, shorebirds and for fish, and the potential for effects beyond Hudson Bay. Huge diversions on this scale, seem unlikely but it should be recognized that implications associated with them go much beyond the scope of existing discussions.

5.1.7 *Hydroelectric development*

Hydroelectric development was once thought environmentally benign but it is now known to exert several forms of environmental stress, particularly as a result of construction and operation of reservoirs. Flooding valleys and lake shore areas changes terrestrial to aquatic habitat (Rosenberg et al., 1987). Depending on the operational requirements of the power facility, water levels and their draw-down may severely limit the usefulness of newly formed reservoirs to fish, birds and mammals (Hecky et al., 1984; Newbury et al., 1984). Indeed, vegetation may not be able to establish itself, leaving a bare and unproductive area of extremely low biodiversity in the fluctuating zone of reservoir. Below the reservoir, changes in channel form, erosion, discharge regime, permafrost formation, and other conditions may create more unusable or degraded forms of habitat. Changes in estuarine habitats may be especially detrimental to species dependent on them for breeding, first feeding, larval and juvenile development, and specific behavioral needs.

Changes in the natural flow regime can alter surface water temperatures below impoundment sites and, depending on whether top or bottom water is released, there may be changes in the concentrations of dissolved oxygen and nutrients (particularly nitrogen and phosphorus). During the early years after formation, nutrients concentrations often increase within reservoirs as a result of their release from the flooded soils and vegetation. This is usually reflected in increased biological productivity. Over a period of a decade or more, fish production may peak and then begin to return to more stable but reduced long term levels related to the modified drainage area. Where reservoir configuration results in a significant increase in water depth, cooler temperatures and less vertical mixing may also result in lower productivities than in the original aquatic system. Both changes in water temperature and discharge velocity may influence ice formation and its stability in areas below impoundment, including plume areas in the coastal zone. Increased freshwater discharge during the winter period may influence the thickness and extent of ice formation of both land fast and offshore ice. Reduced spring discharge may decrease surface flow and thus nutrient upwelling in the nearshore (depending on coastal configuration).

In response to an abundant supply of nutrients (carbon) in flooded soils and vegetation, microbial processes are often enhanced. Many processes occur simultaneously (methylation and demethylation, for example; Jackson, 1987, 1989) and, depending upon the degree to which flooded areas remain oxygenated or become anoxic (Oveden, 1989), reservoirs may become a significant source of methane and carbon dioxide. Both of these are "greenhouse" gases and may, cumulatively, contribute to global warming (Rudd. In: House of Commons, 1991). The most immediate concern, however, relates to the production of methyl mercury and its incorporation within the food web where it bioconcentrates in top predators, including humans (CM-MA, 1987).

Methyl mercury is a naturally occurring but toxic compound and the dominant dissolved form in water is monomethyl mercury. Under most natural conditions, it occurs in extremely small concentrations and at levels below environmental concern. Often, concentrations of methyl mercury in biota are species dependent and age-related. The production of methyl mercury rises rapidly in newly flooded areas and declines slowly over several decades (Brouard et al., 1990; Makivik Corp., 1991). Increases in methyl mercury in fish, in reservoirs, are empirically related to the extent that organic soils are flooded upstream and within the newly flooded reservoirs. Wide shallow basins produce more methyl mercury than deep narrow valleys. Areas subject to forest fire (for example, parts of the Grande Baleine drainage) may produce less methyl mercury than non-affected areas (such as the La Grande drainage). Based on models of reservoirs in northern Manitoba, and without additional mercury from other sources, it may take 50 years for methyl mercury in top predators (pike) to return to background levels. Also, because of the use of different food sources, methyl mercury concentrations in pike make be as much as 10 times those in lake whitefish (Bodaly & Johnston, 1992).

In parts of the La Grande River drainage, for example, methyl mercury was a health concern even before the development of hydroelectric facilities (A. Penn, Grand Council of the Cree-Quebec, Montreal, Quebec, pers. comm.). Background levels of methyl mercury in the region include increasing contributions from atmospheric sources and may range between 0.1 and 1.0 mg · g^{-1} in sport fish, even in remote lakes (McCormick, Ontario Hydro, Toronto, Ontario, pers. comm.).

5.1.8 *Transmission corridors*
The construction of transmission corridors and oil and gas lines creates soil disturbance, and soil and nutrient loss to adjacent waters. The use of herbicides to control vegetation can result in the movement of toxic chemicals into local water courses. The principal concern, however, relates to their effect on habitat where ground-cover is substantially reduced. It is possible for the corridors to act as a barrier to the transverse movements of some species. More often, however, they act as an enhancement to the longitudinal movement of species (including humans seeking easier access to wildlife resources). These forms of impact may have long lasting effects but good design, construction, and conservation techniques can reduce much of the direct habitat stress.

5.1.9 *Air, shipping, road, rail, and construction site activities*
Construction of hydroelectric facilities (Krotz, 1991) is likely a major cause of habitat stress but a wide range of other activities including air, shipping, road, and rail construction all produce similar effects, differing mostly in the extent rather than the nature of direct stress. They also share many similarities with direct stress effects of transmission corridors. Soil and vegetation disturbance and drainage modification can change local habitat, degrading it and breaking connections between areas of specific habitat use. The presence of people, noise, and heavy construction equipment may also cause avoidance reactions in animal species. In addition, the widespread use of powerful outboard motors for hunting, fishing and travel is also thought to result in avoidance behaviours, particularly in marine mammals and birds with highly developed hearing. With good user practices, some of these effects can be limited in their extent and severity. Because of the permanence of structures (buildings, dams, roads, etc), however, their presence contributes in a major way to cumulative environmental impact.

Of particular concern are stress effects related to increased access. Throughout the region, there is often little to restrict off-road access at any point and extensive off-road access generally leads to exploitation of wildlife resources. Road-kill adds a further component to "exploitation". Other means of transport support an increase in human populations at various destination points and it is from such centres that exploitive activities are most likely to extend. With relatively easy means of bringing goods into bay communities, there is also the means to export products, including country foods. Bulk material are carried by rail, road and ships, and oil spills are not uncommon. Locally, the direct effects of oil spills can be extremely damaging to the environment and may persist for several years. In addition, fire, accidental or intentional, may be extremely damaging and the frequency of burns usually increases with human activity.

5.1.10 *Off-road transport*
Cree and Inuit have readily adapted to new technologies (George & Preston, 1987), and off-road and over-ice/snow vehicles have greatly extended the effective travel distance for hunting, fishing and trapping. They have reduced the need for families to fol-

low a semi-nomadic existence. It is now easier to bring food to a settlement, rather than to move people to the food. If the size and number of communities remain more or less constant, off-road and over-ice/snow transport might reduce local harvest pressures by distributing hunting and fishing activities over wider areas. However, the change to permanent settlement has given rise to major societal changes and often to an expanding population (both resident and non-resident).

5.1.11 *Tourism and recreation*
Tourism and recreation usually require additional accommodation at key sites, provision of fuel and supplies, and they generally increase access both to points of entry and to areas of special biological interest. With small numbers of visitors, damage to habitat and interference with animal behaviour should be minimal. The potential for environmental stress is considerably increased if site activities include hunting and fishing. The need for management control is likely to vary from site to site and, at present, tourism throughout most of the coastal and near coastal areas of the region is restricted by availability of air service and accommodation. In southern parts of the basin, however, Churchill, Moosonee, Moose Factory, and La Grande already have road/rail access. Further, it is likely that more roads will serve Great Whale River, Wemindji, Eastmain, and possibly Waskeganish within the next three or four years (A. Penn, pers. comm.). In eastern Manitoba, access may also improve along the Nelson River. Increased access to the coastal region of eastern James Bay and southeastern Hudson Bay may significantly increase the ecological impact of tourism in the region. Already there are 25 000 to 30 000 visitors a year to the La Grande complex, and adult native hunters are outnumbered by non-native hunters and fishermen, even in winter (A. Penn, pers. comm.).

5.1.12 *Settlements*
The populations of both Inuit and Cree settlements in coastal and near-coast areas of the basin are increasing. Between 1986 and 1991, for example, the population of Quebec Inuit communities increased by about 8 percent and Inuit populations within the Northwest Territories increased by about 17 percent. Over the same period, Cree communities in Manitoba, Ontario and Quebec have increased between 1.5 and 3 percent (Berkes & Freeman, 1986; Statistics Canada, 1986). Most recent population data, however, indicate that the rate of increase is declining in almost all bay communities. A summary of resident Cree and Inuit populations (1991) in bay communities is provided in Table 18. As previously noted, the effects of community harvesting could increase with community size but the potential for increased pressure from hunting activities seems to have eased. Over the past two decades or so, there has been a tendency to incorporate a higher proportion of "Kablunaq" (white man's food) into the northern diet. The relationships between northern populations and their use of country foods are complex, and increases in native populations do not necessarily result in increased demand on local food sources.

In addition, settlements can be a specific source of nutrient and toxic stress derived from domestic sewage and animal wastes, oil spills, and solid wastes (municipal and other). Contamination of surface freshwater and groundwater supplies, and the marine environment is local but significant throughout much of the region.

5.1.13 *Long range transport of atmospheric pollutants*
The Long Range Transport of Atmospheric Pollutants (LRTAP) is a significant contributor to toxic stress in the region. Several persistent organic contaminants for which

Table 18. Resident Populations of Bay Communities, 1991 (Sources: Statistics Canada, 1986; F. Berkes, Pers. Comm.).

Location	Population	Location	Population
(Quebec)		(Manitoba)	
Killiniq	-	*York Factory*	251
Kangiasualujjaq	519	*Gillam*	1893
Kuujjuaq	1405	*Churchill*	1370
Kangirsuk	351		
Quaqtac	236		
Kangiqsujuaq	404	(N.W.T.)	
Purtuniq	-	Arviat	1323
Salluit	823	Whale Cove	235
Akulivik	375	Rankin Inlet	1706
Povungnituk	1091	Chesterfield Inlet	316
Ivujivik	263	Baker Lake	1186
Inukjuak	1044	Repulse Bay	488
Umiujaq	284	Hall Beach	526
Kuujjuarapik	605	Pelly Bay	-
Whapmagoostui	508	Igloolik	936
Radisson	2500*	Arctic Bay	543
Chisasibi	2306	Nanisivik	294
Wemindji	919	Pond Inlet	974
Eastmain	444	Clyde River	565
Waskaganish	1344	Broughton Island	461
		Pangnirtung	1135
(Ontario)		Iqaluit	3552
Moosonee	1213	Lake Harbour	365
Moose Factory	-	Cape Dorset	961
Ft. Albany	1199	Coral Harbour	578
Kashachewan	-	Sanikiluaq	526
Attawapiskat	1041		
Peawanuk	150		
Ft. Severn	335		

Native population *Cree*   * Number approximate (Baie James Corp. 1991)

Native population Inuit

there is no local or regional source of emission have been identified in the food web. These include DDT and its derivatives, PCBs, and other organochlorines (Adam *et al.*, 1987; Gregor, 1990; Gregor & Gummer, 1989; Jensen, 1991). Other substances, such as mercury, occur both as an external source of loading and as a "local" input from within the region (for example, reservoirs). At high latitudes (Welch *et al.*, 1991), atmospheric processes tend to increase fallout of contaminants associated with ice-particles and episodic dust-fall. Contaminants have been identified from distant sources in North America, Asia, and elsewhere (Barrie *et al.*, 1992; Lockhart *et al.*, 1992; Thomas *et al.*, 1992). Bioaccumulation of persistent organic contaminants in long-lived animal species results in high levels of contamination. This may pose a potential health risk to both the animals themselves (Twitchell, 1991) and humans who depend on country food (Davies, 1991b). Toxic substances incorporated lower in the food web and taken up by summer feeding migratory species, particularly birds, could be of particular concern if migrants already carry high contaminant burdens from their wintering grounds.

The presence of ozone-depleting gases, particularly CFCs, from widespread global sources is a special concern at high latitudes (UNEP, 1989). Loss of high level ozone results in decreased adsorption of UV radiation and an increase in the amount received at the Earth's surface. Increasing levels of UV-B radiation (mid part of the UV band) are a hazard to human health and to the health of most species exposed to direct sunlight. Reflected radiation from ice, water and snow surfaces adds to the degree of exposure. Recent studies on the effects of UV-B indicate that this radiation can affect marine and freshwater phytoplankton and may reduce productivity at the base of the food web (Smith, 1989; Smith et al., 1992). This stress effect could be felt throughout the food web and may affect the health of many larger animals, including humans.

5.1.14 Global warming

Global warming is induced by a wide range of "greenhouse" gases, and by particulate emissions at low levels within the atmosphere. Carbon dioxide and methane are key greenhouse gases (Rudd. In: House of Commons, 1991). Both of these are released by biochemical processes (respectively, aerobic and anaerobic processes) occurring naturally in the environment. However, the rates at which these processes take place are increased by the effects of flooding soils. This is especially true for methane which is produced by shallow wetland (Oveden, 1989) and reservoir areas (Rudd. in: House of Commons, 1991). Feed-back mechanisms from atmospheric warming are also likely to increase rates of natural microbial degradation and the further release of greenhouse gases.

Climate models predict that the effects of global warming are likely to be greatest at high latitudes such as the Hudson Bay region (Roots, 1987). A wide range of stress effects are possible (McAllistair, 1991a; Dunbar, 1954a) but, in particular, delayed ice formation, less stable ice cover and earlier break-up, and increased wave activity due to extended open water conditions are anticipated. A small sea-level rise is probable, and a large rise is possible though less likely in the near future. Northern inland lakes may also provide a sensitive indicator of some of the effects to be expected from climatic change. A range of effects which already include increased water temperature, longer ice free periods, higher evaporation rates, increased water residence times, increased chemical concentrations, and deepening of the thermocline have been reported by Schindler et al. (1990).

The importance of water supply in high demand areas of the Hudson Bay watershed has been demonstrated in Table 3, and the critical balance between supply and demand is most evident in the interior drainage of the North and South Saskatchewan rivers, and the Red and Assiniboine rivers. Climate modelling projects that soil moisture will decrease and that agricultural water use could increase significantly. Predictions about changes in precipitation, snowpack, and runoff are uncertain and depend heavily on headwater conditions (Cohen, 1991). The likelihood that climate change would significantly alter discharge from this part of the watershed into Hudson Bay is considerable but, also, it is very uncertain. Reduced flow into the Nelson River would affect hydro-electric generating capacity. It would also further modify hydraulic conditions in the lower river and estuary. Future discharge reductions in both the Churchill and Nelson rivers are possible, most particularly in the latter.

5.2 Understanding the principal concerns

Based on these comments, it is likely that the activities or conditions which cause the most significant forms of stress in the basin, and which are important now or are likely to be in the relatively near future include: Reservoir development (including diversions); any activity which is likely to result in loss of productivity, or exploitation; and the long range transport of atmospheric pollutants.

5.2.1 *Reservoir development*

Recently, Bunch & Reeves (1992) reported on a workshop on the potential for cumulative impacts in the Hudson Bay region. They focused particularly on the likely effects of hydroelectric development and parts of the following section are summarized from their workshop proceedings. Causes and effects are presented as hypotheses, which are followed by comments and examples.

Hypotheses Related to Hydraulic Change

Reduced tributary discharge during the summer will have several effects:

(1) Vertical stratification will weaken while vertical mixing in the bay will increase. In turn, this will increase nutrient addition from deeper waters.

(2) Horizontal circulation in the bay will be affected, but only locally.

(3) Offshore ice breakup will be delayed and nearshore ice conditions altered.

In winter, there will be an increased discharge, with the following effects:

(4) Stratification will increase and vertical mixing will lessen, thereby decreasing nutrient renewal in surface waters.

(5) The extent of under-ice freshwater plumes will increase.

(6) Ice will be thicker and form earlier.

(7) The natural regime of freshwater outflow from Hudson Bay will alter.

Comments

Reduced summer discharge might create cooler temperatures at the sea surface but with colder water, the air/sea interface should be more stable. Prinsenberg (1991) suggests this greater stability would enhance heat flux and that Hudson Bay is likely buffered against forced temperature change. However, summer warming may be delayed. Changes are likely to be greatest in winter when the freshwater inflow to James Bay would double, and that to Hudson Bay will increase by 50 percent (Prinsenberg, 1991). Winter surface water salinity will be expected to decrease by about 1.5 °/oo in James Bay, but surface currents are expected to increase. The effects of change are likely to be most noticeable in estuaries and coastal areas, and may extend a considerable distance from source. Hypotheses 1 and 4 are expected to be locally significant. Also, because summer circulation in the bay is both density-driven and wind-driven, the effects of reduced river discharge may be more local than regional (hypothesis 2). Ice thickness could increase in the offshore, and with reduced spring runoff would lead to delayed ice break-up in the bay (hypothesis 3), but it is unclear what may happen in the nearshore. Likely, ice will thicken but if rates of discharge are high enough break-up could be accelerated.

If freshwater plumes overlap under the ice, this could add to the effect of hypothesis 5. This includes displacement of zones in which primary production occurs under optimal conditions (ice algae and phytoplankton). The extent to which these changes (and those suggested in hypothesis 6) would be apparent in the bay, within the normal present and predicted future global climatic warming, is uncertain. Hypothesis 7 refers to the suggestions by Sutcliffe *et al.* (1983) that seasonal salinity changes off Newfoundland correspond to seasonal runoff and melt in Hudson Bay, and that there could be a negative effect on primary production and fish stocks in Atlantic coastal waters. However, it is now thought likely that most seasonal salinity changes off Newfoundland are related to ice-melt along the Labrador coast (Myers *et al.*, 1990).

Hypotheses related to nutrient and sediment loadings

(8) Relative to marine waters, freshwater discharge is less important as a source of inorganic nutrients in the nearshore zone.

(9) Changes in the timing and location of freshwater discharge will significantly influence nutrient availability in surface waters.

(10) Over the long term, impoundments will reduce the input of organic carbon to the bay, and this will reduce nearshore and estuarine productivities.

(11) There will be a temporary increase in nutrient concentrations downstream of impoundments, immediately after construction.

(12) Impoundments will retain much of the sediment load normally carried into the nearshore zone by tributaries. This and changes in flow regime will alter channel and bank configurations downstream.

Comments

On a regional scale, hypothesis 8 may not be significant. Although nutrient levels in discharge to the basin are often less than occur in the receiving marine waters (Legendre & Simard, 1978), freshwater discharge can be an influence on productivity in James Bay, in some estuaries of the basin (for example the Nelson River), and it may influence productivity in the large embayment of Hudson Bay north of Cape Jones. The effects of diversion on the Eastmain River and of increased flow in the La Grande River system are useful indicators of anticipated change (Messier *et al.*, 1986). Between July 1979 and July 1980, about 810 $m^3 \cdot s^{-1}$ or 90 percent of the flow of the Eastmain River was diverted into the La Grande system, and between 1980 and 1984 diversion from the upper watershed of the Koksoak River system provided an additional 776 $m^3 \cdot s^{-1}$ of inflow to the La Grande system. Eventually, the average discharge of the La Grande will increase from about 1800 $m^3 \cdot s^{-1}$ to 3400 $m^3 \cdot s^{-1}$. This represents an increase of only 3 percent in the total annual discharge of freshwater to James Bay but an increase of 50 percent in the winter discharge.

Changes evident in the lower part of Eastmain River (Ingram *et al.*, 1986; Messier *et al.*, 1986) include: lower summer water level, intrusion of the salt-wedge, and an increase in the freshwater residence time. River erosion has increased with increased hydraulic carrying capacity below the diversion, even though flow is much reduced. Erosion of post glacial marine silts and clays and peat from areas below the diversion has increased turbidity and water colour. Sediment is accumulating in the lower river and much of the material is being brought in from the adjacent bay (d'Anglejan, 1982; Ingram *et al.*, 1986). Although the total nutrient load carried by the river has been reduced, nutrient concentrations appear to have increased. In part, this may be due to greater water residence time in the estuary. Increased nutrient levels in the outer estuary are derived from marine sources. The concentrations of chlorophyll a have increased slightly in both the estuary and lower river. Under stable conditions, algal blooms have been observed in the thin photic layer of estuarine water (Ingram et al., 1985). The distributions of biota reflect changes in salinity. Grenon (1982) suggests that benthic macrofauna which utilize a pelagic mode of reproduction will tend to colonize new substrates most quickly, and this may be indicated by the marine bivalve *Macoma balthica*. This mollusc is now more abundant and widespread in the inner estuary, and chironomids are restricted to the remaining area of freshwater. These changes will likely favour the benthic feeding patterns of lake whitefish relative to other local anadromous species.

The size of the La Grande winter plume has increased (Ingram & Larouche, 1987b; Messier et al., 1986) from about 750 km^2 (flow 500 m$^3 \cdot$ s^{-1}) to about 2300 km^2 (flow 3000 m$^3 \cdot$ s^{-1}). The winter water temperature has increased by about 1°C and ice cover has been reduced by four to six weeks. Because of naturally large interannual variations in ice cover and break-up, changes in ice algal productivity and the spring phytoplankton bloom are difficult to establish. Summer flow has slightly decreased, summer water temperature has dropped by about 3°C, some nutrients (such as dissolved silica) have decreased, but organic carbon has increased by about 30 percent. The decrease of inorganic nutrients reflects conditions after the productivity surge frequently apparent in newly formed reservoirs. The increase of organic carbon and fish yield, particularly of cisco in the lower river, is thought to be related to overspill through the turbines and not to increased productivity. Levels of chlorophyll a near the front of the river plume appear to be elevated. This may be due to the enhancement of photosynthesis by freshwater.

Hypothesis 9 is likely to be locally significant. However, the extent to which it may influence under-ice primary production over large areas of the bay is unclear. Conditions in the bay are expected to reflect hypothesis 10. An early but temporary increase of organic carbon may be expected (hypothesis 11), tied to the productivity surge within the reservoir. Levels should decrease over time with the decrease of nutrient availability within the reservoir and the effects of greater water depth within it. Levels of organic carbon may eventually stabilize below the original background level. Estimates suggest that 10,000-12,000 t of particulate organic carbon and perhaps 20 times as much dissolved organic carbon are discharged into Hudson Bay by the Great Whale River annually.

Maximum bacterial activities and abundance occur during spring break-up and chlorophyll concentrations are high at the river mouth, relative to surface waters elsewhere in the bay. Evidence suggests that estuarine productivity is largely heterotrophically driven. Therefore, any reduction by impoundment of the naturally contributed organic load could have significant consequences for estuarine productivity. Offshore waters are not likely to be much affected since they are largely enriched by vertical mixing of Arctic waters within the bay.

The productivity surge following reservoir formation is a well known phenomenon (Hecky et al., 1984) and in boreal reservoirs it may last a few decades (hypothesis 11). If the outflow from a reservoir is greater than the natural flow, or occurs under different seasonal conditions, it is likely that bank and bed erosion will increase downstream. Downstream erosion can also occur, even if flows are less than before impoundment, as a river seeks to meet its increased sediment carrying capacity (hypothesis 12). Major flow reductions, caused by diversion and impoundment (for example, the Churchill River has lost 85 percent its flow), usually result in significant channel and estuarine infilling. Habitat changes due to both erosion and deposition will result in displacement and colonization of new areas by benthos and vegetation. If river discharge is reduced, saltwater-tolerant species will penetrate more deeper into the estuary.

Hypotheses related to mercury and other contaminants

(13) The creation of reservoirs leads to increased levels of methyl mercury and other forms of mercury in downstream (estuarine) biota, including resident and anadromous fish, fish-eating birds, and marine mammals.

(14) As a component of the total mercury loading to estuarine areas, mercury derived from atmospheric sources is as important as that derived from reservoirs.

(15) Increased body burdens of mercury and other contaminants can effect the health of fish and other marine mammals.

(16) Reservoirs are unlikely to have a significant effect on the estuarine loadings of organochlorines, PAHs, or other heavy metals (excluding mercury).

(17) Reservoirs will increase the production of CO_2 and CH_4.

Comments

Although some uncertainty may be associated with early monitoring data from the region due to sampling practices and analytical methods, it is clear that impoundment can lead to increased concentrations of methyl mercury (hypothesis 13), as reported by CM-MA (1987). Mercury levels in fish, for example, from the La Grande 2 reservoir, have been reported at 3 ug $\cdot$ g^{-1} (Brouard *et al.*, 1990), which is 15 times the guideline for long term human consumption (Perusse, 1990). Because of further bioaccumulation, mercury in reservoirs is also a threat to the health of loons, bald eagles (*Haliaetus leucocephalus*), Osprey (*Pandion haliaetus*), and other top predator species (Rimmer, 1992). Although the presence, mobilization and transformation of mercury in freshwater systems has been well documented (Bjornberg *et al.*, 1988; Håkanson *et al.*, 1988; Jackson, 1987, 1989; James Bay Mercury Committee, 1990, 1992), it requires further examination in the marine and estuarine environment (Makivik Corp., 1991). Models have not yet been developed to predict what may happen in these conditions. Methyl mercury moves downstream dissolved in water and attached to sediment and particulate organic matter. Fish that "spill" over the dam or which pass through the turbines and enter the food web of downstream organisms also carry methyl mercury. From the La Grande River (Bodaly & Johnston, 1992), there is evidence that levels of methyl mercury in fish downstream from the dam are, in fact, higher than those in fish from the reservoir.

Atmospheric loadings of mercury are an additional component to the total loading of mercury (hypothesis 14) but the relative proportions of internal and external source loadings to the reservoir are not known. This is important because mercury loading, globally, have been increasing over the past two centuries. The present rate of increase is about 1.5 percent per year (Slemr & Langer, 1992). The addition of external loading may, therefore, extend the period over which body burdens remain elevated in fish within and below a reservoir. Mercury (hypothesis 15) is potentially toxic to fish, birds and mammals (including humans). Its effects may be exacerbated by some contaminants and reduced by the presence of others (in particular, selenium; Nriagu & Wong, 1983; Wren & Stokes, 1988). There is no evidence to suggest that reservoirs significantly effect the downstream concentration of contaminants other than mercury (hypothesis 16).

Reservoirs probably do increase the production of carbon dioxide and methane above pre-impoundment levels at the same location (hypothesis 17). The significance of this in terms of existing regional sources or in relation to gas emissions from thermal generation remains uncertain. Assuming that 60 percent of the organic carbon stored in flooded soils is released over a 50 year period, the release of carbon dioxide from two reservoirs has been estimated and compared with emissions from thermal power plants. The upper and lower range carbon dioxide equivalents for the Grand Rapids reservoir in Manitoba and the Churchill/Nelson development are 0.3-0.5, and 0.04-0.06 Tg $CO_2 \cdot TWhr^{-1}$, respectively (B. J. McCormick, Manitoba Hydro, Winnipeg, Manitoba, pers. comm.). Thermal power production from combustion of coal or natural gas is in the range of 0.4-1.0 Tg $CO_2 \cdot TWhr^{-1}$ (Bodaly, In: Bunch and Reeves, 1992).

Hypotheses Related to Impacts on Biota

(18) Changes in flow, siltation, and obstructions will effect resident and non-resident fish species (including anadromous fish), and other biota downstream from impoundments.

(19) Specifically, flow reduction in the Nastapoka estuary will have a negative impact on the only Atlantic salmon population in Hudson Bay.

(20) Specifically, there will be a negative impact on freshwater harbour seals in impounded watersheds due to habitat reduction and to the seals' increased accessibility to hunters.

(21) Salinity and temperature regimes will be modified in estuaries below impoundments. These will alter the distribution and survival of fish (larval, juvenile, and adult stages) and dependent predators. Other effects are possible.

(22) Changes in salinity patterns may result in changes in the use of the coastal zone by fish and dependent predators (including humans).

(23) Changes in ice conditions and the time of break-up will alter the use of affected areas by marine mammals and, consequently, modify harvesting opportunities.

(24) Altered freshwater flow will change the distribution and possibly the composition of benthic macrophytes and dependent fauna.

(25) In areas affected by modification of the freshwater plume, the balance between ice algae and planktonic primary production will be altered, as will the energy

transfer to higher levels of the food web through zooplankton.
Comments

Impoundments can affect water use, resident species (CM-MA, 1987) and downstream biota (hypothesis 18). Evidence from studies of the effects of the Churchill-Nelson diversion in northern Manitoba (Baker & Davies, 1991; CM-MA, 1987), indicate that regime manipulation and draw down have killed fish eggs. Flooding and draw down have decreased shore stability, increased erosion, and greatly increased turbidity in some areas. High levels of turbidity have restricted water for drinking supplies and may have limited habitat for fish dependent on sight-feeding. Sedimentation and standing debris have also degraded habitat. In some parts of the diversion, populations of resident lake whitefish have greatly declined.

Conditions below the La Grande complex represent some of the most extreme flow modifications within the region. The discharge of the La Grande River now reaches 5000 to 6000 $m^3 \cdot s^{-1}$ in mid-winter, with typical flows in the order of 4000 to 5000 $m^3 \cdot s^{-1}$. These are seven to eight times greater than pre-impoundment flows (A. Penn, pers. comm.). Flow manipulation of the La Grande results in weekly-pulsed discharge during mid-winter, equivalent to conditions of a major natural spring flood. The extent to which this change in regime may influence over-wintering anadromous stocks in the river, downstream, and in the estuary and near-coastal waters is unknown, but likely significant. The effects may result in displacement of stocks and may also affect seasonal metabolic and reproductive capabilities. Other forms of impact could occur.

In west and northwest Quebec, fish that will be most negatively affected include Atlantic salmon, coregonids (lake whitefish, for example), and lake sturgeon. Suckers, perch and pike, however, are likely to benefit from the expanded lacustrine habitat created by reservoir formation. Merging of freshwater plumes in the coastal zone might reduce their value as biological markers and disorient anadromous species whose reproduction depends on spawning in natal habitat.

The small population of Atlantic salmon in the Nastapoka River may be a relict of postglacial colonization. In some scenarios of hydroelectric development, flow reduction in this river has been considered. Since there is only about 1.5 km of river below the first natural barrier (falls) and a salt wedge extends well into the Nastapoka River, a projected 15 percent reduction of flow could be sufficient to cause significant loss of the limited habitat available to this stock. Flow in this river is highly variable and environmental sensitivity is likely to depend on how the flow regime might be modified more than the average reduction of discharge. Reduction of low flow might be more damaging to the freshwater environment than reduction of high flow. The habitat of this stock could be very sensitive to regime change (hypothesis 19). In addition, the change in habitat could favour competitive species such as brook trout. Changes in the flow regime of the Koksoak River flowing into Ungava Bay, as a result of headwater manipulation of Lake Caniapiscau, may already affect another local stock of Atlantic salmon.

Very little is known, with certainty, about the distribution, ecology and behaviour of the freshwater harbour seals (hypothesis 20). It is likely that any increase in access for hunters or modification of unique habitat would endanger these animals.

Impoundments are likely to influence downstream estuarine conditions (hypothesis 21). Although potentially of major importance, little is known of the estuarine communities in Hudson Bay. The extent of change is uncertain. For example, the Nelson estuary is subject to a relatively large tidal range and the estuarine plume is compressed with each tidal cycle. Unlike many of the east shore estuaries, the plume is well mixed

(Baker, 1989), there is no salt wedge, and maximum productivity (defined by chlorophyll a) occurs in the nearshore zone rather than in the river or its estuary. Tidal cycles produce rapid variations in depth, temperature, salinity and currents. Warm water may enhance estuarine productivity but the extent to which changes have been brought about by the Churchill-Nelson diversion are uncertain. Temperature records of natural estuarine conditions before diversion are not available. The Nelson River is regulated for power production and this results in diurnal and weekly cycles, and low summer flow. The extent to which discharge and regime changes have influenced productivity in the Nelson and Churchill River estuaries is uncertain. Likely, the plume of the Churchill River has significantly decreased (now 25 % of natural flow), and that of the Nelson River increased (post-diversion flow has increased by about one-third). Because of the prevalence of drought conditions, however, the present flow of the Nelson River is generally less than historic maximum.

The proposed Conawapa dam (low on the Nelson River) would limit the use of spawning habitat by migratory lake whitefish and this could result in a large decline in the estuarine population of this species. Species-dependent activities such as the moulting of beluga whales in the Churchill estuary may be affected. Their requirements are thought to include warm temperatures, low salinity and substrate upon which to rub. Based on limited information, there appears to be no impairment of this particular activity, so far. Nevertheless, the lag between cause and effect may be greater than the time elapsed since diversion of the Churchill River in 1976. Impoundments may also influence salinity in the coastal zone but the extent to which this may affect use of the zone are uncertain (hypothesis 22). Hypothesis 23 refers mainly to ringed and bearded seals hunted by the Inuit, on or near sea-ice. Ringed seals have a relatively long lactation period (42 days) and rely on stable sea-ice for pupping and moulting in the spring. Thus, any degradation of this habitat could reduce their reproductive success. In addition, poor spring sea-ice may limit successful predation by polar bears, thereby affecting not just their food supply but also their reproductive health. Poor sea-ice would also create a major hazard for Inuit hunters and limit the availability of a major food supply.

Changes in freshwater flows may have an effect on aquatic macrophytes and on the distribution of dependent species (hypothesis 24). The degree and extent of effect are site-specific and related to the time of ice break-up. Hypotheses 23 and 24 may be linked through the role of kelp. Kelp covers extensive areas of shallow water in the bay and it stores nitrate in winter, before the growth of ice algae. Since reservoir discharge will increase during the winter (relative to the natural flow) a greater proportion of the available nitrate may be taken up by kelp and less may be available to the ice algae for late winter/early spring production. The importance and the role of kelp in the ecology of the bay is not understood. Also, the role of eelgrass meadows and tidal marshes is not clearly defined. Ice algae may represent 10-50 percent of the primary production of Hudson Bay and the timing of ice break-up could be critical to the effectiveness of grazing invertebrates upon which larval fish depend (hypothesis 25). Under various changes in discharge regime, impacts might result in either a productivity loss or gain, or no net change.

5.2.2 *Excessive harvest and exploitation*

Hypotheses Related to Growth and Development

(1) Population growth in northern communities and increasing access to wilderness areas (including sea ice) is leading to resource exploitation.

(2) With increasing levels of human activity and habitat modification, the range and population of many native species is being affected (and often reduced). An additional threat is posed by sport demand. This increases harvest pressure on alternative species and stocks, and increases the threat of extinction or extirpation of unique or endangered species or stocks.

(3) Increasing interest in northern resources as specialty foods will raise harvesting levels to meet "export" demand.

Comments

Northern populations have grown but it is not known if the number of Cree and Inuit in the Hudson Bay region equal or exceed their historical maximum (hypothesis 1). Generally, the Cree include a much greater proportion of non-marine species in their diet than the Inuit and, therefore, they tend to exert less pressure on aquatic food resources. However, at several locations within the basin and most noticeably in parts of the Nelson-Churchill River diversion and the James Bay hydroelectric development, it has been necessary for Indian communities to shift hunting and fishing practices away from traditional areas. This need has risen because of habitat change, redistribution of wildlife, and, in particular, mercury contamination of the food-chain. In some areas these shifts have led to more intensive harvesting at alternative sites where there have been reductions in the fish and wildlife resource.

In addition to the use of aquatic resources by resident populations, significant numbers of people now move into the region for temporary employment (for example, government, and civil and military construction and maintenance). They may place additional demand on local resources, particularly Arctic char, lake trout, northern pike and walleye, and in the western part of the Hudson Bay drainage, grayling (*Thymallus arcticus*). Northern stocks are very sensitive to harvest pressures (Arctic char, for example, Ayles, 1987; McCart & Den Best, 1979) and fly-in fishing may have depleted stocks in inland lakes and streams. The number is unknown. Non-resident hunting has also depressed the populations of other species including ducks and geese (in-turn, increasing the use of remaining aquatic stocks by resident Inuit and Cree). Within areas covered by the JBNQA, however, wildlife stocks do not appear to be depressed. In part, this may be due to more effective management practices, but also to a decrease in dependence on traditional foods (noted previously). Similar shifts in dependence are less evident in Inuit communities.

Habitat modification is most evident in association with reservoir construction, and the creation of power line corridors and roads (hypothesis 2). Post-construction environmental assessments are available from few sources in the region but it is clear that habitat changes have had a particularly significant impact on native communities dependent on natural resource harvesting (Usher & Weinstein, 1991). In more recent developments, the effects of major changes in river discharge (due to reservoir construction and diversion) are not fully evident and may not be for several decades. It can take a long time for induced physical and chemical changes to become established. To a large extent, many effects are potential rather than actual. The sensitivity of Arctic char to even small changes in stream flow during migratory runs (Power & Barton, 1987) is a particular reminder of the importance of connectedness between habitat types and the extent to which even slight modification may effect the availability of critical resources.

Some changes are known. For example, variations in ice formation and condition influence survival of sea mammals and birds, and the hunting abilities of surface predators (including humans). The lack of stable ice in some rivers affected by hydro-electric development has limited access and ice travel for native people. Openings in ice cover caused by the movement of ships may also limit access to ice edge hunting sites for Inuit.

Air transportation now provides a ready means of export for fresh and frozen specialty foods, and there is potential, if not effectively regulated, for stocks of high quality fish and shellfish to be rapidly depleted (hypothesis 3). The small size of beluga populations in eastern Hudson Bay are thought to be, in part, a reflection of past hunting practices and, may indicate long recovery time. The near extirpation of the bowhead whale (between 1850 and 1915) and the slow recovery of this species attests to the sensitivity of Arctic marine systems to exploitation, and there are examples throughout the Northwest Territories and elsewhere around the bay (Keith et al., 1987) where inland and coastal fisheries have severely depleted fish stocks (many of which have not recovered). In the Ungava District, Boivin et al. (1989) noted significant declines in Arctic char populations in proximity to Inuit communities, where about 35 percent of the fishery was for non-subsistence use. In the Keewatin District, lake whitefish, lake trout, and Arctic char form the basis for continued interest as commercial fisheries (McCart & Den Beste, 1979; Topolniski et al., 1987). The relatively high productivity of parts of the Hudson Strait is known to support large but local populations of seabirds and marine mammals, and may be capable of supporting a specialized shellfish fishery. Mariculture may also offer a specialized capability in other selected areas of the nearshore.

As with all predator-prey relations, human use of selected species will result in a compensatory response by the species affected, and the balance between human demand and resource production can be maintained. Over-harvesting occurs only when compensatory mechanisms are exceeded. Unfortunately, the extent of such carrying capacities is not adequately known. Evidence suggests that many problems associated with over exploitation in the region owe their origins to management and commercial practices of non-native and non-resident groups., This strongly implies the need to accommodate northern development and recreational use of resources only after the needs of the native people have been met (Usher, 1981).

Hypotheses Related to Information and Management

(4) Traditional environmental knowledge (TEK) should play a valuable role in resource management. Often there is insufficient knowledge to support application of standard harvest/yield models for management of many northern species. Even where sufficient information exists, a lack of holistic insight may result in poor allocation and conservation. TEK offers complementary understandings and a powerful guide to practical decision-making in resource management.

(5) There is a need to address synergisms associated with indirect and negative effects of human activities in the basin with growing harvest pressures. This need is not yet understood or recognised as an essential part of long term strategies in sustainable resource development.

Comments

For the most part, standard catch and stock data are not adequate to provide the basis for effective resource management (hypothesis 4). Survey and monitor information have certainly defined areas where resource exploitation has occurred (McCart & Den Beste, 1979; Freeman, 1981; Topolniski *et al.*, 1987), but data are neither sufficiently comprehensive nor compiled and interpreted frequently enough to provide reliable guidance within realistic time frames. There is no unified management objective for basin resources. Studies on appropriate management techniques for resource recovery or enhancement are, also, largely lacking (hypothesis 5). Opportunities exist for significant improvements in resource management through the linkage of scientific method and TEK.

At present, five governments (and many more agencies) carry various levels of responsibility for resource management within and around the basin (Federal Government, Quebec, Ontario, Manitoba, and Government of the Northwest Territories). In addition, there are increasing levels of responsibility for renewable resource management within both the Inuit and Cree communities. There is no evidence of a comprehensive approach to basin-wide sustainable resource management, nor to the integration of near-bay terrestrial and freshwater wildlife and fisheries resource management with marine resources management. The successful use of reindeer (raised locally) to supplement the marine food source in the Belcher Islands is just one example of what may be required to provide a sustainable, local, food supply for communities around the bay.

5.2.3 Long Range Transport of Atmospheric Pollutants (LRTAP) and health

Hypotheses Related to Organic Contaminants

(1) LRTAP is the principal contributor of persistent organic contaminants to the bay.

(2) Increased levels of persistent organic contaminants pose a potential threat to the health of top predator species, including humans.

(3) At present, benefits to human health derived from the traditional use of country foods outweigh a shift to alternative food sources (imported "southern" foods).

Comments

Within the bay area, there are few point sources of persistent contaminants and most are associated with military and construction wastes and storage. Releases are thought to be small (most often PCBs). There are no known major emissions from agriculture or forestry or manufacturing sources around the bay and direct loadings from tributary sources are small. Spatial distributions and temporal trends of Arctic contaminants are not adequately defined, and unknown "hot spots" may exist. For many contaminants reported in biota from the bay and adjacent shoreline and tributary areas, there are no known sources within several thousands of kilometres (hypothesis 1).

Weather event back-casting, the identity of dry-fall material, and the types and "finger-printing" of contaminants present in the region strongly imply atmospheric transport from distant sources in North America, Europe, and Asia (Welch *et al.*, 1991). This includes loadings from the Arctic haze which appears to reach maximum development during February and March (Anon., 1990; Muir *et al.*, 1992). The Arctic haze, which is concentrated in the lowest 5 km of the atmosphere, was not reported prior to the 1950s (Lockhart *et al.*, 1992). Dust fall (brown snow) in the Keewatin district is thought to comprise loess from western China and soot from coal combustion in either China or Russia, or both (Gregor & Gummer, 1989; Welch *et al.*, 1991). The composition of PAHs in snow samples is compatible with emissions from fuel combustion in mid-latitudes of the northern hemisphere. Increasing concentrations of PAHs in Arctic soil, snow and lake bed samples parallel global trends of increasing fuel combustion.

River loadings from sources in northern Russia are also thought to be important (though of lesser scale). Seven of the eight largest rivers entering the Arctic Ocean flow from northern Russia and Siberia. The top 50 m of the polar mixed layer of Arctic Ocean water is least dense and most stable. With very low particulate concentrations, contaminant scavenging and settlement in Arctic waters is minimal. Residence times are long (up to 20 years) and with a general west to east circulation, the polar mixed layer can act as a significant source of contaminants for northern Canada (Barrie *et al.*, 1992). Seasonal volatilization provides a mechanism for contaminant entry into atmospheric transport, and ice rejection may act as a contaminant pump. Collectively, precipitation/deposition, vapour phase transfer and warm/cold transfer via volatilization may result in a systematic build up of contaminants in the Arctic. For example, relatively high concentrations of HCH have been found in many northern freshwaters. Concentrations of both alpha and gamma isomers are near Canadian guidelines; and alpha HCH exceeds the guideline of 10 ng $\cdot$ l^{-1} in some northern areas. Concentrations of HCH are latitudinally associated.

Organic contaminants in the Canadian Arctic include DDT and its degradation products, PCBs, dieldrin, heptachlor epoxide, chlordane, toxaphene, hexachlorocyclohexane (HCH), chlorobenzenes, chlorinated dioxines and chlorinated dibenzofurans. Many of these compounds are pesticides (Gregor, 1990; Gregor & Gummer, 1989). Many of the polyaromatic hydrocarbons (PAHs) present in the Arctic environment are thought to be derived from mid-latitude combustion (power generation from fossil fuels, and vehicular emissions). The highest levels of total organochlorine contamination in the Canadian Arctic occur in Hudson Bay (Norstrom & Muir, 1988). The loadings of PCBs are greatest from dry fall (Barrie *et al.*, 1992). Levels of tris (4-chlorophenyl)-methanol (TCP) and dieldrin are relatively high in polar bears from Hudson Bay, and are likely derived from widespread atmospheric transport (Jarman *et al.*, 1992). Levels of alpha-HCH are elevated in polar bears in areas of the bay receiving high surface runoff and the distributions of TCDD and OCDD (polychlorinated dibenzo-p-dioxins) in marine mammals suggest derivation from combustion sources in Eurasia (Norstrom *et al.*, 1990). Levels of DDT metabolites and PCBs are higher in biota from Hudson Bay than other areas of the Canadian Arctic and may reflect atmospheric transport from Mexico and Central America (Norstrom & Muir, 1988).

Reports of persistent organic contaminants in many species of animals (Muir *et al,.* 1986, 1990) within the Hudson Bay basin are limited and records of these contaminants in humans (Davies, 1991b) are even fewer. It is therefore difficult to address hypothesis 2 with quantitatively adequate data. Qualitatively, the information base is nevertheless significant. Chlordane and PCBs account for about 80 percent of the organochlorine present in the adipose tissue of marine mammals (Norstrom *et al.*, 1988). Generally, Arctic marine and freshwater fishes contain levels of organochlorine contaminants which are similar to or lower than levels in marine and freshwater fishes from southern Canada. More specifically, concentrations are generally an order of magnitude or two lower than in comparable levels of the food web in Lake Ontario (Norstrom & Muir, 1988). Fish do not bioconcentrate PAHs and concentrations of them are similar to those from relatively uncontaminated marine and freshwater areas in southern Canada. Relatively high concentrations of HCH and toxaphene have been reported in Pacific herring and Arctic cod.

Throughout the Arctic, levels of toxaphene (a mixture of polychlorinated camphenes or PCCs) are several times higher than levels of PCBs in fish, but toxaphene has not been found in polar bears (Norstrom & Muir, 1988). This may reflect selective metabolism. Beluga and narwhal likely have the ability to selectively metabolise TCDD-like substrates (Norstrom *et al.*, 1992). Although PCBs and hexachlorobenzene (HCB) biomagnify in the food-chain between seals and polar bears, there is no evidence of a similar increases in TCDD, OCDD, or TCDF (a polychlorinated dibenzofuran) and DDT (Norstrom & Simon, 1990).

Arctic cod is a principal source of food for sea birds (thick-billed murres), ringed seals, and beluga; and ringed seals, themselves, are an important source of food for polar bears and humans (Muir *et al.*, 1988). Contaminant levels and their long term trends in the ringed seal are a useful indicator of ecosystem health (including human health). Both DDT and PCBs were first reported in Arctic wildlife (including ringed seals) in the late 1960s. Since then, periodic sampling has established that levels of total PCB have declined by about 50 percent since the early 1970s (Environment Canada 1991, Indicators Task Force, 1991; Nettleship & Peakall, 1987). The decline has continued, less rapidly, since then. Concentrations of total DDT have shown no comparable decline but they are generally less now than two decades ago. However, concentrations of chlordane in polar bears from Hudson Bay have increased by about four times between 1969 and 1984 (Norstrom *et al.*, 1988).

Recent comparisons of contaminant levels in beluga (Muir *et al.*, 1990) suggest that total DDT and total PCB concentrations are slightly less in animals from west Hudson Bay than east Hudson Bay. Concentrations of total DDT and PCBs in beluga are 25 to 30 times lower in Arctic animals than St. Lawrence animals (concentrations in males are higher than females at almost all sites). At present, there is no indication that either ringed seals or beluga are subject to health problems in Hudson Bay. In Beluga, total DDT concentrations are generally between 0.3 and 6.3 ug $\cdot g^{-1}$, and total PCB levels are generally between 0.3 and 3.7 ug $\cdot g^{-1}$. Contaminant effects are thought to be an important factor affecting the health and reproductive success of the St. Lawrence beluga whale stock. Therefore, although contaminant concentrations appear to be decreasing in the marine mammals of Hudson Bay, any increase to early 1970 contaminant levels, or above, could give rise to concern (regarding both animal health and human food supply).

The principal pathway by which persistent organic contaminants enter human populations is through the food supply. The Inuit derive a substantial portion of their diet from country foods and these meet important nutritional needs. A high proportion of country foods are consumed raw (hypothesis 3). In addition, even if alternative foods were to be acceptable both in terms of nutrition and preference, they would be exceedingly costly to transport into the region (in sufficient quantities) and to store. Also, it is necessary to consider the potential of adverse social and spiritual effects in shifting to other sources of food. The most important components of Inuit diet include sea mammals, fish, birds, and small mammals. In the western part of the basin, the Inuit make use of caribou; some muskox hunting is permitted, and reindeer are kept on the Belcher Islands. Cree populations are more dependent on terrestrial mammals, fish and birds, and consume foods after they have been cooked. As a result of differences in both diet and the preparation of food, concentrations of organochlorines tend to be lower in the Cree. Nevertheless, Indian communities are subject to the effects of significant contaminant exposure (Young, 1988).

Concentrations of persistent organic contaminants have been measured in a wide range of human tissues and, in general, contaminant levels are highest in tissues with a high lipid content. The presence of organochlorines in adipose tissue suggests exposure over an extended period of time whereas the presence of organochlorines in blood samples suggests a more immediate exposure. Concentrations of contaminants in humans have the potential to reach high levels as a result of bioaccumulation and biomagnification through the food web. Organic contaminant concentrations are not the same in all tissues, concentrations in liver are usually less than in adipose tissue and progressively decrease in kidney, gonad and brain tissues. Concentrations of total PCBs and DDT and its derivatives in mothers' milk from northern Quebec Inuit are among the highest values recorded in humans and about five times greater than in Caucasian women from southern Quebec (Dewailly et al., 1989). The mean level of PCBs in breast milk from the northern mothers was about 111 ug $\cdot$ L^{-1} and the concentration of PCBs in milk fat was 3.6 ug $\cdot$ L^{-1}, comparable values from the southern mothers were about 28 ug $\cdot$ L^{-1} and 0.8 ug $\cdot$ L^{-1}, respectively.

Despite these high levels, it is not possible to make detailed comparisons between Inuit and other population data because of a lack of consistent reporting. It is also hard to establish cause and effect; rather it is necessary to depend on the "weight of evidence". No significant differences were observed within Inuit settlements that were studied; they were heavily depended on freshwater fish (18 meals/mo), marine mammals (10 meal/mo), and marine fish (9 meals /mo).

In the Great Lakes region, health studies were undertaken on a group of mothers in Michigan (Fein et al., 1984; Jacobson et al., 1985, 1990) who consumed a modest quantity of lake fish in their diet (about 7 kg $\cdot$ a^{-1}, for at least 16 years). Concentrations of PCBs in the fat of breast milk form these women averaged 0.75 ug $\cdot$ L^{-1}, similar to those reported in women from southern Quebec (Dewailly et al., 1989). In another study of women from North Carolina, Rogan et al. (1986) reported the median level of PCBs in breast milk fat as 1.74 ug $\cdot$ L^{-1}. Although no significant negative health effects were observed in the adult women of either the Michigan or North Carolina studies, there were indications of subtle behavioural differences in their infants, especially those having high concentrations of PCBs. The behavioural differences, however, were not necessarily related to breast feeding since transfer of contaminants in blood serum between mother and embryo may have been a more important means of transfer. Whether or not the subtle behavioural differences reported by Jacobson et al. (1985, 1990) can be directly related to the presence of PCBs, DDT, and or other contaminants remains unclear, nevertheless there is potential for greatest harm to occur in the embryonic stage of a population.

As a result of the Broughton Island study (Baffin Island) on Inuit food consumption and contaminant levels, it was found that blood levels of PCB exceeded the tolerable levels defined by Health and Welfare Canada. Dewailly et al. (1989) concluded that some Arctic residents could be at risk from their intake of PCBs. Kinloch et al. (1992) reported that 73 percent of the population tested exceed the recommended maximum total daily intake for PCBs. Most of the PCBs were derived from consumption of narwhals and this species was also the single greatest source of DDT, chlordane, HCH and dieldrin. Seasonally, contaminant intake varied with the food source. The highest contaminant intake levels were in September, and the lowest in January.

There may be some parallel between this example of PCB contamination in humans and that reported by Jacobson et al. (1985, 1990). However, the value of country food must be weighed carefully against other concerns. At Broughton Island, traditional foods have a high protein content and they are a good source of omega-3 fatty acids which are thought to provide protection against cardiovascular disease (Kinloch et al., 1992). As well, there appear to be advantages to breast feeding which supplies infants with a natural and traditional diet. Recent medical studies in other parts of Canada suggest that juvenile diabetes may be most prevalent in children not fed on breast milk (Juvenile Diabetes Foundation of Canada, Richmond Hill, Ontario). Ways need to be found to reduce the intake of contaminants while keeping the nutritional benefits of traditional diet.

Hypotheses Related to Metal Contaminants

(4) Mercury is the most important metal contaminant brought into the Hudson Bay basin by LRTAP. With few local exceptions, it is the only metal present in sufficient quantities to be considered a threat to environmental health.

(5) Over the long term, LRTAP loadings will significantly raise the quantity of mercury present in the northern environment above natural background levels. Thus, although quantities of methyl mercury associated reservoir construction will decrease, they may not return to pre-impoundment levels.

(6) The presence of selenium, particularly in marine mammals, tends to reduce the toxic effects of mercury which is present at relatively high concentrations in several top predators.

(7) Similar antagonistic relations between selenium and mercury may have a beneficial effect on human health. There is a potential for negative health effects from synergisms among other contaminants that are present.

Comments

Since the 1960s, accumulations of radionuclides from long range atmospheric transport have declined in northern Canada (north of 60 °) and atmospheric sources are not presently of concern in the Hudson Bay region (Anon., 1989). However, a number of metals are present in both particulate and dissolved form in Arctic and sub-Arctic air masses. The most important sources are probably mineral smelting and the combustion of fossil fuels (particularly coal). The long range transport of pollutants likely contributes significant quantities of mercury and to a lesser extent zinc, cadmium, lead and other metals (hypothesis 4).

Mercury is widespread as a global contaminant (Andersson et al., 1991; Muir et al., 1986; SEPA, 1991; Slemr & Langer, 1992; Wagemann et al., 1990) and, like cadmium, its concentration in biota appears to be related to loading rates. Mercury in marine mammals tends to be lower in Hudson Bay than many other areas of Arctic Canada and this is probably as an expression of atmospheric loading (Braune et al., 1991; Renozi & Norstrom, 1990). Elevated levels of cadmium are known to occur in narwhals, and also in bearded seals and polar bears (Braune et al., 1991). Most of the cadmium is thought to be of natural origin but the significance of atmospheric loading is not known. There is no evidence of cadmium toxicity in narwhals. Lead is also an important contaminant. Often, it is most concentrated close to the general area of its emission. Lead emissions are probably too distant to be of much significance to the aquatic environment of Hudson Bay. Concentrations of copper and zinc in biota do not appear to be related to their emissions (Johnson, 1987).

Hypothesis 5 draws attention to the selective nature of mercury bioconcentration and the importance of food web structures. Biota, in which mercury levels tend to be high under natural conditions are most likely to respond to increased global loadings. Slemr & Langer (1992) have demonstrated that global loadings of mercury, particularly mercury derived from combustion sources (coal), have been increasing for the past two centuries. Mercury concentrations are increasing about 1.5 percent annually, in the atmosphere of the northern hemisphere. These observations are supported by dated soil, peat, and lake sediment core profiles which suggest that there has been a doubling in such concentrations since the start of the 19th Century. The concentrations of other metals have also increased as a result of human activities in the environment but mercury is most widespread.

Johnson (1987) demonstrated that there is a close relationship between the concentration of mercury in fish tissue and the rate of mercury loading to the environment, but no comparable relationship exists between concentrations of mercury in fish and ambient waters or bottom sediments. This is evident, for example, from comparisons between Quebec and Ontario lake surveys. Mercury concentrations in flooded sediments from the La Grande area average 0.164 ug $\cdot$ g^{-1} and those from the flooded Caniapiscau River system average 0.085 ug $\cdot$ g^{-1} (Brouard et al., 1990). These levels are similar to the background levels in sediments from 14 Ontario lakes described by Johnson (1987). However, the most recent sediments (the top few centimetres of the core profile) in the Ontario lakes usually have higher concentrations of mercury (0.2-1.5 ug $\cdot$ g^{-1}) and this increase is ascribed to the effects of human activity over the past two centuries. Some increase would be expected in the upper part of the La Grande and Caniapiscau sediments but it appears, generally, that the concentrations of mercury in them are lower than in the Ontario data.

Despite the differences in sediment mercury concentration, the concentrations of mercury in northern pike, walleye, and lake trout from the Quebec sites were higher than in the Ontario sites. Johnson (1987) reported the average values for mercury (wet weight) as 0.1 ug $\cdot$ g^{-1} (lake whitefish), 0.13 ug $\cdot$ g^{-1} (walleye), 0.18 ug $\cdot$ g^{-1} (lake trout), and 0.2 ug $\cdot$ g^{-1} (northern pike). Comparable values before reservoir flooding at the Quebec sites were about 0.15 ug $\cdot$ g^{-1} (lake whitefish), and about 0.6 ug $\cdot$ g^{-1} for lake trout, walleye, and northern pike (Brouard et al., 1990). The higher values in fish from the Quebec sites, in part, probably reflect their slower rates of growth and greater age (due to lower ambient water temperatures and lower rates of productivity).

It seems probable that metal emissions to the atmosphere will continue to rise globally over the next several decades and, at present rates of increase, this would amount to a doubling of the mercury loading from global emissions over the next 50 years. With improved emission controls, release of mercury (especially from coal combustion) may be reduced, and this should limit the rate at which mercury loadings increase. However, it is most probable that while production of methyl mercury in the flooded areas of reservoirs will decrease towards background conditions (over the next 50 years or so), background levels will continue to rise. Perusse (1990) noted that, even before formation of the La Grande reservoir, many fish important to the diet of Cree and Inuit (for example, lake trout, walleye, and northern pike) already had mercury close to tolerable levels defined by Health and Welfare Canada for sport fish consumption (0.5 ug $\cdot$ g^{-1}) and exceeded the guideline for unlimited consumption (0.2 ug $\cdot$ g^{-1}). It is also possible that other species (for example, longnose sucker and lake whitefish) may be placed on limited consumption in the future. A similar situation exists in northern Manitoba where flooding for the Churchill-Nelson diversion has raised mercury concentrations in pike and walleye well above consumption guidelines (CM-MA, 1987). Concentrations in lake whitefish remain, generally, within the guidelines.

Hypothesis 6 addresses the presence of both selenium and mercury in biota. Selenium is naturally present in the environment, usually at low concentrations (Nriagu & Wong, 1983). As an emission from some industrial sources, also, it can be present in atmospheric loadings (Wren & Stokes, 1988). Global average values in near surface ocean waters are about 50 ng $\cdot$ L^{-1}, and in river water they are about 200 ng $\cdot$ L^{-1}. Based on work by Harrison & Klaverkamp (1990), Klaverkamp et al. (1983) and Salki et al. (1985), selenium is an important micro-nutrient at low concentrations. Experimental data and field observations indicate there is little evidence of its toxicity within the freshwater aquatic food web up to levels of at least 100 ug $\cdot$ L^{-1} in water. Behavioral impairment is apparent in freshwater fish at about 250 ug $\cdot$ L^{-1} and acute mortalities occur at the low mg $\cdot$ L^{-1} level. Toxic effects appear to be similar in several species of freshwater fish. Concentrations in plankton appear to be related to the concentrations in water. At concentrations in water below about 200 ng $\cdot$ L^{-1}, the principal means of selenium uptake in fish is through ingestion of food. At higher concentrations in water, both direct uptake and uptake from food contribute progressively more equal amounts of selenium to the body burden.

Although, on its own, selenium is considered to be a toxic substance at relatively low concentrations, it can have an antagonistic relationship with many heavy metals. This is most apparent between selenium and cadmium, and selenium and mercury. The relationships between selenium and metals, however, are complex and differ with concentrations and, sometimes, with species (Pelletier, 1985; 1986a; Srikumar & Akesson, 1992; Wren et al., 1986). At low concentrations in water, selenium appears to limit bioaccumulation of mercury and, therefore, toxic effect (C-OSC, 1983; Speyer, 1980; Turner & Swick, 1983). At higher concentrations, selenium and mercury may both accumulate but the presence of selenium still seems to limit the toxic effect of mercury. Synergistic effects are also known. Concentrations of mercury and selenium are usually greatest in liver tissue, less in kidney tissue, and least in muscle tissue.

The concentrations of selenium tend to be higher in most marine fish than freshwater fish, but levels of mercury are generally lower in most marine fish. High levels of mercury and selenium have been reported from many marine mammals in Canadian waters, especially ringed seals and beluga (Wagermann et al., 1990) and elsewhere (Kari & Kauranen, 1978). There is no evidence of detriment to health by metals in beluga populations of the eastern Arctic (including Hudson Bay).

Concentrations of cadmium in the muscle tissue of beluga from the eastern Arctic are generally less than 0.5 ug $\cdot$ g^{-1} and mercury is generally less than 3.5 ug $\cdot$ g^{-1}. Comparable levels of selenium in muscle tissue are about 1.5 ug $\cdot$ g^{-1}. Concentrations of cadmium are greatest in kidney tissue. Beluga from both east and west Hudson Bay and from the east coast of Baffin Island have high cadmium levels in kidney tissue (69-106 ug $\cdot$ g^{-1}) but these values are not related to the age of the animals (Wagermann et al., 1990). Although global loadings of cadmium have increased with human activities, there is no clear indication that high levels of cadmium in beluga from the eastern Arctic are related to the effects of long range contaminant transport. Most of these beluga are thought to spend part of the year feeding in the Hudson Strait but it is not known if or how elevated levels of cadmium may be associated with this. Concentrations of mercury and selenium in beluga liver tissue range between 25-38 ug $\cdot$ g^{-1} and 16-17 ug

· g^{-1}, respectively, for the east and west shore waters of the bay. High levels also occur in beluga caught from Baffin Island.

Antagonistic relations between selenium and mercury have been reported in a wide range of mammalian species and it has been suggested that selenium might be used as a supplement to reduce the toxic effects of mercury in human diets (hypothesis 7). Regrettably, sufficient information is not available to make an objective assessment of such a strategy. Interactions between selenium and metals (particularly mercury) need to be better understood in terms of real environmental effects and at concentrations typical of actual conditions. Most particularly, the effects of selenium in human diets needs to be better understood (Valentine et al., 1992) and the potential for other less controversial supplements explored.

These issues are of particular importance to northern populations that continue to depend on country foods, even though concentrations of mercury and selenium may well increase. The intentional addition of selenium to human diet is not new. It was first applied in agriculture about three decades ago (Oldfield, 1992). Most recently, sodium selenate has been applied in Finland as a crop fertilizer with the result that the calculated daily intake of selenium has risen from 20-40 ug · d^{-1} to about 100 ug · d^{-1}. This practice began in 1985 and the addition of selenium appears to have been generally beneficial (Hultunen et al., 1992). There has been a significant change in the dietary source of selenium in which fish provided about one-third of the selenium present in the diet during the 1970s. The daily intake of selenium by Inuit populations largely dependent on marine foods is likely higher than Cree populations that derive more of their diet from terrestrial mammals. A comparison of the diet, health, and metal levels in these populations may give further insight into the value selenium enrichment.

Hypotheses Related to UV-B Radiation and Climatic Effects

(8) LRTAP distributes emissions of ozone-depleting gases. In high latitude regions, atmospheric processes result in the greatest loss of high level ozone.

(9) Ozone depletion results in an increase in UV-B radiation received at the Earth's surface. This may negatively affect the health of many terrestrial and aquatic species, and may reduce primary production in aquatic systems.

(10) LRTAP redistributes greenhouse gases, globally. There are complex interactions between different gases, including ozone, which can cause both heating and cooling of the atmosphere. The greatest effect of resulting climatic change is expected to occur at high latitudes, including the Hudson Bay basin.

Comments

Ozone occurs both at high (stratospheric) and low levels in the atmosphere. Ozone is produced naturally as a result of photo-oxidation (particularly in the stratosphere) but one of the effects of atmospheric pollution (particularly from urban sources) is, also, to increase the amount of ozone present at low levels. Excessive quantities of ozone at ground level act as a pollutant and affect both plant and animal health, but in the stratosphere ozone provides global protection against part of the spectrum of solar radiation.

Atmospheric circulation and long range transport redistribute gaseous and fine particulate emissions and re-emissions, globally (including those in Arctic regions). There is a continual movement of air-masses from low to high latitude in both hemispheres of the globe. General circulation carries a number of ozone-depleting gases (mostly from mid and low latitude regions) to high latitudes as part of the pollutant load, and a number of persistent chlorinated and other halogenated compounds are of particular concern (hypothesis 8). These include CFCs 11, 12, and 113 (lifetimes between about 75 and 110 years, UNEP 1989), carbon tetrachloride (lifetime about 67 years), methylchloroform (lifetime about 8 years), and halon 1301 (lifetime 110 years). Nitrous oxides (mostly from sources of combustion and the application of fertilizers), also, can destroy ozone but mostly breakdown before being carried into the stratosphere.

In the stratosphere, chlorine (from CFCs and related long-life man-made chemicals) undergoes a catalytic reaction with ozone in the presence of sunlight, forming oxygen and chlorine monoxide. The chlorine monoxide disassociates and the process repeats with chlorine splitting many more molecules of ozone. The two processes of ozone formation and ozone depletion occur simultaneously but, because of the increasing emission of ozone-depleting gases and their persistence, rates of depletion exceed ozone formation. Ozone depletion is occurring, globally, but the process is most severe at high latitudes.

During each winter, atmospheric circulation tends to separate the polar air mass from the rest of global circulation (polar vortex) and clouds of ice-particles form at very low temperatures in the high latitude stratosphere. Because of landmass configuration, the polar vortex over the northern hemisphere is less stable than over the southern hemisphere. As the ozone-depleting gases breakdown, chlorine becomes attached to the ice particles in the stratosphere and this process results in the concentration of ozone-depleting substances over each pole during its winter period. With the advance of polar spring, sunlight triggers photochemical reactions in the stratosphere and the cycle of ozone depletion begins again. Largely because of the greater stability of the south polar vortex, ozone depletion is most severe over the Antarctic. Although ozone depletion appears to be 4-5 times more severe over the south pole than the north pole there has been a decrease of about 2.5 percent in stratospheric ozone cover between 1978 and 1987, in regions north of latitude 50° N. Ozone depletion is not permanent and, over time, natural regeneration will replace lost ozone. However, there will be a lag time of several decades between the effective control and reduction of critical emissions and increasing quantities of stratospheric ozone.

Based on selected literature (Rampino & Etkins, 1990; Ryan, 1992; Smith, 1989; Smith & Baker, 1979; Smith et al., 1992) stratospheric ozone in a layer 15 to 50 km above the Earth, shields life-forms from ultraviolet (UV) radiation. Solar radiation, within the UV band, is received as short (C), medium (B) and long wave (A) radiation. The UV-C is lethal to life-forms but it is almost entirely absorbed by the ozone layer. Very little of the UV-A is absorbed by the ozone layer but this long wave radiation is relatively harmless. Although UV-B radiation (280-320 nm) is somewhat less severe in its effect than UV-C, more of this radiation penetrates the ozone layer. Depletion of the stratospheric ozone is therefore of particular significance to the quantity of UV-B received at the Earth's surface (hypothesis 9). Increased UV-B (as a result of ozone depletion) may give rise to a wide range of health related problems in plants and animals (for example, sunburn, cataracts, skin cancer, dysfunction of the immune system such as reduced resistance to infectious disease, growth impairment and loss of fecundity).

A reduction of more than 50 percent has already occurred during the spring in the Antarctic ozone layer, with a significant increase in the penetration of UV-B radiation.

In clear water, UV-B penetrates deeply into the photic zone and photosynthesis by marine algae appears to have decreased by 6 to 12 percent. About 10 percent of the received UV-B will penetrate 2 m of ice and photosynthesis by ice algae appears to have decreased by about 5 percent. Although plants and animals (including humans) produce pigments to absorb UV-B radiation, the extent to which increased radiation can be tolerated is uncertain. Humans produce melanin, and many species of invertebrates, algae, and fish taken from Antarctic waters have mycosporine-like amino acids (MAA) that absorb UV-B and which may provide radiation shielding (Karentz et al., 1991). Comparable data from Arctic and sub-Arctic Canada are not available, but strong similarities are likely to exist between north and south high latitude environments.

A number of factors influence the temperature of the global atmosphere and, hence, climate. Eccentricities in the Earth's orbit around the sun and a slight wobble in the Earth's axis result in small but significant changes in received radiation, especially at the poles. These cycles (Milankovitch cycles) have periodicities which extend over extremely long time frames (from about ten thousand to one hundred thousand years). Based on astronomical observations, the Earth is shortly expected to enter a cooling phase (or to have already entered it). In addition, the loss of ozone is also thought to lead to stratospheric cooling. Ozone warms as it absorbs UV radiation and this raises the temperature of the stratosphere. With ozone depletion, this effect is reduced and the polar vortex may become more persistent (through feedback, this mechanism might also further increase the rate of ozone depletion and reduce temperature).

A counter effect is also occurring. As a result of human activities, a number of greenhouse gases are being released into the atmosphere and it is thought that their influence will override any immediate tendencies towards global cooling (hypothesis 10). The most important greenhouse gases include water vapour, carbon dioxide, ozone, methane, CFCs, and nitrous oxides. These gases tend to be transparent to short wave solar radiation but block long wave radiation re-emitted from the Earth's surface. The gases differ greatly in the degree of effect. For example, CFC 12 is about 10,000 times more effective as a greenhouse gas than carbon dioxide, and methane is 60 times more effective than carbon dioxide. Estimates suggest that carbon dioxide is responsible for more than half of the present trend in global warming, that CFCs (before breakdown) could contribute about 20 percent of the effect, and that methane may account for nearly 15 percent of the effect. Over the past century, human activities have led to a global warming of about 0.5°C and models predict that concentrations of atmospheric carbon dioxide will double by 2050. There will be a commensurate rise in the mean annual global temperature of between 1.5 and 4.5°C. Most of the increase in carbon dioxide will come from combustion of fossil fuels, especially coal. The rise in global temperature is expected to be uneven, with the greatest increase occurring at high latitudes.

It is expected that the release of methane will increase from peatlands and wetlands as a result of temperature induced microbial activity, and that rising temperature will also affect gas hydrates. Estimates suggest that levels of methane have increased considerably over the past three centuries and that during interglacial periods this gas contributed about one-quarter of the greenhouse effect of carbon dioxide. Methane breaks down in the atmosphere to produce water vapour. For every 1°C increase in temperature there will be a 6 percent increase in atmospheric water vapour.

One of the major uncertainties in estimating the degree of global warming which may result from the release of greenhouse gases is what will happen to the increasing amount of water vapour, how much will enter the stratosphere and how much will remain at lower levels in the atmosphere (where it would have a warming effect). It is unclear whether present warming trends are likely to develop conditions similar to those of the late Tertiary Period of geological history (Matthews, 1989). Potentially, more water vapour could result in the formation of more ice clouds in the stratosphere. If this happens, there could be significant stratospheric cooling and a loss in total received solar radiation at the Earth's surface. In relation to this, dimethyl sulphide (DMS) is produced by phytoplankton and may be an important precursor of cloud condensation nuclei in the marine atmosphere. Geological records indicate DMS was relatively high during glacial periods.

Increased precipitation is projected at high latitudes, based on analogy with palaeoclimatic information. If the precipitation occurred as snow, its high albedo would increase the amount of reflected radiation (McAllister, 1991a; 1991b). A result of this would be to further increase the effect of surface cooling. Smith (1990) suggests that changes could be expected to large areas of permafrost which are presently within 1 to 3°C of melting. Because permafrost conditions are influenced by vegetation, soil moisture and snow cover, it is difficult to predict the general response of the active layer or the extent to which feed-back mechanisms may occur. Surface organic layers can buffer cryogenic substrates from increasing air temperature and increased snow cover can insulate the ground from geothermal heat loss. Soils with low moisture contents warm more quickly than those with high moisture contents. Changes in the active layer will result in surface disruption and are likely to evolve over periods of decades to centuries. Responses may include increased rates of erosion, slope failure, ponding, soilifluction, creep and slumping.

Considerable uncertainties exist with regard to the extent of global warming that may occur and the stability of associated climatic conditions. Within recorded history, large variations in winter and summer temperature are already known to exist within the Hudson Bay basin (Wang *et al.*, 1991), as are considerable variations in precipitation. Prinsenberg (1991) notes that year to year differences in ice cover and thickness vary by about 30 percent about the mean.

6. Cumulative impact evaluation

At present, there is neither a generally accepted definition of cumulative impact evaluation nor a generally accepted means of quantifying cumulative impact (Beanlands *et al.*, 1985; Davies, 1991a, 1991b; Bunch & Reeves, 1992; Spaling & Smit, 1993). Essential components include a description of the current state of environment, a review of the extent to which past effects have led to the present, determination of thresholds (often not apparent until exceeded), and priorities for sustainable development (Peterson, 1992). Thermodynamics may help to provide a conceptual view of how ecosystems evolve and Kay (1991) and Schneider & Kay (1992) have discussed relations between thermodynamic branching and ecological thresholds. Meaningful indicators are required and focus should be given to environmental and socio-economic factors that are likely to be most significant (Schindler, 1987; CGRM, 1991). Here, cumulative impact evaluation is considered only at the conceptual level.

6.1 *Cumulative impact*

Cumulative impact describes the collective effects of many individual, multiple, and interactive forms of stress over time, on the ecosystems and environments of some geographically defined region. Stresses can be associated with many different forms of development (such as power production, manufacturing, resource extraction and agriculture) and domestic sources, including demands imposed by expansion of human populations. Growing populations do not necessarily impose additional stress. This is particularly evident where technological improvements and better understanding of the environment and ecosystem processes have been used to reduce both cause and effects of stress. With respect to the Hudson Bay region, concerns so far relate mostly to the additive effects of multiple development decisions rather than population growth.

6.2 *Ecosystem, environment and habitat*

The ecosystem describes interactions among living things and between them and their surroundings (biotic and abiotic components). It includes microbial forms, plants and animals, and it also includes human beings. The environment describes the totality of surroundings within which organisms or communities of organisms exist. Some organisms are found only in water (aquatic environment) and some only on land (terrestrial environment), and some are further specialized, for example to the freshwater aquatic environment or to even smaller microenvironments. Environment is descriptive of the physico-chemical media (air, land and water) and the nature of their interfaces.

Habitat is used to describe some smaller part of the environment in which specific characteristics are dominant, and current terminologies and the significance of structural groupings are discussed by Bailey *et al.* (1985). Physical features largely dominate habitat (for example, rivers, lakes, mountains, plateaux, wetlands, or ice-fields) and within each broadly defined habitat there may be numerous subtleties of form and exposure that provide specialized conditions which support particular forms of biological organization. Thus wetlands and estuaries for example, which because physico-chemical conditions usually favour high primary productivity, are frequently associated with breeding and nursery sites for birds, fish, and aquatic mammals.

6.3 *Sustainable development*

Following usage adopted by the United Nations World Commission on Environment and Development (Bruntland Commission; WCED 1987), sustainable development means that human activities and resource use should not diminish the abilities of future generations to sustain themselves. While simple in concept, this is an enormously complex issue to address in a meaningful way. Fig. 13 provides an illustration of the link between cumulative impact and sustainable development. If sustainable development is thought of as equivalent to the long term upper limit of sustainable cumulative impact, due to the combined effects of population growth and technological ability (demotechnic growth), three major scenarios arise:

1) Cumulative impact increases to the intercept with level (A). Here, there is sufficient reserve in the environment and ecosystem to accommodate variations imposed by natural forms of stress (such as that induced by climatic or seismic events) and this level of development is sustainable over the long term.

2) At level (B), which represents the impact of a further increase in population growth and technology, it may be possible to sustain the stress effects of occasional events. However, it is not possible to sustain this level of stress on a continuing basis.

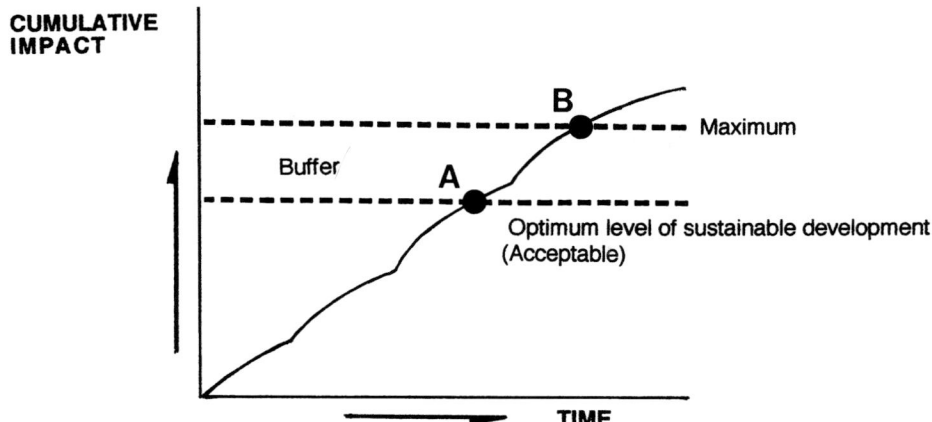

CUMULATIVE IMPACT AND SUSTAINABLE DEVELOPMENT

Fig. 13. Forms of Cumulative Impact (Source: Peterson, 1992).

3) If the effects of demotechnic growth result in a cumulative impact which exceeds level (B), over the long term, there will be some form of collapse or change since, from thermodynamic principles, the system can no longer sustain itself.

In presenting these scenarios it must be recognized that it is necessary to define whether we are concerned only with sustainable development at the global scale, or whether the idea is to be applied to some region of the globe such as a country or a continent. For example, by sustainable development, do we mean that the Hudson Bay ecosystem should remain entirely self-sufficient (including its human component)? As discussed further, it must be noted, also, that ecosystems are both resilient and adaptive to stress. This dynamic state means, therefore, that what we deem to be sustainable development based on the characteristics of the present environment and its ecosystem may not be reflected by some future state (of environment).

6.4 *System properties*

The evolution and behaviour of ecosystems conform to fundamental laws of science. Ecosystems may be described in terms of non-equilibrium thermodynamics. For example, if no development stress is present and the system functioned under natural conditions, system conditions would remain within a limited range. Likely, this would be characterized by strong biological integrity (Karr, 1991). However, with the application of some form of continued stress, there would be a shift along a thermodynamic branch. If driven past some threshold, a catastrophic event could occur and this would cause ecosystem components to change and follow another path along a new thermodynamic branch. The system would seek to reorganize itself to maximize the use of available energy and self-organization would occur within the food web.

6.5 *Hierarchical structures*

Relationships between organisms and environment may be described at different levels (Ryder & Kerr, 1989), for example at the level of an individual, a species, or a community (Fausch et al., 1990; Miller et al., 1988). Likewise, relationships can be described at various levels of habitat complexity (Busch & Sly, 1992), or across the aquatic environment of a large region such as Hudson Bay. In all ecosystems, long term sustainable productivity is less than the theoretical maximum calculated from the sum of the parts. This is due to the effects of many forms of natural temporal and geographic variation, such as climate, connectedness between habitat components, and patchiness in the distributions of both food and predators. In effect, it is necessary to insert some form of probability function between sets of information that describe the details of habitat and productivity at low levels of hierarchy, and those above. The differences between the deterministic models for lake trout and walleye production based on habitat suitability (McMahon et al., 1984; Marcus et al., 1984), and the empirical Morphoedaphic Index model for fish yield (Ryder, 1965, 1982) for example, substantiate this point.

Several recent developments in understanding regional morphology and the effect of loadings and loading dynamics may also lead to a better appreciation of cumulative impact model requirements at high levels in environmental hierarchy (Håkanson &Wallin, 1991; Håkanson et al., 1986; Keup, 1985; Nilsson & Håkanson, 1992; and Wallin, 1991). A recent attempt to link environmental and socio-economic factors in a large scale ecosystem model was recently reported by Sonntag et al. (1991). This model approach was developed for the Great Lakes basin and attempts to provide linkages between different spatial (and temporal) scales, within hierarchical structure.

6.6 *Environmental impact assessment*

Assessments of environmental impact and cumulative impact differ, although the distinction is not always clearly made, and of course the larger the project the more complex the assessment (Larkin, 1984). Assessments of environmental impact often occur at lower levels in the hierarchical structure of the global environment than cumulative impact. Also, by analogy, the sum of individual environmental impacts is not equal to their cumulative effect (which could well be greater). Figure 14 illustrates the general approach to environmental impact assessment which usually seeks to determine cause and the extent of effect (often as a basis for mitigative action).

Beanlands et al. (1885) raised points which may serve to further differentiate between assessments of environmental and cumulative impact. The study suggested several areas of concern. Environmental impact assessments, for example, do little to set regional objectives that reflect broad societal goals. Nor do these assessments fully recognize their role as precedent-setters. In addition, assessments must not only consider additive effects, but must also deal with the non-linear response of ecosystems.

While no single approach has general support, one example of a recent conceptual approach is given by Irwin & Rodes (1992), and a more detailed synoptic approach by Liebowitz et al. (1992). Emerging conceptual frameworks have been reviewed, most recently, by Spaling & Smit (1993) and these authors refer to two broad approaches, one scientific and the other planning-oriented. As noted by Lawrence (pers. comm., North/South Consultants Inc., Winnipeg, Manitoba), evaluation of environmental impact is applicable to projects and cumulative impact to systems. In large measure, both thoughts also reflect the level at which decision-making is taking place within the hierarchical structure of social, economic and environmental factors.

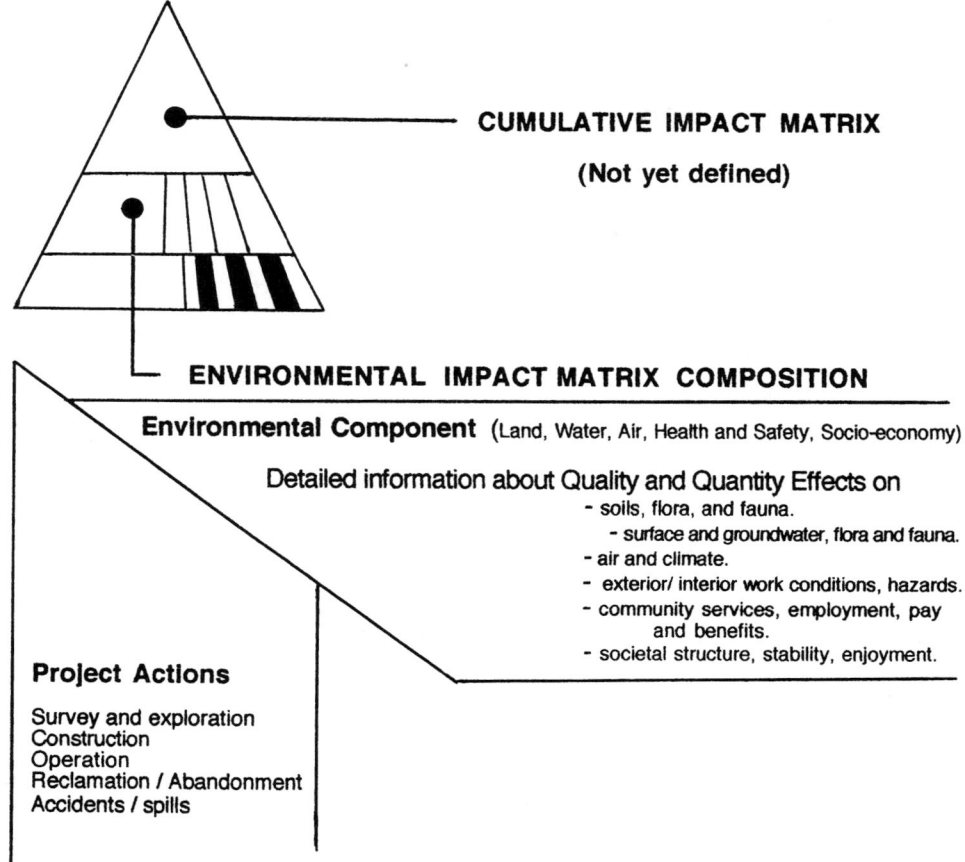

Fig. 14. Environmental and Cumulative Impact Assessment (Source: Wright & Greene, 1987).

6.7 *State of environment*

Ecological monitoring provides a basis for establishing the state of environment (Beanlands & Duinker, 1983; EC-DFO, 1992) but, to be effective, the scope of monitoring must be clearly defined and scaled. Without clear focus, the collection of information is likely to expand to consume any and all resources available. Figure 15 provides a summary of environmental trends in relation to some initial state. In terms of the natural environment, restoration is never truly possible since we can not return to the temporal origin. Through remediation, however, it may be possible to achieve something similar to the initial state. Enhancement describes any form of improvement, including that at previously degraded sites. Conservation describes a condition which is maintained over the long term but at something less than its original value. Preservation retains the present value which might, in some circumstances, be close to the original state.

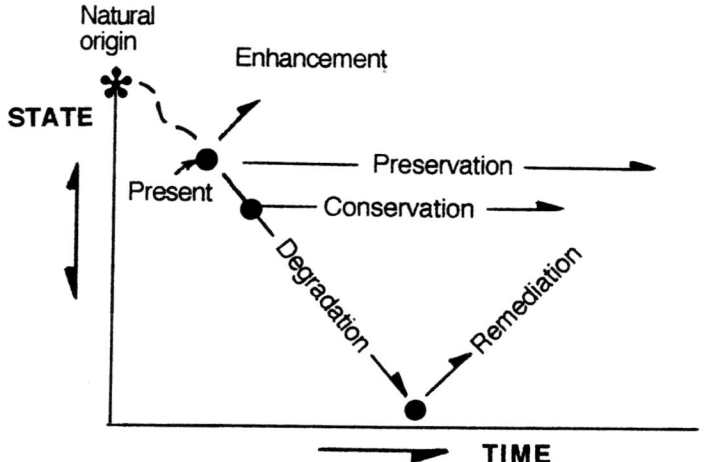

CHANGES IN STATE

Fig. 15. Changes in State (Source: Francis et al., 1979).

6.8 *Vision*

State of environment reporting is in its infancy and, so far, it is largely a qualitative tool. However, by using integrative measures appropriate to different levels of environmental hierarchy, it should be possible to achieve quantitative expressions against which the effects of cumulative impact can be assessed. Can this approach define and quantify sustainable development? On its own, clearly the answer is no. However, that is largely because the state of conditions in the environment, to be associated with sustainable development, has not been established. This may be termed a vision of the state of environment which is desired in various areas or regions. Obviously, the closer the vision to the natural environment ,the less cumulative impact that can be allowed. The establishment of a vision is an essential criterion for measuring sustainable development.

In the North American Great Lakes, for example, it was not until the concept of most sensitive use was applied that ambient water quality standards were defined and could be used to establish allowable nutrient loadings. The 1972 Great Lakes Water Quality Agreement and subsequent amendments have not only provided an early form of vision but caused the application of technology to reduce the effects of cumulative environmental impact. Application of emission control technology and improved process technology has significantly reduced many forms of environmental stress while, at the same time, the human population in the Great Lakes Basin has continued to increase. In short, it is necessary to know where "we" wish to be, before it is possible to assess how closely present conditions and trends match that goal (the vision).

The concept of a public charter in which not only environmental quality but desired/ undesired activities were defined would take the concept of vision closer to its prime application (RAAS, 1989).

7. Acknowledgements

This contribution has been prepared for the independent Hudson Bay Program of the Canadian Arctic Resources Committee (CARC), Ottawa, the Rawson Academy of Aquatic Science (RAAS), Ottawa, and the Environmental Committee of the Municipality of Sanikiluaq, NWT. The program is funded by donations, grants, contracts, and in-kind contributions of a number of agencies, including:

The Richard and Jean Ivey fund
The Harold Crabtree Foundation
Walter and Duncan Gordon Charitable Foundation
The McLean Foundation
George Cedric Metcalf Charitable Foundation
Helen McCrea Peacock foundation
Murphy Foundation Incorporated
John D. and Catherine T. MacArthur Foundation
The Molson Family Foundation
The Low-Beer Foundation
Government of the Northwest Territories
Indian and Northern Affairs Canada
Environment Canada
Environmental Innovation Programme
Grand Council of the Crees (Quebec)
Mushkegowuk Tribal Council
Manitoba Hydro
Ontario Hydro

Their interest and support for the program, and particularly this overview of existing scientific knowledge is gratefully acknowledged.

The preparation of this overview has greatly benefited from information, comments, and detailed reviews provided by many authorities with close experience and knowledge of the Hudson Bay region and its people. I am indebted to the valuable assistance of: Jim Bunch (Fisheries and Oceans Canada, Ottawa, Ontario), Milton Freeman (Circumpolar Institute, University of Alberta, Edmonton, Alberta), Murray Clamen (Conservation and Protection, Environment Canada, Ottawa, Ontario), Ian Dickson (Environmental Affairs, Manitoba Hydro, Winnipeg, Manitoba), Harvey Feit (Department of Anthropology, McMaster University, Hamilton, Ontario), Michaela Huard (Director, Habitat Operations, Fisheries and Oceans Canada, Ottawa, Ontario), Robbie Keith (University of Waterloo, Waterloo, Ontario), Mike Lawrence (North/South Consultants Ltd., Winnipeg, Manitoba), Pat LeBlanc (Director of Research, Federal Environmental Assessment and Review Office, Ottawa, Ontario), Louise Legendre (University of Laval, Ste Foy, Quebec), Danielle LeMessier (Quebec Hydro, Montreal, Quebec), Don McAllister (Canadian Museum of Nature, Ottawa, Ontario), Everett Peterson (President, Western Ecological Services Ltd., Victoria, British Columbia), Dick Preston, (Department of Anthropology, McMaster University, Hamilton, Ontario), Bruce Stewart (President , Arctic Biological Consultants, Winnipeg, Manitoba), Fred Roots (Senior Scientist, Emeritus, Environment Canada, Ottawa, Ontario), Paul Wilkinson (President, P. F. Wilkinson and Associates Inc., Montreal, Quebec), Ted Yuzyk (Conservation and Protection, Environment Canada, Ottawa, Ontario)

In particular, the views and knowledge contributed by the following reviewers have been most valuable in bringing the state of knowledge as up-to-date as possible, and in trying to present a balanced interpretation of available information and understandings: Fikret Berkes (Director, Natural Resources Institute, University of Manitoba, Winnipeg, Manitoba), Max Dunbar (Professor Emeritus, Department of Oceanography, McGill University, Montreal, Quebec), Miriam Fleming (Environmental Committee, Municipality of Sanikiluaq, Belcher Islands, Northwest Territories), Peter Martini (Department of Geology, Guelph University, Guelph, Ontario), Bob Milko (Biological Resources, Indian and Northern Affairs, Canada, Ottawa, Ontario), Ross Norstrom (Head Toxicology Research, Canadian Wildlife Service, Ottawa, Ontario), Alan Penn (Environment Project Manager, Grand Council of the Cree-Quebec, Montreal, Quebec), Simon Prinsenberg (Fisheries and Oceans Canada, Bedford Institute of Oceanography, Dartmouth, Nova Scotia), Chris Rimmer (Director Research, Vermont Institute of Natural Science, Woodstock, Vermont), Buster Welch (Fisheries and Oceans Canada, Resolute, Northwest Territories).

Without the willingness of so many expert and knowledgeable researchers, scientists and interested residents, this contribution would not have been possible. I would also like to express my appreciation to Glen OKrainetz, Manager of the Hudson Bay Program, whose particular skills which have done much to ensure financial support, to access community and expert resources, and to advance the preparation of this overview. The contributions by staff of CARC, RAAS, and other members of the Environmental Committee of Sanikiluaq and are also grateful acknowledged. It is rare that programs of this sort receive the close support of all levels of government, key industrial stakeholders, academia, and resident interests to the extent that has been achieved in this program.

References

Abraham, K. F. & G. H. Finney, 1986. Eiders of the eastern Canadian Arctic. In : A. Reed (ed.), Eider ducks in Canada, pp.55-73. Can. Wildl. Serv. Rept. Ser. 47. Ste-Foy, Quebec.

Adam, M., W. D. McKone & H. Shear, 1987. Report from a Workshop on Chemical Hazards to Fish and Fisheries. Can. Tech. Rept. Fish. Aquat. Sci. 1525, Ottawa, Ontario, 105p.

Adams, N. J., G. Power, & D. R. Barton, 1989. Diet and Daily Ration of Anadromous Arctic Charr in Ungava Bay. Physiol. Ecol. Japan, Spec. Vol. 1: 253-264.

Allan, R. J., M. Clamen, R. J. Daley, J. Moenig & B. Sadler, 1992. Toward an Ecosystem Approach to Research on the Hudson Bay Bioregion. National Water Research Institute Contribution 92-77, Burlington, Ontario, 19p.

Anderson, J. T. & J. C. Roff, 1980a. Seston Ecology of the Surface Waters of Hudson Bay. Can. J. Fish. Aquat. Sci. 37: 2242-2253.

Anderson, J. T. & J. C. Roff, 1980b. Subsurface Chlorophyll a Maximum in Hudson Bay. Naturaliste Can. 107: 207-213.

Andersson, A., A. Nilsson & L. Håkanson, 1991. Metal Concentrations of the Mor Layer. Swedish Environ. Protect. Agency Rept 3990, Solna, Sweden, 85p.

d'Anglejan, B., 1982. Patterns of Recent Sedimentation in the Eastmain Estuary, Prior to River Cut-off. Naturaliste Can. 109: 362-374.

Anon. 1989. Contaminants in Northern Ecosystems and Native Diets. Summary Rept. Evaluation Meeting, February 28-March 2, 1989, Ottawa. Dept. Indian and Northern Affairs, Ottawa, Ontario. 7p.

Anon. 1990. Arctic Pollution: How Much is Too Much? Northern Perspectives 18 (3): 1-8.

Anon. 1991, Madison's Canadian Lumber Directory. Madison's Canadian Lumber Reporter, Vancouver, B.C.

Anon. 1992. Lockwood-Post's Directory of the Pulp and Paper and Allied Trades. Miller Freeman Publications, San Francisco, California.

Arnold, C. D. 1982. Archeology in the Northwest Territories. Northern Perspectives 10 (6): 1-4.

Arragutainaq, L. & B. Fleming, 1991. Community-Based Observations on Sustainable Development in Southern Hudson Bay. Alternatives 18(2): 9-11.

Atkinson, E. G. & J. W. Wacasey, 1989a. Benthic Invertebrates Collected from Hudson Bay, Canada 1953-

1965. Can. Data Rept. Fish. Aquat. Sci. 744. Ottawa, Ontario. 125p.
Atkinson, E. G. & J. W. Wacasey, 1989b. Benthic Invertebrates Collected from Hudson Strait, Foxe Channel and Foxe Basin, Canada 1949-1970. Can. Data Rept. Fish. Aquat. Sci. 746. Ottawa, Ontario. 102p.
Ayles, G. B. 1987. Arctic Fisheries Research in Canada: An Informal Perspective. Northern Perspectives 15 (4): 13-15.
Bailey, R. G., S. Zoltai & E. B. Wiken, 1985. Ecological Regionalization in Canada and the United States. Geoforum 16: 265-275.
Baker, R. F. 1989. An Environmental Assessment and Biological Investigation of the Nelson River Estuary. Unpubl. Rept., prepared for Manitoba Hydro, North/South Consultants Ltd., Winnipeg, Manitoba, 179p.
Baker, R.F. & S. Davies, 1991. Physical, Chemical and Biological Effects of the Churchill River Diversion and Lake Winnipeg Regulation on Aquatic Ecosystems. Can. Tech. Rept. Fish. Aquat. Sci. 1806, 53p.
Barrie, L.A., D. Gregor, B. Hargrave, R. Lake, D. Muir, R. Shearer, B. Tracey & T. Bidleman, 1992. Arctic Contaminants: Sources, Occurrence and Pathways. Sci. Total Environ. 122: 1-74.
Beals, C. S. & D.A. Schenstone (eds), 1968. *Science, History and Hudson Bay*. Department of Energy Mines and Resources, Ottawa, Ontario, 2 Vols., 1050p.
Beanlands, G. E. & P. N. Duinker, 1983. An Ecological Framework for Environmental Assessment in Canada. Instit. Resource and Environ. Studies, Dalhousie University, Halifax, Nova Scotia.
Beanlands, G. E., W. J. Erkmann, G. H. Orians, J. O'Riordan, D. Policansky, M. H. Sadar & B. Sadler (eds), 1985. *Cumulative Environmental Effects: A Bi-national Perspective*. Cumulative Environmental Assessment Research Council (CEARC) and U. S. National Science Council Board on Basic Biology. Supply and Services Canada, Ottawa, Ontario. 175p.
Beattie, C. E. & K. R. Greenway, 1986. Offering Up Canada's North. Northern Perspectives 14(4): 5-8.
Bellrose, F. C., 1980. *Ducks, geese and swans of North America*. Stackpole Books, Harrisburg Pa. 544pp.
Berkes, F., 1990. Native Subsistence Fisheries: A Synthesis of Harvest Studies in Canada. Arctic 43: 35-42.
Berkes, F. & M. M. R. Freeman, 1986. Human Ecology and Resources Use. In: I. P. Martini (ed.) *Canadian Inland Seas*, Elsevier Oceanographic Ser. 44, Elsevier, New York, p 425-455.
Berkes, F., P.George & R. Preston, 1991a. Co-Management: The Evolution of the Theory and Practice of Joint Administration of Living Resources. TASO Research Rept. Second Ser., No. 1. McMaster University, Hamilton, Ontario.
Berkes, F., P.George & R. Preston, 1991b. Co-Management. Alternatives 18(2): 12-18.
Berkes, F., P.George, R. Preston & J. Turner, 1992a. The Cree View of Land and Resources. TASO Rept., Second Ser. No. 8. McMaster University, Hamilton, Ontario. 28p.
Berkes, F., P. George, R. Preston, J. Turner, A. Hughes, B. Cummins & A. Haugh, 1992b. Wildlife Harvests in the Mushkegowuk Region. TASO Rept. Second Ser. No. 6. McMaster University, Hamilton, Ontario.
Bjornberg, A., L. Håkanson & K. Lundbergh, 1988. A Theory on the Mechanisms Regulating the Bioavailability of Mercury in Natural Waters. Environ. Poll. 49: 53-61.
Bodaly, R.A. & T.A. Johnson, 1992. The Mercury Problem in Hydro-Electric Reservoirs with Predictions of Mercury Burdens in Fish in the Proposed Grande Baleine Complex, Québec. Hydro-Electric Development: Environmental Impacts. Paper No. 3. North Wind Information Services, Inc. 15p.
Bodaly, R. A., J. D. Reist, D. M. Rosenberg, P. J. McCart & R. E. Hecky, 1989. Fish and Fisheries of the Mackenzie and Churchill River Basins, Northern Canada. In: D. P. Dodge (ed.) Proceedings of the International Large Rivers Symposium (LARS), pp 128-144. Can. Spec. Publ. Fish. Aquat. Sci. 106.
Boivin, T. G., G. Power & D. R. Barton, 1989. Biological and Social Aspects of an Inuit Winter Fishery for Arctic Charr (*Salvelinus alpinus*). Physiol. Ecol. Japan, Spec. Vol. 1: 653-672.
Bousefield, E. L., 1955. The Cirripede Crustacea of the Hudson Strait Region, Canadian Eastern Arctic. J. Fish. Res. Bd. Canada. 12: 762-767.
Boyd, H., G. E. J. Smith & F. G. Cooch, 1982. The Lesser Snow Geese in the eastern Canadian Arctic. Can. Wildl. Serv. Occasional Paper No.46. 23pp.
Braune, B. M., R. J. Norstrom, M. P. Wong, B. T. Collins & J. Lee, 1991. Geographical Distribution of Metals in Livers of Polar Bears From the Northwest Territories, Canada. Sci. Total Environ. 100: 283-299.
Brouard, D., C. Demers, R. Lalumiere, R. Schetagne & R. Verdon, 1990. Evolution of Mercury Levels in Fish of the La Grande Hydroelectric Complex, Québec. Summary Report, Hydro Québec (Montreal) and Schooner Inc., (Québec), P.Q., 97p.
Brousseau, C. S. & G. A. Goodchild, 1989. Fisheries and Yields in the Moose River Basin, Ontario. In : D. P. Dodge (ed.) Proceedings of the International Large Rivers Symposium (LARS), pp 145-158. Can. Spec. Publ. Fish. Aquat. Sci. 106.
Bunch, J. N. & R. R. Reeves, 1992. Proceedings of a Workshop on the Potential Cumulative Impacts of Development in the Region of Hudson and James Bays, 17-19 June 1992. Can. Tech. Rept. Fish. Aquat. Sci. 1874, Ottawa, Ontario, 39p.
Burnham, C. D., 1990. Annual Environmental Performance Report. Ontario Hydro. Toronto, Ontario, 78p.
Busch, W.-D. N.& P. G. Sly (eds), 1992. *The Development of an Aquatic Habitat Classification System for Lakes*. CRC Press. Boca Raton, Florida, 225p.

(CM-MA) Canada-Manitoba Mercury Agreement, 1987. Summary Report on the Study and Monitoring of Mercury in the Churchill River Diversion. Governments of Canada and Manitoba, Ottawa, and Winnipeg.

(C-OSC) Canada-Ontario Steering Committee, 1983. Mercury Pollution in the Wabigoon-English River System of Northwestern Ontario, and Possible Remedial Measures. Summary of the Tech. Rept., Canada and Ontario, Ottawa and Toronto, 18p.

Canadian Mines Handbook, 1992. Energy Mines and Resources, Ottawa, Ontario.

(CPPA) Canadian Pulp and Paper Association, 1990. Pulp and Paper from Canada. Trade Directory, CPPA, Montreal, Québec, 42p.

Chenier, M., 1991. James Bay. Earthkeeper 1(5): 30-33, 36.

Clancy, P., 1987. The Making of Eskimo Policy in Canada, 1952-62: The Life and Times of the Eskimo Affairs Committee. Arctic 40: 191-197.

Cohen, S.J., 1991. Possible Impacts of Climatic Warming Scenarios on Water Resources in the Saskatchewan River Sub-Basin, Canada. Climatic Change 19: 291-317.

Comeau, N., 1915. Report on the Fisheries Expedition to Hudson Bay in the Auxiliary Schooner "Burleigh" 1914. Dept. Naval Service, Ottawa, Ontario.

Conservation and Protection, 1992. Detailed Envirodat Data Report (Water Quality of Tributaries to Hudson Bay). Unpubl. data summary, Environment Canada, Ottawa, Ontario.

Conservation et Protection, (Undated). Carte: Aménagement Hydro-Electriques, Baie-James - Baie d'Hudson. Region du Québec, Environnement Canada, Québec, P. Q.

(CGLRM) Council of Great Lakes Research Managers, 1991. A Proposed Framework for Developing Indicators of Ecosystem Health for the Great Lakes Region. Internat. Joint Comm., Windsor, Ontario, 47p.

Crossman, E. D. & D. E. McAllister, 1985. Zoogeography of Freshwater Fishes of the Hudson Bay Drainage, Ungava Bay and Arctic Archiplego. In : E. Wiley and C. H. Hocutt (eds). *Zoogeography of the Freshwater Fishes of North America*, pp 53-104. J. Wiley and Sons, New York, New York.

Davies, K., 1991a. Cumulative Environmental Effects: A Compendium. Unpubl. Rept. for Federal Environmental Assessment Review Office by Ecosystems Consulting Inc., Orleans, Ontario.

Davies, K., 1991b. Health and Environmental Impact Assessment in Canada. Can. J. Public Health 82: 19-21.

Dewailly, E., A. Nantel, J.-P. Weber & F. Meyer, 1989. High Levels of PCBs in Breast Milk of Inuit Women from Arctic Québec. Bull. Environ. Contam. Toxicol. 43: 641-646.

Dickason, O. P., 1992. *Canada's First Nations: A History of Founding Peoples from Earliest Times.* McClelland & Stewart. Toronto, Ontario.

Dignard, N., R. Lalumière, A. Reed & M. Julien, 1991. Habitats of the northeast coast of James Bay. Can. Wildl. Serv. Occasional Paper No. 70. 27p.

Donaldson, J., 1984. 1982 Wildlife Harvest Statistics for the Baffin Region, N. W. T. Baffin Region Inuit Assoc., Tech. Rept. 2, 64p.

Douglas, R. J. W. (ed.), 1969. *Geology and Economic Minerals of Canada.* Economic Geology Rept. 1, Geological Survey of Canada, Ottawa, Ontario, 838p.

Drolet, R., L. Fortier, D. Ponton & M. Gilbert, 1991. Production of Fish Larvae and Their Prey in SubArctic Southeastern Hudson Bay. Mar. Ecol. Prog. Ser. 77: 105-118.

Drinkwater, K. F., 1986. Physical Oceanography of Hudson Strait and Ungava Bay. In :I. P. Martini (ed.) *Canadian Inland Seas*, pp 237-264. Elsevier Oceanographic Ser. 44, Elsevier, New York.

Dunbar, M. J., 1954a. A Note on Climatic Change in the Sea. Arctic 7 (1): 27-30.

Dunbar, M. J., 1954b. The Amphipod Crustacea of Ungava Bay, Canadian Eastern Arctic. J. Fish. Res. Bd. Canada. 11: 709-798.

Dunbar, M. J., 1958. Physical Oceanographic Results of the Calanus Expeditions in Ungava, Frobisher Bay, Cumberland Sound, Hudson Strait and Northern Hudson Bay, 1947-55. J. Fish. Res. Bd. Canada, 15: 155-174.

Dunbar, M.J., 1988. Marine Regional Boundaries in the Canadian North: Foxe Basin, Hudson Strait, Hudson Bay and James Bay. Unpubl. Manuscript Rept. prepared for the Canadian Parks Service, Ottawa. 43p.

Dunbar, M.J. & H. H. Hildebrand, 1952. Contribution to the Study of Fishes in Ungava Bay. J. Fish. Res. Bd. Canada. 9: 83-128.

(EWG) Ecoregions Working Group, 1989. Ecoclimatic Regions of Canada: First Approximation. Conservation and Protection, Environment Canada, Ottawa, Ontario, 118p.

Elmgren, R., 1989. Man's Impact on the Ecosystem of the Baltic Sea: Energy Flows Today and at the Turn of the Century. Ambio 18: 326-332.

Environment Canada, 1991. *The State of Canada's Environment.* Supply and Services Canada, Ottawa, Ontario.

(EC-DFO) Environment Canada and Department of Fisheries and Oceans, 1992. Federal Ecological Monitoring Program. Summary and Final Rept. Ottawa, Ontario. 2 Vol.

Evans, D. O., J. Brisbane, J. M. Casselman, K. E. Coleman, C. A. Lewis, P. G. Sly, D. L. Wales & C. C. Wilcox, 1991. Anthropogenic Stressors and Diagnosis of Their Effects on Lake Trout Populations in Ontario Lakes. Lake Trout Synthesis Response to Stress Working Group. Ontario Min. Nat. Res. Toronto. 115p.

Fausch, K. D., J. Lyons, J. R. Karr & P. L. Angermeier, 1990. Fish Communities as Indicators of Environmental Degradation. Amer. Fish. Soc. Symp. 8: 123-144.

Fein, G. G., J. L. Jacobson, S. W. Jacobson, P. M. Schwartz & P. M. Dowler, 1984. Prenatal Exposure to Polychlorinated Biphenyls: Effects on Birth Size and Gestational Age. J. Pediatrics 105: 315-320.
Feit, H. A., 1988. Self-Management and State Management: Forms of Knowing and Managing Northern Wildlife. In : M. M. R. Freeman & L. N. Carbyn (eds), *Traditional knowledge and renewable resource management in northern regions.* pp.72-91. Boreal Institute for Northern Studies, Edmonton, Alberta.
Forestry Canada, 1991. The State of Canada's Forests. Forestry Canada, Ottawa, Ontario, 85p.
Francis, G. R., J. J. Magnuson, H. A. Regier & D. R. Talhelm, 1979. Rehabilitating Great Lake Ecosystems. Great Lakes Fishery Comm. Tech. Rept. 37, 99p.
Francis, D. & T. Morantz, 1983. *A History of the Fur Trade in Eastern James Bay 1600-1870.* McGill-Queen's University Press, Kingston and Montreal. 203p.
Freeman, M. M. R. (ed.), 1981. Renewable Resources and the Economy of the North, Proc. First Internat. Symp. Renewable Resources and the Economy of the North, May 1981, Banff, Alberta. Assoc. Can. Univ. for Northern Studies, Ottawa, Ontario.
Freeman, M. M. R., 1988. Environment, Society and Health Quality of Life Issues in the Contemporary North. Arctic Med. Res. 47 (Supp. 1): 54-59.
Freeman, M. M. R. 1989. Graphs and Gaffs: A Cautionary Tale in the Common Property Resource Debate. In : F. Berkes (ed.) *Common Property Resources: Ecology and Community-based Sustainable Development.* pp 92-109. Belhaven Press, London.
Gamble, R. L., 1984. A Preliminary Study of the Native Harvest of Wildlife in the Keewatin Region, North West Territories. Can. Tech. Rept. Fish. Aquat. Sci. 1282, Ottawa, Ontario, 48p.
Gaston, A.J. ,1985a. Energy Invested in Reproduction by Thick-Billed Murres (*Uria lomvia*). Auk 102: 447-458.
Gaston, A.J., 1985b. The Diet of Thick-Billed Murre Chicks in the Eastern Canadian Arctic. Auk. 102: 727-734.
(GSC) Geological Survey of Canada, 1991. Map 900A, Principal Mineral Areas of Canada. Department of Energy Mines and Resources, Ottawa, Ontario.
George, P.J., 1989. Native Peoples and Community Economic Development in Northern Ontario. British Canadian Studies 4: 58-73.
George, P.J. & R.J. Preston, 1987. "Going in Between": The Impact of European Technology on the Work Patterns of the West Main Cree of Northern Ontario. Economic History 47: 447-460.
Gilbert, M., L. Fortier, D. Ponton & R. Drolet, 1992. Feeding Ecology of Marine Fish Larvae Across the Great Whale River Plume in Seasonally Ice-covered Southeastern Hudson Bay. Mar. Ecol. Prog. Ser. 84: 19-30.
Gosselin, M., L. Legendre, S. Demers & R.G. Ingram, 1985. Response of Sea Ice Microalgae to Climatic and Fortnightly Tidal Energy Inputs (Manitounuk Sound, Hudson Bay). Can. J. Fish. Aquat. Sci. 42: 999-1006.
Gosselin, M., L. Legendre, J.-C. Therriault, S. Demers & M. Rochet, 1986. Physical Control of the Horizontal Patchyness of Sea-Ice Microalgae. Mar. Ecol. Prog. Ser. 29: 289-298.
Gouvernement du Québec, 1984. Les Régions Naturelles du Québec. Unpubl. Map. Ministère du Loisir, de la Chasse et de la Pêche, Québec, P.Q.
Gouvernement du Québec, 1992. La Nature en Hértage: Plan d'Action sur les Parcs. Ministère du Loisir, de la Chasse et de la Pêche, Québec, P.Q., 22p.
Government of Manitoba, 1986. A Systems Plan for Manitoba's Provincial Parks. Technical Report, Department of Natural Resources. Winnipeg, Manitoba, 63p.
Government of Ontario, 1992a. Endangered Spaces Action Plan: Protecting Natural Heritage Areas. Ministry of Natural Resources, Provincial Parks and Natural Heritage Policy Branch, Peterborough, Ontario, 8p.
Government of Ontario, 1992b. Ontario Provincial Parks Guide. Ministry of Natural Resources, Toronto, Ontario.
Government of Ontario, 1992c. Ontario Provincial Parks Statistics. Report by Ministry of Natural Resources, Toronto, Ontario, 69p.
Government of Ontario, (undated). Ontario Provincial Parks. Unpubl. Map. Ministry of Natural Resources, Toronto, Ontario.
Government of the Northwest Territories, 1989. Explorers' Guide and Map. Travel Arctic and Public Works and Highways, Yellowknife, NWT.
Grainger, E.H., 1959. The Annual Oceanographic Cycle at Igloolik in the Canadian Arctic. 1. The Zooplankton and Physical and Chemical Observations. J. Fish. Res. Bd. Canada 16: 453-501.
Grainger, E.H. 1963. Copepods of the Genus *Calanus* as Indicators of Eastern Canadian Waters. In : M.J. Dunbar (ed.), *Marine Distributions*, pp.68-94. Roy. Soc. Canada. Spec. Publ. 5., University of Toronto Press, Toronto.
Grainger, E.H., 1982. Factors Affecting Phytoplankton Stocks and Primary Productivity at the Belcher Islands, Hudson Bay. Naturaliste Can. 109: 787-791.
Grant, S. D., 1991. A Case of Compound Error: The Inuit Resettlement Project, 1953, and the Government Responce, 1990. Northern Perspectives 19 (1): 3-29.
Gregor, D. J., 1990. Water Quality Research. In : T. D. Prowse and C. S. L. Omanney (eds), Northern Hydrol-

ogy: Canadian Perspectives. Science Rept. 1, National Hydrology Insitiute, Environment Canada, Saskatoon, Sask., p163-186.

Gregor, D. J. & W. D. Gummer, 1989. Evidence of Atmospheric Transport and Deposition of Organochlorine Pesticides and Polychlorinated Biphenyls in Canadian Arctic Snow. Environ. Sci. Technol. 23: 561-565.

Greig, L. A., J. K. Pawley, C. H. R. Wedels, P. Bunnell & M. J. Rose, 1992. Hypotheses of Effects of Development in the Moose River Basin: Workshop Summary. Unpubl. Rept. for the Department of Fisheries and Oceans, ESSA Environmental and Social Systems Analysts Ltd., Richmond Hill, Ontario.

Grenon, J. F., 1982. The Macrobenthic Fauna of the Eastmain Estuary (James Bay, Quebec). Naturaliste Can. 109: 793-802.

Guilbault, R. A., W. D. Gummer & V. T. Chacko, 1979. The Churchill Diversion: Water Quality Changes in the Lower Churchill and Burntwood Rivers. Water Quality Interpretative Rept. 2, Inland Waters Directorate, Environment Canada, Regina, Sask., 9p.

Hachey, H. B., 1931. Biological and Oceanographic Conditions in Hudson Bay. 2, Report on the Hudson Bay Fisheries Expedition of 1930. A. Open Water Investigations With the S. S. Loubryne. Contrib. Can. Biol. Fish., N. S., 6: 465-471.

Håkanson, L. & M. Wallin, 1991. An Outline of Econometric Analysis to Establish Load Diagrams for Nutrients/Eutrophication. Environmetrics 2: 49-68.

Håkanson, L., H. Kvarnas & B. Karlsson, 1986. Coastal Morphometry as Regulator of Water Exchange - a Swedish Example. Estur. Coast. Shelf. Sci. 23: 873-887.

Håkanson, L., A. Nilsson & T. Andersson, 1988. Mercury in Fish in Swedish Lakes. Environ. Poll. 49: 145-162.

Harrington, F.H., 1991. Caribou Populations in the James Bay Region. Hydro-Electric Development: Environmental Impacts. Paper No. 1. North Wind Information Services, Inc. Montreal. 15p.

Harrison, S. E. & J. F. Klaverkamp, 1990. Metal Contamination in Liver and Muscle of Northern Pike (*Esox lucius*) and White Sucker (*Catosomus commersoni*) and in Sediments from Lakes Near the Smelter at Flinn Flon, Manitoba. Environ. Toxicol. Chem. 9: 941-956.

Hearnden, K.W., S.V. Millson & W.C. Wilson, 1992. A Report on the Status of Forest Regeneration. Independent Forest Audit Committee. Ontario Ministry of Natural Resources, Forest Audit Secretariat, Sault Ste. Marie, Ontario, 117p.

Hecky, R. E. & S. J. Guildford, 1984. Primary Productivity of Southern Indian Lake Before, During, and After Impoundment and Churchill River Diversion. Can. J. Fish. Aquat. Sci. 41: 591-604.

Hecky, R. E., R. W. Newbury, R. A. Bodaly, K. Patalas & D. M. Rosenberg, 1984. Environmental Impact Prediction and Assessment: The Southern Indian Lake Experience. Can. J. Fish. Aquat. Sci. 41: 720-732.

House of Commons, 1991. Pursuant to Standing Order 108 (2), a Study on Sustainable Energy and Mineral Development: A Realistic Response to the Environmental Challenges. Minutes of Proceedings and Evidence of the Standing Committee on Energy, Mines and Resources. Ottawa, Ontario, Third Session, Thirty-fourth Parliament (November 26-28, 1991), Issue 8.

Hultunen J., P. Varo, G. Aifthan & A. Aro, 1992. Nationwide Selenium Supplementation in Finland. Abstr. Fifth Internat. Symp. Selenium in Biology and Medicine, July 20-23, 1992, Vanderbilt University, Nashville, Tennessee.

Hunter, J. G., S. Leach, D. E. McAllister & M. Steigerwald, 1984. A Distributional Atlas of Records of the Marine Fishes of Arctic Canada in the National Museums of Canada and Arctic Biological Station. Syllogeus National Museums of Canada 52. Ottawa, Ontario. 35p.

Indicators Task Force, 1991. A Report on Canada's Progress Towards a National Set of Environmental Indicators. Environment Canada, State of Environment Rept. 91-1, Ottawa, Ontario, 98 p.

Ingram, R. G. & P. Larouche, 1987a. Changes in the Under-Ice Characteristics of La Grande Riviére Plume Due to Discharge Variations. Atmos. Ocean. 25: 242-250.

Ingram, R. G. & P. Larouche, 1987b. Variability of an Under-Ice River Plume in Hudson Bay. J. Geophys. Res. 92(C9): 9541-9547.

Ingram, R. G., B.F. d'Anglejan, S. Lepage & D. Messier, 1986. Changes in Current Regime and Turbidity in Response to a Freshwater Pulse in the Eastmain Estuary. Estuaries 9(4B): 320-325.

Ingram, R. G., L. Legendre, Y. Simard & S. Lepage, 1985. Phytoplankton Response to Freshwater Runoff: The Diversion of the Eastmain River, James Bay. Can. J. Fish. Aquat. Sci., 42: 1216-1221.

Ingram, R. G., J.C. Osler & L.Legendre, 1989. Influence of Internal Wave Induced Vertical Mixing on Ice Algal Production in a Highly Stratified Sound. Estuarine Coast. Shelf Sci. 29: 435-446.

Irwin, F. & B. Rodes, 1992. Making Decisions on Cumulative Environmental Impacts: A Conceptual Framework. World Wildlife Fund, Washinton, D. C.

Irwing, C., 1989. Lords of the Arctic: Wards of the State. Northern Perspectives 14(4): 2-12.

Jackson, T. A., 1987. Methylation, Demethylation, and Bio-accumulation of Mercury in Lakes and Reservoirs of Northern Manitoba, with Particular Reference to Effects of Environmental Changes Caused by the Churchill-Nelson River Diversion. In : Summary Rept., Canada-Manitoba Agreement on the Study and Monitoring of Mercury in the Churchill River Diversion. Tech. Appendices, Vol. 2, Governments of Canada and Manitoba, Ottawa, and Winnipeg.

Jackson, T. A., 1989. The Influence of Clay Minerals, Oxides, and Humic Matter on the Methylation and Demethylation of Mercury by Micro-organisms in Freshwater Sediments. Applied Organometal. Chem. 3: 1-30.

Jacobson, J., S. G. Fein, J. Jacobson, P. Schwartz & J. Dowler, 1985. The Effect of Intrauterine PCB Exposure on Visual Recognition Memory. Child Devel. 56: 853-869.

Jacobson, J., S. Jacobson & H. Humphrey, 1990. Effects of in Utero Exposure to Polychlorinated Biphenyls and Related Contaminants on Cognitive Functioning in Young Children. J. Pediatrics, 116: 38-45.

James Bay Mercury Committee, 1990. Report of Activities 1988-1989. Montreal, Québec, 16p.

James Bay Mercury Committee, 1992. Report of Activities 1990-1991. Montreal, Québec, 16p.

Jarman, W. M., M. Simon, R. J. Norstrom, S. A. Burns, C. A. Bacon, B. R. T. Simonelt & R. W. Risebrough, 1992. Global Distribution of Tris (4-Chlorophenyl) Methenol in High Trophic Level Birds and Mammals. Environ. Sci. Technol. 26: 1770-1774.

Jensen, J., 1991. State of the Arctic Environment Report on Organo-chlorines. Unpubl. Rept. Dept. Indian Affairs and Northern Development, Ottawa, Ontario, 31p.

Johnson, M. G., 1987. Trace Element Loadings of Sediments of Fourteen Ontario Lakes and Correlations with Concentrations in Fish. Can. J. Fish. Aquat. Sci. 44: 3-13.

Johnson, R. D., F. R. Joubin, S. J. Nelson & E. Olsen, 1986. Mineral Resources. In : I. P. Martini (ed.) *Canadian Inland Seas*, pp. 387-402. Elsevier Oceanographic Ser. 44, Elsevier, New York.

Karentz, D., F. S. Mc Euen, M. C. Land & W. C. Dunlap, 1991. Survey of Mycosporine-like Amino Acid Compounds in Antarctic Marine Organisms: Potential Protection from Ultraviolet Exposure. Mar. Biol. 108: 157-166.

Kari, T. & P. Kauranen, 1978. Mercury and Selenium Contents of Seals from Fresh and Brackish Waters in Finland. Bull. Environ. Contam. Toxicol. 19: 273-280.

Karr J. R., 1991. Biological Integrity: A Long-Neglected Aspect of Water Resource Management. Ecological Applications 1: 66-84.

Kay, J. J., 1991. A Nonequilibrium Thermodynamic Framework for Discussing Ecosystem Integrity. Environ. Management 15: 483-495.

Keith, R. F., A. Kerr & R. Vles, 1981. Mining in the North. Northern Perspectives 9(2): 1-7.

Keith, R. F., T. Fenge, P. Jacobs & S. J. Woods, 1987. Arctic Fisheries: New Approaches for Troubled Waters. Northern Perspectives 15(4): 1-5.

Keup, L. E., 1985. Flowing Water Resources. Water Resources Bull. 21: 291-296.

Kinloch, D., H. Kuhnein & D.C.G. Muir, 1992. Inuit Foods and Diet: A Preliminary Assessment of Benefits and Risks. Sci. Total Environ. 122: 247-278.

Klaverkamp, J. F., D. A. Hodgins & A. Lutz, 1983. Selenite Toxicity and Mercury-Selenium Interactions in Juvenile Fish. Arch. Environ. Contam. Toxicol. 12: 405-413.

Krotz, L., 1991. Dammed and Diverted. Canadian Geographic (Feb/Mar): 36-44.

Larkin, P. A., 1984. A Commentary on Environmental Impact Assessment for Large Projects Affecting Lakes and Streams. Can. J. Fish. Aquat. Sci. 41: 1121-1127.

Lawrence, M.J., M. Paterson, R.F. Baker & R. Schmidt, 1992. Report on the Workshop Examining the Potential Effects of Hydroelectric Development on Beluga of the Nelson River Estuary, Winnipeg, Manitoba, November 6 and 7, 1990. Can. Tech. Rept. Fish. Aquat. Sci. 1838. 39p.

Legendre, L., R. G. Ingram & M. Poulin, 1981. Physical Control of Phytoplankton Under Sea Ice (Manitounuk Sound, Hudson Bay). Can. J. Fish. Aquat. Sci. 38: 1385-1392.

Legendre, L., R. G. Ingram & Y. Simard, 1982. Aperiodic Changes of Water Column Stability and Phytoplankton in an Arctic Coastal Embayment, Manitounuk Sound, Hudson Bay. Naturaliste Can. 109: 775-786.

Legendre, L. & Y. Simard, 1978. Dynamique Estivale du Phytoplankton dans l'Estuaire de la Baie de Rupert (Baie James). Naturaliste Can. 105: 243-258.

Legendre, L. & Y. Simard, 1979. Oceanographie Biologique et Phytoplancton dans le Sud-est de la Baie d'Hudson. Mar. Biol. 52: 11-22.

Leibowitz, S. G., B. Abbruzzese, P.-R. Adamus, L. E. Hughes & J. T. Irish, 1992. A Synoptic Approach to Cumulative Impact Assessment: A Proposed Methodology. U. S. Environ. Protect. Agency, Rept USEPA/600/R. 92/167. Washinton, D. C.

Lepage, S. & R.G. Ingram, 1986. Salinity Intrusion in the Eastmain River Estuary Following a Major Reduction of Freshwater Input. J. Geophys. Res. 91(C1): 909-915.

Lepage, S. & R.G. Ingram, 1991. Variation of Upper Layer Dynamics During Breakup of the Seasonal Ice Cover in Hudson Bay. J. Geophys. Res. 96 (C7): 12,711-12,724.

Lindsey, A., 1991. Manitoba Megaproject, the Conawapa Dam, Electric Northland. Earthkeeper 2(1): 34-38.

Lockhart, W.L., R. Wagemann, B. Tracey, D. Sutherland & D.J. Thomas, 1992. Presence and Implications of Chemical Contaminants in the Freshwaters of the Canadian Arctic. Sci. Total Environ. 122: 165-245.

Lower, A. R. M., 1915. A Report on the Fish and Fisheries of the West Coast of James Bay. Ann. Rept. Can. Dept. Naval Serv. 1913-1914. Appendix, pp. 29-67.

Lubinsky, I., 1980. Marine Bivalve Molluscs of the Canadian Central and Estern Arctic: Faunal Composition

and Zoogeography. Can. Bull. Fish. Aquat. Sci. 207. 111p.
McAllister, D. E., 1990. List of Fishes of Canada. Syllogues National Museums of Canada 64. Ottawa, Ontario. 320p.
McAllister, D. E., 1991a. Measuring Arctic Ocean Icepack Shrinkage and Warming: The Planetary Solar Mirror and Global Thermostat. Sea Wind 5(4): 19-24.
McAllister, D. E., 1991b. Questions on Ocean Impacts of James and Hudson Bay Hydro Projects. Sea Wind 5(3): 22-30.
McAllister, D. E. & M. B. Steigerwald, 1986. Bibliography of the Marine Fishes of Arctic Canada 1771-1985. Can. Manuscript Rept Fish. Aquat. Sci. 1909. Ottawa, Ontario. 108p.
McCart, P. J. & J. Den Beste, 1979. Aquatic Resources of the Northwest Territories. Rept. for the Science Advisory Board of the Northwest Territories, Yellowknife, Northwest Territories, 56p.
McCrea, R. C. & G. M. Wickware, 1986. Organochlorine Contaminants in Peatlands and Selected Estuary Sites of the Moose River. Water Poll. Res. J. Can., 21: 251-256.
McGhee. R. ,1987. Climate and People in the Prehistoric Arctic. Northern Perspectrives 15 (5): 13-15.
McMahon, T. E., J. W. Terrell & P. C. Nelson, 1984. Habitat Suitability Information: Walleye. FWS/OBS-82/10.56. U. S. Fish and Wildlife Service, Washington, D. C.
Makivik Corporation, 1991. Contaminants in the Marine Environment of Nunavik. Proc. Conf. on Contaminants in the Marine Environment of Nunavik, September 12-14, 1990, Montreal. Makivik Corp., Montreal, Canada, 101p.
Manitoba Hydro, 1992. The Hydro Province. Fact Sheet 92-08, Manitoba Hydro, Winnipeg, Manitoba.
Mansfield, A. W., 1990. Marine Mammals. In : J. A. Percy (ed.) Proceedings of a Workshop: Marine Ecosystem Studies in Hudson Strait. November 9-11, 1989, pp 134-137. Can. Tech. Rept. Fish. Aquat. Sci. 1770. Ottawa, Ontario.
Marcus, M. D., W. A. Hubert & S. H. Anderson, 1984. Habitat Suitability Index Models: Lake Trout (Exclusive of the Great Lakes). FWS/OBS-82/10.84. U. S. Fish and Wildlife Service, Washington, D. C.
Markham, W. E., 1986. The Ice Cover. In : I. P. Martini (ed.) *Canadian Inland Seas*, pp.101-116. Elsevier Oceanographic Ser. 44, Elsevier, New York.
Martini, I. P. (ed.), 1986a. *Canadian Inland Seas*. Elsevier Oceanographic Ser. 44, Elsevier, New York, 494p.
Martini, I. P., 1986b. Coastal Features of Canadian Inland Seas. In : I. P. Martini (ed.) *Canadian Inland Seas*, Elsevier Oceanographic Ser. 44, Elsevier, New York, p 117-142.
Matthews, J.V., 1989. Late Tertiary Arctic Environments: A Vision of the Future. Geos 18(3): 14-18.
Maxwell, J. B., 1986. A Climate Overview of the Canadian Inland Seas. In : I. P. Martini (ed.) *Canadian Inland Seas*, pp.79-99. Elsevier Oceanographic Ser. 44, Elsevier, New York.
Melville, C. D., 1915. Report on the East Coast Fisheries of James Bay. Ann. Rept. Can. Dept. Naval Serv. 1913-1914. Appendix, pp. 3-29.
Messier, D., R. G. Ingram & D. Roy, 1986. Physical and Biological Modifications in Response to La Grande Hydroelectric Complex. In : I. P. Martini (ed.) *Canadian Inland Seas*, pp. 403-424. Elsevier Oceanographic Ser. 44, Elsevier, New York.
Mikhail, M.Y. & H.E. Welch, 1989. Biology of Greenland Cod *Gadus ogac*, at Saqvaqjuac, Northwest Coast of Hudson Bay. Envrion. Biol. Fishes, 26: 49-62.
Milko, R. ,1986. Potential Ecological Effects of the Proposed GRAND Canal Diversion Project on Hudson and James Bays. Arctic, 39: 316-326.
Miller, D. L., P. M. Leonard, R. M. Hughes, J. R. Karr, P. B. Moyle, L. H. Schrader, B. A. Thompson, R. A. Daniels, K. D. Fausch, G. A. Fitzhugh, J. R. Gammon, D. B. Halliwell, P. L. Angermeier & D. J. Orth, 1988. Regional Applications of an Index of Biotic Integrity for Use in Water Resource Management. Fisheries 13: 12-20.
Morin, R., & J. J. Dodson 1986. The Ecology of Fishes in James Bay, Hudson Bay and Hudson Strait. In : I. P. Martini (ed.) *Canadian Inland Seas*, pp. 293-325. Elsevier Oceanographic Ser. 44, Elsevier, New York.
Morrison, R. I. G. & A. J. Gaston, 1986. Marine and Coastal Birds of James Bay, Hudson Bay and Foxe Basin. In : I. P. Martini (ed.) *Canadian Inland Seas*, pp. 355-386. Elsevier Oceanographic Ser. 44, Elsevier, New York.
Mortsch, L. (ed.) ,1990. Eastern Canadian Boreal and Sub-Arctic Wetlands: A Resource Document. Environment Canada and the Canadian Institute for Research in Atmospheric Chemistry, Climatological Studies 42. Downsview, Ontario, 169p.
Mowat, F., 1967. *Canada North*. The Canadian Illustrated Library, McClelland and Stewart Ltd, Toronto, Ontario, 128p.
Muir, D. C. G., C. A. Ford, R. E. A. Stewart, T. G. Smith, R. F. Addison, M. E. Zinck, & P. Beland. 1990. Organochlorine Contaminants in Beluga, *Delphinapterus leucas*, from Canadian Waters. In : T. G. Smith, D. J. St. Aubin, and J. R. Geraci (eds) Advances in Research on the Beluga Whale, *Delphinapterus leucas*, pp. 165-189. Can. Bull. Fish. Aquat. Sci. 224.
Muir, D. C. G., R. J. Nostrom & M. Simon, 1988. Organochlorine Contaminants in Arctic Marine Food Chains: Accumulation of Specific Polychlorinated Biphenyls and Chlordane-related Compounds. Environ. Sci. Technol. 22: 1071-1079.

Muir, D. C. G., R. Wagemann, W. L. Lockhart, N. P. Grift, B. Billeck & D. Mether, 1986. Heavy Metal and Organic Contaminants in Arctic Marine Fishes. Environ. Studies 42, Indian and Northern Affairs Canada, Ottawa, Ontario, 64p.

Muir, D.C.G., R. Wageman, B.T. Hargrave, D.J. Thomas, D.B. Peakall & R.J. Nostrom, 1992. Arctic Marine Ecosystem Contamination. Sci. Total Environ. 122: 75-134.

Mulvihill, P.R. & P. Jacobs, 1991. Towards New South/North Development Strategies in Canada. Alternatives 18(2): 34-39.

Murray, J. L. & R.G. Shearer (eds), 1992. Synopsis of Research Conducted Under the 1991/92 Northern Contaminants Program. Environmental Studies 68. Dept. Indian Affairs and Northern Development, Ottawa. 213p.

Myers, R. A., S. A. Akenhead & K. F. Drinkwater, 1990. The Influence of Hudson Bay Runoff and Ice Melt on the Salinity of the Inner Newfoundland Shelf. Atmos. - Ocean 28: 241-256.

Nakashima, D. J. & D. J. Murray, 1998. The Common Eider (*Somateria mollissima sedentaria*) of Eastern Hudson Bay: A Survey of Nest Colonies and Inuit Ecological Knowledge. Environmental Studies Revolving Funds Report, No.102, Ottawa, xxiv + 174p.

(NHR) Native Harvest Research, 1982a. Harvests of the James Bay Cree 1977-1978 and 1978-79. James Bay and Northern Quebec Harvesting Research Committee, Montreal, Quebec.

(NHR) Native Harvest Research, 1982b. The Wealth of the Land. Wildlife Harvests by the James Bay Cree 1972-73 to 1978-79. James Bay and Northern Quebec Harvesting Research Committee, Montreal, Quebec.

Natural Resources Manitoba, 1991. Five Year Report on the Status of Forestry. Winnipeg, Manitoba. 2Vol.

(NSL) Natural Science of Canada Ltd, 1970. *The Arctic Coast*. Illustrated Natural History of Canada, McClelland and Stewart Ltd, Toronto, Ontario, 160p.

Nettleship, D. N. & D. B. Peakall, 1987. Organochlorine Residue Levels in Three High Arctic Species of Colonially Breeding Sea Birds from Prince Leopold Island. Mar. Poll. Bull. 18: 434-438.

Newbury, R. W., G. K. McCullough, & R. E. Hecky, 1984. The Southern Indian Lake Impoundment and Churchill River Diversion. Can. J. Fish. Aquat. Sci. 41: 548-557.

Nilsson, A. & L. Håkanson, 1992. Relationships Between Drainage Area Characteristics and Lake Water Quality. Environ. Geol. Water Sci. 19: 75-81.

Norris, A. W., 1986. Review of Hudson Platform Palaeozoic Stratigraphy and Biostratigraphy. In : I. P. Martini (ed.) *Canadian Inland Seas*, pp. 17-42. Elsevier Oceanographic Ser. 44, Elsevier, New York.

Norstrom, R. J. & D. C. G. Muir, 1988. Long-Range Transport of Organochlorines in the Arctic and Sub-Arctic: Evidence from Analysis of Marine Mammals and Fish. In : N. W. Schmidtke (ed.), *Chronic Effects of Toxic Contaminants in Large Lakes*, pp.83-111. Lewis Publ. Inc., Chelsea, Michigan.

Norstrom, R. J., M. Simon & D. C. G. Muir, 1990. Polychlorinated Dibenzo-p-Dioxins and Dibenzofurans in Marine Mammals in the Canadian North. Environ. Poll. 66: 1-19.

Norstrom, R. J., D. C. G. Muir, C. A. Ford, M. Simon, C. R. Macdonald & P. Béland, 1992. Indications of P450 Monooxygenase Acitivities in Beluga (*Delphinapterus leucas*) and Narwhal (*Monodon monoceros*) From Patterns of PCB, PCDD and PCDF Accumulation. Mar. Environ. Res. 34: 267-272.

Norstrom, R. J., M. Simon, D. C. G. Muir & R. E. Schweinsburg, 1988. Organochlorine Contaminants in Arctic Marine Food Chains: Identification, Geographical Distribution, and Temporal Trends in Polar Bears. Environ. Sci. Technol. 22: 1063-1071.

Nriagu, J. O. & H. K. Wong, 1983. Selenium Pollution of Lakes Near the Smelters at Sudbury, Ontario. Nature 301: 55-57.

Okrainetz, G., 1992. Towards a Sustainable Future in Hudson Bay. Northern Perspectrives 20(2): 12-16.

Oldfield, J. E., 1992. Risks and Benefits in Agricultural Uses of Selenium. Abstr. Fifth Internat. Symp. Selenium in Biology and Medicine, July 20-23, 1992, Vanderbilt University, Nashville, Tennessee.

Oveden, L., 1989. Peatlands: A Leaky Sink in the Global Carbon Cycle. Geos 18 (3): 19-24.

Pearse, P. H., F. Bertrand & J. W. MacLaren, 1985. Currents of Change: Final Report of the Inquiry on Federal Water Policy. Ottawa, Ontario, 222p.

Pelletier, E., 1985. Mercury-Selenium Interactions in Aquatic Organisms: A Review. Mar. Environ. Res. 18: 111-132.

Pelletier, E., 1986a. Modification de la Bioaccumulation du Selenium chez *Mytilus edulis* en Présence du Mercure Organique et Inorganique. Can. J. Fish. Aquat. Sci. 43: 203-210.

Pelletier B. R., 1986b. Seafloor Morphology and Sediments. In : I. P. Martini (ed.) *Canadian Inland Seas*, pp.143-162. Elsevier Oceanographic Ser. 44, Elsevier, New York.

Percy, J. A. (ed.), 1990. Proceedings of a Workshop: Marine Ecosystem Studies in Hudson Strait. November 9-11, 1989. Can. Tech. Rept. Fish. Aquat. Sci. 1770. Ottawa, Ontario.

Perusse, M. 1990. Grande Baliene: Mercury in the Natural Environment. Serv. Rech. Envir. Sante Publique. Unpubl. Rept. Hydro-Québec, Montreal, Québec.

Peterson, E.B., 1992. Cumulative Effects Aspects of Ontario Hydro's Proposed Hydroelectric Potential in the Moose River Basin, Ontario. Unpubl. Rept. prepared for Moose River/James Bay Coalition Ontario Hydro Demand/Supply Plan Hearing Panel 3e6, Cumulative Impacts. Western Ecological Services Ltd., Victoria,

British Columbia. 49p.

Power, G. & D. R. Barton, 1987. Some Effects of Physiographic and Biotic Factors on the Distribution of Anadromous Arctic Charr (*Salvelinus alpinus*) in Ungava Bay, Canada. Arctic 40: 198-203.

Prinsenberg, S. J., 1980. Man-Made Changes in the Freshwater Input Rates of Hudson and James Bays. Can. J. Fish. Aquat. Sci. 37: 1101-1110.

Prinsenberg, S. J., 1986a. Salinity and Temperature Distributions of Hudson Bay and James Bay. In : I. P. Martini (ed.) *Canadian Inland Seas*, pp.163-186. Elsevier Oceaonographic Ser. 44, Elsevier, New York.

Prinsenberg, S. J., 1986b. The Circulation Pattern and Current Structure of Hudson Bay. In : I. P. Martini (ed.) *Canadian Inland Seas*, pp.187-204. Elsevier Oceanographic Ser. 44, Elsevier, New York.

Prinsenberg, S. J., 1986c. On the Physical Oceanography of Foxe Basin. In : I. P. Martini (ed.) *Canadian Inland Seas*, pp. 217-236. Elsevier Oceanographic Ser. 44, Elsevier, New York.

Prinsenberg, S. J., 1991. Effects of Hydro-Electric Projects on Hudson Bay's Marine and Ice Environments. Hydro-Electric Development: Environmental Impacts. Paper No. 2. North Wind Information Services, Inc., Montreal. 8p.

Prinsenberg, S. J. & N. G. Freeman, 1986. Tidal Heights and Currents in Hudson Bay and James Bay. In : I. P. Martini (ed.) *Canadian Inland Seas*, pp. 205-216. Elsevier Oceanographic Ser. 44, Elsevier, New York.

Prinsenberg, S. J. & R.J. Ingram, 1991. Under-Ice Physical Oceanographic Processes. J. Mar. Systems, 2: 143-152.

Quigley, N.C. & N.J. McBride, 1987. The Structure of an Arctic Microeconomy: The Traditional Sector in Community Economic Development. Arctic 40: 204-210.

Rampino, M. R. & R. Etkins, 1990. The Greenhouse Effect, Stratospheric Ozone, Marine Productivity, and Global Hydrology: Feedbacks in the Global Climate System. In : R. Paepe, R. W. Fairbridge, and S. Jejgersma (eds), *Greenhouse Effect, Sea Level and Drought*, pp. 3-20. Kluwer Acad. Publ., Netherlands.

Rapport, D. J., 1989. Symptoms of Pathology in the Gulf of Bothnia (Baltic Sea): Ecosystem Response to Stress from Human Activity. Biol. J. Linnean Soc. 37: 33-49.

(RAAS) Rawson Academy of Aquatic Science, 1989. Towards an Ecosystem Charter for the Great Lakes - St. Lawrence. Rawson Academy of Aquatic Science, Occasional Paper 1, Ottawa, Ontario, 112p.

(RAAS) Rawson Academy of Aquatic Science, 1993. Annotated Bibliography on the Biophysical Environment of the Hudson Bay and James Bay Regions. Unpubl. Rept. for Environment Canada; Rawson Academy of Aquatic Science, Ottawa, Ontario.

Ray, A.J., 1990. *The Canadian Fur Trade in the Industrial Age*. University of Toronto Press, Toronto. 283p.

Reed, A., 1991. Subsistence Harvesting of Waterfowl in Northern Quebec: Goose Hunting and the James Bay Cree. Trans. 56th. N. Amer. Wildl. & Nat. Res. Conf. pp.344-349.

Reed, A. & A. J. Erskine, 1986. Population of the Common Eider in Eastern North America: Their Size and Status. In : A. Reed (ed.), Eider Ducks in Canada, pp.156-162. Can. Wildl. Serv. Rept. Ser. 47. Ste-Foy, Quebec.

Reed, A., P. Dupuis, & G. E. J. Smith, 1987. A Survey of Lesser Snow Geese on Southampton and Baffin Islands, NWT, 1979. Can. Wildl. Serv. Occasional Paper No.61. 24p.

Reeves, R. R. & E. Mitchell, 1988. Status Report on the White Whales, *Delphinapterus leucas*, in Ungava Bay and Eastern Hudson Bay. COSEWIC, 41p.

Renozi, A. & R. J. Norstrom, 1990. Mercury in the Hairs of Polar Bears (*Ursus maritimus*). Polar Record 26: 326-328.

Rimmer, C.C. 1987. A Literature Review on the Significance of James Bay to Migratory Birds. New England/ Hydro-Québec +/-450 kV Transmission Line Interconnection, Phase II. U.S. Dept. Energy Final Environmental Impact Statement. C52-C69.

Rimmer, C.C., 1992. James Bay: Birds at Risk. Amer. Birds 46: 216-218.

Roff, J. C. & L. Legendre, 1986. Physico-chemical and Biological Oceanography of Hudson Bay. In : I. P. Martini (ed.) *Canadian Inland Seas*, pp. 265-291. Elsevier Oceanographic Ser. 44, Elsevier, New York.

Rogan, W., B. Gladden, J. McKinney, N. Carreras, P. Hardy, J. Thullen, J. Tinglestad & M. Tully, 1986. Neonatal Effects of Transplacental Exposure to PCBs and DDE. J. Pediatrics 109: 335-341.

Roots, E. F., 1987. Developing Policies for Responding to Climate Change: High Latitude Regions. Keynote Address, Workshop on Developing Policies for Responding to Climate Change, Sept. 28 - Oct. 2, 1987, Villach, Austria, Unpubl. Manuscript 38p.

Rosenberg, D. M., R. A. Bodaly, R. E. Hecky & R. W. Newbury, 1987. The Environmental Assessment of Hydroelectric Impoundments and Diversions in Canada. In: M. C. Healy and R. R. Wallace (eds) Canadian Aquatic Resources, pp.71-104. Can. Bull. Fish. Aquat. Sci. 215.

Rougerie, J. F., 1990. James Bay Development Project: Hydroelectric Development in Northwestern Québec. Canadian Water Watch 3:56-58.

Roy, D., 1989. Physical and Biological Factors Affecting the Distribution and Abundance of Fishes in the Rivers Flowing into James Bay and Hudson Bay. In : D. P. Dodge (ed.) Proceedings of the International Large Rivers Symposium (LARS), pp.159-171. Can. Spec. Publ. Fish. Aquat. Sci. 106.

Runge, J.A. & R.G. Ingram, 1991. Under-Ice Feeding and Diel Migration by Planktonic Copepods *Calanus glacialis* and *Pseudocalanus minutus* in Relation to the Ice Algal Production Cycle in Southeastern Hudson

Bay, Canada. Mar. Biol. 108: 217-225.
Ryan, K. G., 1992. UV Radiation and Photosynthetic Production in Antarctic Sea Ice Microalgae. J. Photochem. Photobiol. B: Biol. 13: 235-240.
Ryder, R. A., 1965. A Method for Estimating the Potential Fish Production of North-Temperate Lakes. Trans. Am. Fish. Soc. 94: 214-218.
Ryder, R. A., 1982. The Morphoedaphic Index - Use, Abuse, and Fundamental Concepts. Trans. Am. Fish. Soc. 111: 154-164.
Ryder R. A. & S. R. Kerr, 1989. Environmental Priorities: Placing Habitat in Heirarchic Perspective. Can. J. Fish. Aquat. Sci. Special Publ. 105, pp 2-12.
St. Aubin, D. J. & J. R. Geraci, 1989. Seasonal Variation in Thyroid Morphology and Secretion in the White Whale, *Delphinapterus leucas*. Can. J. Zool. 67: 263-267.
St. Aubin, D. J., T. G. Smith & J. R. Geraci, 1990. Seasonal Epidermal Moult in Beluga Whales, Delphinapterus leucas. Can. J. Zool. 68: 359-367.
Salki, A., M. Turner, K. Patalas, J. Rudd & D. Findlay, 1985. The Influence of Fish-Zooplankton-Phytoplankton Interactions on the Results of Selenium Toxicity Experiments Within Large Enclosures. Can. J. Fish. Aquat. Sci. 42: 1132-1143.
Savard J.-P.L., 1988. Memorandum sur les Developments Hydro-Electriques a la Baie James et leurs Impacts Possibles sur l'Aviefaune: Identification du Problème et des Lacunes dans nos Connaisances. Unpubl. Rept. Canadian Wildlife Service, Ste-Foy, Québec. 55p.
Savard J.-P.L. & P. Lamothe, 1991. Distribution, Abundance, and Aspects of Breeding Ecology of Black Scoters, *Melanitta nigra*, and Surf Scoters *M. perspicillata*, in Northern Quebec. Can. Field-Nat. 105: 488-496.
Schindler, D. W., 1987. Detecting Ecosystem Responses to Anthropogenic Stress. Can. J. Fish. Aquat. Sci. 44 (Suppl 1): 6-25.
Schindler, D. W., K. G. Beaty, E. J. Fry, D. R. Cruikshank, E. R. DeBruyn, D. L. Findlay, G. A. Linsey, J. A. Shearer, M. P. Stainton & M. A. Turner, 1990. Effects of Climatic Warming on Lakes of the Central Boreal Forest. Science, 250: 967-970.
Schneider, E. D. & J. J. Kay, 1992. Life as a Manifestation of the Second Law of Thermodynamics. Unpubl. Working Paper Series, Department of Environment and Resource Studies, University of Waterloo, Waterloo, Ontario, 32p.
Schuurman, L., R. Preston, F. Berkes & P. George, 1992. Cultural-Historical Reconstruction of New Post. TASO Rept. Second Ser. No.7, McMaster University, Hamilton, Ontario.
Scott, C., 1986. Hunting Territories, Hunting Bosses and Communal Production Among Coastal James Bay Cree. Anthropologia (N.S.) 28: 163-173.
Scott, C., 1992. Political Spoils or Political Largesse? Regional Development in Northern Quebec, Canada and Australia's Northern Territory. Centre for Aboriginal Economic Policy Research Discussion Paper 27. Australian National University, Canberra, Australia.
Sergeant, D. E., 1973. Biology of White Whales (*Delphinapterus leucas*) in Western Hudson Bay. J. Fish. Res. Bd. Canada, 30: 1065-1090.
Sergeant, D. E., 1986. Sea Mammals. In : I. P. Martini (ed.) *Canadian Inland Seas*, pp. 327-340. Elsevier Oceanographic Ser. 44, Elsevier, New York.
Shearer, R.G. (ed.), 1991. Synopsis of Research Conducted Under the 1990/91 Northern Contaminants Program. Unpubl. Rept., Depart. Indian Affairs and Northern Development, Ottawa. 84p.
Shilts, W. W. 1986. Glaciation of the Hudson Bay Region. In: I. P. Martini (ed.) *Canadian Inland Seas*, pp. 55-78. Elsevier Oceanographic Ser. 44, Elsevier, New York.
Sinclair W. F., 1990. Controlling Pollution from Canadian Pulp and Paper Manufacturers: A Federal Perspective. Conservation and Protection, Environment Canada, Ottawa, Ontario, 360p.
Slemr, F. & E. Langer, 1992. Increase in Global Atmospheric Concentrations of Mercury Inferred from Measurements Over the Atlantic Ocean. Nature 335: 434-437.
Sly, P. G., 1992. Ecosystems Health: An Example of Implications Arising from Regulations Under the Canadian Environmental Protection Act (CEPA), and the Importance of Setting Precedent. J. Aquat. Ecosyst. Health 1: 39-48.
Sly, P. G.& W. -D. N. Busch, 1992. Introduction to the Process, Procedure, and Concepts Used in the Development of an Aquatic Habitat Classification System for the Great Lakes. In: W. -D. N. Busch, and P. G. Sly (eds), *The Development of an Aquatic Habitat Classification System for Lakes*, pp. 2-13. CRC Press, Boca Raton, Florida.
Sly, P. G., 1994. Human Impacts on the Hudson Bay Bioregion, its Present State and Future Environmental Concerns. Science Rept. to the Hudson Bay Programme, Rawson Academy of Aquatic Science, Ottawa, Ontario.
Smith, M., 1990. Potential Responses of Permafrost to Climatic Change. J. Cold Regions Engng. 4: 29-37.
Smith, R. C., 1989. Ozone, Middle Ultraviolet Radiation and the Aquatic Environment. Photochem. Photobiol. 50: 459-468.

Smith, R. C. & K. S. Baker, 1979. Penetration of UV-B and Biologically Effective Dose-rates in Natural Waters. Photochem. Photobiol. 29: 311-323.
Smith, R. C., B. B. Prezelin, K. S. Baker, R. R. Bidigare, N. P. Boucher, T. Coley, D. Karentz, S. MacIntyre, H. A. Matlick, D. Menzies, M. Ondrusek, Z. Wan & K. J. Waters, 1992. Ozone Depletion: Phytoplankton Biology in Antarctic Waters. Science 255: 952-959.
Sonntag, N. C., L. A. Greig, J. D. Meisner, & J. Koonce, 1991. Development of a Great Lakes - St. Lawrence Ecosystem Model. ESSA Environmental and Social Systems Analysts Ltd., Toronto, Ontario. Report for the International Joint Commission Regional Office, Windsor, Ontario.
Spaling, H. & B. Smit, 1993. Cumulative Environmental Change: Conceptual Frameworks, Evaluation Approaches and Institutional Perspectives. Environmental Management (in press).
Speyer, M. R., 1980. Mercury and Selenium Concentrations in Fish, Sediments, and Water of Two Northwestern Québec Lakes. Bull. Environ. Contam. Toxicol. 24: 427-432.
Srikumar, T. S. & B. Akesson, 1992. Occurrence of Low and High Molecular Weight Selenium Compounds in Fish. Abstr. Fifth Internat. Symp. Selenium in Biology and Medicine, July 20-23, 1992, Vanderbilt University, Nashville, Tennessee.
Statistics Canada, 1986. *Human Activity and the Environment. Statistics Canada*, Ottawa, Ontario, 374p.
Stephenson, K., 1991. The Community of Moose Factory: A Profile. TASO Report, Second Series, No. 2, McMaster University, Hamilton, Ontario, 101p.
Stewart, D.B., L.M.J. Bernier & M.J. Dunbar, 1991. Marine Natural Areas of Canadian Significance in the Hudson Bay Marine Region. Unpubl. Rept. for Canadian Parks Service, Ottawa; Arctic Biological Consultants, Winnipeg, Manitoba, 241 p.
Stirling, I. & H. Cleator (eds), 1981. Polynyas in the Canadian Arctic. Can. Wildl. Serv. Occasional Paper No.45. 70p.
Stirling, I. & M. A. Ramsay, 1986. Polar Bears in Hudson Bay and Foxe Basin: Present Knowledge and Research Opportunities In: I. P. Martini (ed.) *Canadian Inland Seas*, Elsevier Oceanographic Ser. 44, Elsevier, New York, p 341-354.
Sutcliffe, W. H. Jr., R. H. Loucks, K. F. Drinkwater & A. R. Coote, 1983. Nutrient Flux on the Labrador Shelf from Hudson Strait and its Biological Consequences. Can. J. Fish. Aquat. Sci. 40: 1692-1701.
(SEPA) Swedish Environmental Protection Agency, 1991. Mercury in the Environment: Problems and Remedial Measures in Sweden. Swedish Environ. Protect. Agency, Solna, Sweden, 36p.
Symington, F., 1978. *The First Canadians*. Canada's Illustrated Heritage, McClelland and Stewart Ltd, Toronto, Ontario, 128p.
Thomas, D. J., B. Tracey, H. Marshall & R. J. Nostrom, 1992. Arctic Terrestrial Ecosystem Contamination. Sci. Total Environ. 122: 135-164.
Thomas, V. G. & J. P. Prevett, 1982. The Roles of James Bay and Hudson Bay Lowland in the Annual Cycle of Geese. Naturaliste Can. 109: 913-925.
Topolniski, D., S. Kerwan & A. Kristofferson, 1987. Economic Opportunities in the Commercial Fisheries of the Northwest Territories. Northern Perspectives 15(4): 8-10.
(TFN) Tungavik Federation of Nunavut, 1989. An Inuit Response. Northern Perspectives 17(1): 13-18.
Turner, M. A. & A. L. Swick, 1983. The English-Wabigoon River System: IV. Interaction Between Mercury and Selenium Accumulated from Waterborne and Dietary Sources by Northern Pike (*Esox lucius*). Can. J. Fish. Aquat. Sci. 40: 2241-2250.
Twitchell, K., 1991. The Not-so-Pristine Arctic. Can. Geographic (Feb/Mar), p 53-60.
(UNEP) United Nations Environment Programme, 1989. Action on Ozone. UNEP, Nairobi, Kenya,16p.
Upper Lakes Reference Group, 1977. The Waters of Lake Huron and Lake Superior: Lake Superior, Vol. 3 (Parts A and B). Internat. Joint Comm., Windsor, Ontario, 575p.
Urbatique (Inc.), 1991. Charactérisation Socio-Economique Démographique et Urbanistique des Villages de Whapmagoostui et Kuujjuarapik. Final Rept. for Vice President Hydro-Québec, Montreal. 55p.
Usher, P. J., 1981. Sustenance or Recreation? The Future of Native Wildlife Harvesting in Northern Canada. In : M. M. R. Freeman (ed.). Renewable Resources and the Economy of the North, Proc. First Internat. Symp. Renewable Resources and the Economy of the North, May 1981, Banff, Alberta. Assoc. Can. Univ. for Northern Studies, Ottawa, Ontario.
Usher, P. J., & M. S. Weinstein, 1991. Towards Assessing the Effects of Lake Winnipeg Regulation and Churchill River Diversion on Resource Harvesting in Native Communities in Northern Manitoba. Can. Tech. Rept Fish. Aquat. Sci. 1794. Ottawa, Ontario, 69p.
Valentine, J. L., M. E. Cebrian, B. Faraji, J. Kuo & P. A. Lachenbruch, 1992. Daily Selenium Intake Estimates for Residents of Arsenic Edemic Areas. Abstr. Fifth Internat. Symp. Selenium in Biology and Medicine, July 20-23, 1992, Vanderbilt University, Nashville, Tennessee.
Veilleux, L., R.G. Ingram & A. van der Boaren, 1992. A Description of Summer Physical Oceanographic Conditions in Rupert Bay (James Bay, Canada). Arctic 45: 258-268.
Wagemann, R., R. E. A. Stewart, P. Beland & C. Desjardines, 1990. Heavy Metals and Selenium in Tissues of Beluga Whales, Delphinapterus leucas, from the Canadian Arctic and the St. Lawrence Estuary. In : T. G.

Smith, D. J. St. Aubin & J. R. Geraci (eds). Advances in Research on the Beluga Whale *Delphinapterus Leucas*, pp.191-206. Can. Bull. Fish. Aquat. Sci. 224.

Wallin, M., 1991. Nutrient Loading Models for Coastal Waters. Department of Physical Geography, Uppsala University. UNGI Rept # 80, Uppsala, Sweden, 38p.

Wang, J., A.L. Mysak & R.G. Ingram, 1991. Interannual Variability of the Atmospheric Circulation and Sea Ice Cover in the Hudson Bay-Baffin Bay-Labrador Sea Region 1953-88. Proc. Fifth Conf. on Climatic Variations, October 14-18, 1991, pp.358-361. Denver, Colorado. Amer. Meteorol. Soc.

Welch, H. E. & M.A. Bergmann, 1989. Seasonal Development of Ice Algae and its Prediction from Environmental Factors Near Resolute, Northwest Territories, Canada. Can. J. Fish. Aquat. Sci. 46: 1793-1804.

Welch, H. E. & J. Kalff, 1975. Marine Metabolism at Resolute Bay, Northwest Territories. Circumpolar Conference on Northern Ecology, Vol. II, 69-75.

Welch, H. E., M.A. Bergmann, T.D. Siferd & P.S. Amarvalik, 1991. Seasonal Development of Ice Algae Near Chesterfield Inlet, Northwest Territories, Canada. Can. J. Fish. Aquat. Sci. 48: 2395-2402.

Welch, H. E., D. C. G. Muir, B. N. Billeck, W. L. Lockhart, G. J. Brunskill, H. J. Kling, M. P. Olson & R. M. Lemoine, 1991. Brown Snow: A Long-range Transport Event in the Canadian Arctic. Environ. Sci. Technol. 25: 280-286.

Wickware, G. M. & C. D. A. Rubec, 1989. Ecoregions of Ontario. Ecological Land Classification Series 26, Environment Canada, Ottawa, Ontario, 37p.

(WCED) World Commission on Environment and Development (Chair: G. H. Bruntland), 1987. *Our Common Future*. Oxford University Press, Oxford, U. K.

World Wildlife Fund Canada, 1992. Endangered Spaces. Open Report to Canadian Ministers of the Environment, World Wildlife Fund, Canada, Ottawa, Ontario, 42p.

Wren, C. D. & P. M. Stokes, 1988. Depressed Mercury Levels in Biota from Acid and Metal Stressed Lakes Near Sudbury, Ontario. Ambio, 17: 28-30.

Wren, C. D., P. M. Stokes & K. L. Fischer, 1986. Mercury Levels in Ontario Mink and Otter Relative to Food Levels and Environmental Acidification. Can. J. Zool. 64: 2854-2859.

Wright D. S. & G. D. Greene, 1987. An Environmental Impact Methodology for Major Resource Developments. J. Environ. Management 24: 1-16.

Young, T. K., 1988. Are SubArctic Indians Undergoing the Epidemiologic Transition? Soc. Sci. Med., 26: 659-671.

Workshop sessions

Two workshops were arranged in connection with the symposium, to discuss the following subjects:

1. The means of assessing the ecotoxicological effects and risks related to chemicals in Northern conditions for the administrative decision-making procedures
2. Data gaps in research concerning chemicals in the Northern environment and possibilities for co-operation in research

Mr. Alf Lundgren (Sweden), acted as the chairperson for Workshop I, and Mr. Jukka Malm (Finland), as the reporter.

Dr. Lyle Lockhart (Canada), acted as the chairperson and opening speaker for Workshop II, and Dr. Rolv Lundheim (Norway), as the reporter.

The following is a presentation of the themes dealt with at the workshops:

1. Workshop I

1.1 *Hazard identification*

1.1.1 *Is new information on the intrinsic properties of the chemicals needed?*

Hazard identification can be defined as the identification of the adverse effects which a chemical has an inherent capacity to cause. Hence, hazard identification is based on the data of the intrinsic properties of chemicals. For environmental hazard identification, data are needed on physical-chemical properties (*e.g.* water solubility, vapour pressure), on environmental fate (abiotic degradability, biodegradability, bioaccumulation potential), and on toxicity to organisms. In the workshop session, the OECD Premarketing Set of Data was regarded as reference data to provide a basis for the discussion.

There was a general agreement in the group that the current OECD Premarketing Set of Data provides a sufficient basis for administrative screening purposes. Although the need for further data on intrinsic properties of chemicals (*e.g.* toxicity to terrestrial organisms) may be justified, this need is not based on the specific conditions of the Arctic-Boreal environment.

1.1.2 *Is development/modification of test guidelines needed?*

In most cases the testing of chemicals in order to obtain data on the intrinsic properties of chemicals is carried out according to internationally accepted test guidelines. This practice is of crucial importance in order to get reliable and comparable data on chemicals. The most widely accepted guidance is the OECD Guidelines for Testing of Chemicals. The test conditions in the guidelines try to some extent, reflect the situation in the real environment. In some cases, however, the differences between the standardized test conditions and the Arctic-Boreal ecosystems are dramatic.

For the toxicity testing there is, in most cases, some flexibility in the guidelines when choosing the test species. On the other hand, it can be the case that none of the species recommended in the guidelines are present or a key species in the local environment (*e.g.* Arctic-Boreal) for which a risk assessment should be carried out is missing.

Although there may be considerable variation between the susceptibility of different species to the toxic actions of chemicals, species-specific effects were not regarded as a major concern in the administrative hazard identification procedure. However, the use of endogenous species in chemicals testing should not be encouraged.

Although several studies concerning the impact of low temperature, the presence of humic substances, and other relevant factors affecting toxicity already exist, it was the consensus of the group that more data and better understanding on this area are needed.

Concerning fate testing, the best documented and possibly also the most important aspect is the effect of the temperature on the degradation of chemicals. The relatively high temperature (20–25 °C) recommended in the biodegradability test guidelines was identified to be a major problem. So far, there is not enough information for the extrapolation of the biodegradability test results to Arctic-Boreal field conditions.

The impact of temperature as well as different soil and water parameters on the mobility and bioaccumulation potential of chemicals is not sufficiently understood.

Although the need for more detailed and locally applicable information on the hazardous properties of the chemicals is obvious, the workshop participants were clearly aware of the problem that even the present guidelines for chemicals testing are not sufficiently used, and consequently, the lack of basic data on environmental effects of chemicals is one of the main problems in this area.

1.2 *Risk assessment*

1.2.1 *Are present fate models applicable to Arctic-Boreal conditions?*

Regionally or locally modified models are used in many countries for pesticide assessment. Many workshop participants indicated a desire for the promotion of information exchange on the development and use of these models.

Concerning industrial chemicals, the primary obstacle for exposure assessment in most cases is the lack of data on the industrial use and emission of chemical substances. So far the use of fate/exposure models have not been very common. In the near future, however, exposure models are going to be used more often, for example, in the systematic investiga–tion of the existing chemicals. Since risk assessment should be based on a realistic worst case approach, it should be studied because the Arctic-Boreal environmental conditions (low degradation rate), could in some cases lead to underestimation of the exposure of the environment to a certain chemical.

1.2.2 *Are field tests and monitoring data used in the assessment?*

Although field tests are in most situations regarded as valid and a valuable source of information, they are not used very commonly. One major reason for this is the high costs of such testing.

Monitoring programmes exist today in many countries. The programmes may cover the area of a whole country and go on for decades, or they can be focused on a certain site and a limited period of time. The monitored substances are in most cases well known persistent and bioaccumulative toxicants. Especially for industrial chemicals, a careful priority setting is needed when selecting the substances to be monitored. One possibility to use monitoring could be to, in some cases, set a condition for pesticide approval by registration authorities.

Some biological effect monitoring programmes are ongoing. An exchange of the experiences of those programmes between authorities and researchers in different countries is desirable.

2. Workshop II

In Workshop II the participants were asked for their opinions on the research needs and proposals for co-operation in research. The most important concerns that were brought up are presented below. The discussion mostly dealt with research needs, and included short comments on the need for co-operation in research.

2.1 Air-borne pollution

2.1.1 Atmospheric data on chlorohydrocarbons
A possible source of trichloroacetic acid (TCA) and monochloroacetic acid (MCA) in northern ecosystems may be the photochemical transformation of other chlorohydrocarbons. Cold condensation may explain higher levels in the north than in more temperate areas. Data on chlorohydrocarbons in the atmosphere of the Arctic and Boreal areas in Europe are lacking. As a starting point to look into these problems, it was proposed that chlorohydrocarbons in the atmosphere should be monitored. The existing network of air quality stations could be used. It was also proposed that mosses and lichens for these substances should be analized and the air samples measured directly.

There was a general agreement on the need for ecotoxicological studies with regard to the chemicals in order to link the measured levels to the effects on the ecosystem. Conifer seedlings and Arctic fish were proposed for testing.

The work could possibly be organized in some form of co-operation within the EEC, OECD or the Nordic Council. The studies should also be co-ordinated with the Arctic Monitoring and Assessment Programme (AMAP).

2.1.2 Modelling
Most of the physical data on chemicals are measured at room temperature. To make good models, data are needed on the physical properties of chemicals under the conditions prevailing in the atmosphere and in the Arctic. In addition, data are needed on the emission and use of chemicals.

2.2 Features of the environment

2.2.1 Behaviour of chemicals in snow and ice
We know that chemicals are released in pulses during snow melting, but more detailed information on what happens to the chemistry of soil and water during snow melting is needed. Also physical data on the behaviour of chemicals in snow are needed, *e.g.* partition coefficients of snow/air.

2.2.2 Seasonal dynamics of lipid content
Animals often use lipids as an form of energy storage before the cold season. Chemicals which are accumulated in the deposited lipids may be released when the lipids are metabolised and possibly raise the levels of toxic substances in other organs. Studies of chemicals which are accumulated in lipids, and the effect of starvation are therefore relevant.

2.2.3 *Cold hardening mechanisms and effects of low temperature*

The mechanisms of cold hardening and seasonal and thermal acclimation are vulnerable to pollution.

All poikilothermic organisms in Arctic and Boreal areas have to possess special adaptations to cope with low temperatures during the winter. If these adaptations are disturbed, the animals will die as a result of the severe winter temperatures in those areas.

Effects of chemicals on the cold hardening mechanisms of Arctic and Boreal organisms have been shown. This is a problem specific to cold regions which has been given little attention and should be focused on in the future.

The physiological, cellular and molecular mechanisms of the actions of chemicals should be studied in natural low-temperature conditions.

A combination of the approaches, methods and experiences of ecotoxicologists and thermal physiologists is needed.

The impact of low temperature on the uptake, bioconcentration and elimination of chemicals should be studied.

2.3. Development of test methods

2.3.1 *Calibration of biomarkers*

There is a need for calibration of biomarkers concerning levels of pollution. There is also the problem of separating natural variation from pollution-induced changes.

One possible way to find relevant parameters to study may be to aim at finding the mechanisms of action of a given chemical.

2.3.2 *Population size as a monitoring parameter*

The use of population size as a monitoring parameter was criticised because of its limitations. First, the casual relationship between a certain type of pollution and the population size is often unclear. Second, when a decline in the population is registered, the environment may already be severely damaged. Therefore, there is a need for early warning parameters.

2.3.3 *Speciation*

The toxicity of a given chemical may vary with regard to charge, isomers, etc. It is essential to identify such properties when measuring levels of a pollutant.

2.4 Field studies

2.4.1 *Lower trophic levels*

There is little information on the lower trophic levels in limnic and marine systems with regard to the effects of pollution. It was therefore suggested to put more effort into studies of plankton.

2.4.2 *Biodegradation of chemicals in soil*

The biodegradation rates of chemicals in soil have been shown to be different in northern soils as compared to the rates in more temperate areas. It was therefore proposed to study biodegradation in soils from different climatic zones to find the potential unique features of northern soils.

List of Participants

Ingela Andersson
National Chemicals
Inspectorate
P.O.Box 1384
S–171 27 Solna, SWEDEN

Sari Autio
Finnish Environment Agency
Chemicals Control Unit
P.O. Box 140
FIN-00251 Helsinki, FINLAND

Hannu Braunschweiler
Finnish Environment Agency
Chemicals Control Unit
P.O. Box 140
FIN-00251 Helsinki, FINLAND

Kim Dahlbo
Finnish Environment Agency
Chemicals Control Unit
P.O. Box 140
FIN-00251 Helsinki, FINLAND

L.G. Emelyanova
Moscow State University
Geographical Dept.
P.O. Box 148
117574 Moscow B-574,
RUSSIA

Sten Engblom
Water and Environment District of Vaasa
P.O. Box 262
65101 Vaasa, FINLAND

Inger Grethe England
State Pollution Control Authority
P.O. Box 8100 DEP
N-0032 Oslo, NORWAY

Alan E. Etzel
Kaskenpolttajantie 12 A
00670 Helsinki, FINLAND

Kaj Forsius
Helsinki University of
Technology
Otakaari 1
02150 Espoo, FINLAND

Tom Frisk
Water and Environment District of
Tampere
P.O. Box 297
SF-33101 Tampere, FINLAND

Tone Karin Frost
ALLFORSK/AVH
Dept. of ecotoxicology
N-7055 Dragvoll, NORWAY

Björn Ganning
Ambio
Royal Swedish Academy of Sciences
P.O. Box 50005
10405 Stockholm, SWEDEN

Lise Grønlie
ALLFORSK/AVH
Dept. of ecotoxicology
N-7055 Dragvoll, NORWAY

Johannes Hahl
Monsanto Oy
Rautatiekatu 14 B 5
15110 Lahti, FINLAND

Jari Haimi
University of Jyväskylä
Dept. of Biology
P.O. Box 35
40351 Jyväskylä, FINLAND

Taina Hammar
Water and Environment District of
Kuopio
P.O. Box 49
70101 Kuopio, FINLAND

Erkki Haukioja
University of Turku
Dept. of Biology
Laboratory of Ecology and
Animal Systematics
20500 Turku, FINLAND

Kaisa Heinonen
Finnish Environment Agency
P.O. Box 140
FIN-00251 Helsinki, FINLAND

Jussi Hautala
Finnewos Agri Oy
P.O. Box 239
20101 Turku, FINLAND

Anne Hernesmaa
University of Helsinki
Dept. of Applied Chemistry and
Microbiology
P.O. Box 27
Viikki, Building B
00014 Helsinki, FINLAND

Olga Hilmo
University of Trondheim
Dept. of Botany
7055 Dragvoll, NORWAY

Juha-Pekka Hirvi
Finnish Environment Agency
P.O. Box 140
FIN-00251 Helsinki, FINLAND

Satu Huttunen
University of Oulu
Linnanmaa
90570 Oulu, FINLAND

Merja Itävaara
Technical Research Centre of
Finland
Food Research Laboratory
P.O. Box 203
02151 Espoo, FINLAND

Maija Jarva
Finnish Forest Research
Institute
P.O. Box 18
01301 Vantaa, FINLAND

Jouni Jokela
University of Helsinki
Dept. of Applied Chemistry and
Microbiology
P.O. Box 27
Viikki, Building B
00014 Helsinki, FINLAND

Jukka Järvi
Water and Environment district of
Helsinki
P.O. Box 58
00421 Helsinki, FINLAND

Mai Järvinen
Helsinki University of Technology
Otakaari 1
02150 Espoo, FINLAND

Leena Kajaste
Vanha Pisantie 9 F
02280 Espoo, FINLAND

Johanna Kankaanranta
Helsinki University of Technology
Otakaari 1
02150 Espoo, FINLAND

Aarno Karels
Helsinki University of Technology
Otakaari 1
02150 Espoo, FINLAND

Birgit Kemiläinen
Finnish Environment Agency
Chemicals Control Unit
P.O. Box 140
FIN-00251 Helsinki, FINLAND

List of participants

Anu Kettunen
University of Helsinki
Dept. of Applied Chemistry and
Microbiology
P.O. Box 27
Viikki, Building B
00014 Helsinki, FINLAND

Merja Kiljunen
Ministry of the Environment
P.O. Box 399
00121 Helsinki, FINLAND

Pirkko Kivelä-Ikonen
Ministry of the Environment
P.O. Box 399
00121 Helsinki, FINLAND

Jarle Klugsøyr
Inst. of Marine Research
P.O. Box 1870
N-5024 Bergen-Nordnes,
NORWAY

Seppo Knuuttila
Finnish Environment Agency
Chemicals Control Unit
P.O. Box 140
FIN-00251 Helsinki, FINLAND

Jaana Koistinen
University of Jyväskylä
Dept. of Chemistry
P.O. Box 35
40351 Jyväskylä, FINLAND

Marko Kolari
University of Helsinki
Dept. of Applied Chemistry and
Microbiology
P.O. Box 27
Viikki, Building B
00014 Helsinki, FINLAND

Kaija Korhonen
Finnish Environment Agency
P.O. Box 140
FIN-00251 Helsinki, FINLAND

Markku Korhonen
Finnish Environment Agency
P.O. Box 140
FIN-00251 Helsinki, FINLAND

Michail Kozlow
University of Turku
Dept. of Biology
Laboratory of Ecology and
Animal Systematics
20500 Turku, FINLAND

Eszer Kostyål
University of Helsinki
Dept. of Applied Chemistry and
Microbiology
P.O.Box 27
Viikki
00014 Helsinki, FINLAND

Jussi Kukkonen
University of Joensuu
Dept. of Biology
P.O. Box 111
80101 Joensuu, FINLAND

Arto Kultamaa
Finnish Environment Agency
Chemicals Control Unit
P.O. Box 140
FIN-00251 Helsinki, FINLAND

Anne-Mari Kurka
Finnish Forest Research Institute
P.O. Box 18
00131 Vantaa, FINLAND

Juha Kämäri
Finnish Environment Agency
P.O. Box 140
FIN-00251 Helsinki, FINLAND

Kari Lagerspetz
University of Turku
Dept. of Biology
20500 Turku, FINLAND

Minna Laine
University of Helsinki
Dept. of Applied Chemistry and
Microbiology
P.O. Box 27
Viikki
00014 Helsinki, FINLAND

Jaana Lehtimäki
University of Helsinki
Dept. of Applied Chemistry and
Microbiology
P.O. Box 27
Viikki, Building B
00014 Helsinki, FINLAND

Erkki Leppäkoski
Åbo Akademi
Institute of biology
Biocity
P.O. Box 2
20521 Turku, FINLAND

Marja-Liisa Lindell
Finnish Environment Agency
P.O. Box 140
FIN-00251 Helsinki, FINLAND

W. Lyle Lockhart
Freshwater Institute
501 University Crescent
Winnipeg, Manitoba,
CANADA R3T 2N6

Martin Lodenius
University of Helsinki
Dept. of Env. Conservation
P.O. Box 27
Viikki
00014 Helsinki, FINLAND

Alf Lundgren
Swedish National Chemicals
Inspectorate
P.O. Box 1384
S-171 27 Solna, SWEDEN

Rolv Lundheim
ALLFORSK
Dept. of Ecotoxicology
7055 Dragvoll, NORWAY

Marita Luotamo
Institute of Occupational Health
Laboratory of Biochemistry
Arinatie 3
00370 Helsinki, FINLAND

Marja Luotola
Finnish Environment Agency
P.O. Box 140
FIN-00251 Helsinki, FINLAND

Christina Lye
Pesticides Approval Division
National Chemicals Inspectorate
P.O. Box 1384
S-171 27 Solna, SWEDEN

Jukka Malm
Finnish Environment Agency
Chemicals Control Unit
P.O. Box 140
FIN-00251 Helsinki, FINLAND

Merja Manninen
Water and Environment District of
Northern Carelia
P.O.Box 69
80101 Joensuu, FINLAND

Pertti Manninen
Water and Environment District of
Mikkeli
Jääkärinkatu 14
50100 Mikkeli, FINLAND

Jaakko Mannio
Finnish Environment Agency
P.O. Box 140
FIN-00251 Helsinki, FINLANDD

Esko Martikainen
Institute for Environmental
Research
University of Jyväskylä
P.O. Box 35
40351 Jyväskylä, FINLAND

Tuula Matilainen
Finnish Environment Agency
c/o University of Helsinki
Dept. of Limnology
P.O. Box 27
00014 Helsinki, FINLAND

List of participants

Katri Mattila
University of Helsinki
Dept. of Applied Chemistry and
Microbiology
P.O. Box 27
Viikki, Building B
00014 Helsinki, FINLAND

Matti Melanen
Finnish Environment Agency
P.O. Box 140
FIN-00251 Helsinki, FINLAND

Jukka Meriluoto
University of Helsinki
Dept. of Applied Chemistry and
Microbiology
P.O. Box 27
Viikki, Building B
00014 Helsinki, FINLAND

Veijo Miettinen
Finnish Environment Agency
P.O. Box 140
FIN-00251 Helsinki, FINLAND

Arun B. Mukherjee
University of Helsinki
Dept. of Limnology and
Environmental Protection
P.O. Box 27
Viikki
00014 Helsinki, FINLAND

Mohiuddin Munawar
Fisheries and Oceans Canada
Canada Centre for Inland Waters
867 Lakeshore Rd.
P.O. Box 5050
Burlington, Ontario
CANADA L7R 4A6

Irma Mäkinen
Finnish Environment Agency
P.O. Box 140
FIN-00251 Helsinki, FINLAND

Tarja Nakari
Finnish Environment Agency
P.O. Box 140
FIN-00251 Helsinki, FINLAND

Esa Nikunen
Finnish Environment Agency
Chemicals Control Unit
P.O. Box 140
FIN-00251 Helsinki, FINLAND

Eeva Nurmi
Finnish Environment Agency
Chemicals Control Unit
P.O. Box 140
FIN-00251 Helsinki, FINLAND

Beryl C. Nygreen
State Pollution Control Authority
P.O. Box 8100 DEP
N-0032 Oslo, NORWAY

Magnus Nyström
Finnish Environment Agency
Chemicals Control Unit
P.O. Box 140
FIN-00251 Helsinki, FINLAND

Aimo Oikari
Helsinki University of Technology
Otakaari 1
02150 Espoo, FINLAND

Jaakko Paasivirta
University of Jyväskylä
Dept. of Chemistry
P.O. Box 35
SF-40351 Jyväskylä, FINLAND

Sirpa Paattakainen
Finnish Environment Agency
P.O. Box 140
FIN-00251 Helsinki, FINLAND

Jukka Pellinen
University of Joensuu
Dept. of Biology
P.O. Box 111
80101 Joensuu, FINLAND

Rainer Peltola
University of Helsinki
Dept. of Applied Chemistry and
Microbiology
P.O. Box 27
Viikki, Building B
00014 Helsinki, FINLAND

Tiina Petänen
Helsinki University of Technology
Otakaari 1
02150 Espoo, FINLAND

Seppo Peuranen
Finnish Game and Fisheries Research Institute,
Fisheries Division
P.O. Box 202
SF-00151 Helsinki, FINLAND

Petri Porvari
Finnish Environment Agency
P.O. Box 140
FIN-00251 Helsinki, FINLAND

Helena Poutanen
Finnish Environment Agency
P.O. Box 140
FIN-00251 Helsinki, FINLAND

Tarja Pyykkö
Finnish Environment Agency
P.O. Box 140
FIN-00251 Helsinki, FINLAND

Tiina Rantio
University of Jyväskylä
Dept. of Chemistry
P.O. Box 35
40351 Jyväskylä, FINLAND

Jarkko Rapala
University of Helsinki
Dept. of Applied Chemistry and Microbiology
P.O. Box 27
Viikki, Building B
00014 Helsinki, FINLAND

Marko Reinikainen
University of Turku
Dept. of Biology
20500 Turku, FINLAND

Mirja Salkinoja-Salonen
University of Helsinki
Dept. of Applied Chemistry and Microbiology
P.O. Box 27
Viikki, Building B
00014 Helsinki, FINLAND

Janne Salminen
University of Jyväskylä
Dept. of Biology
P.O. Box 35
40351 Jyväskylä, FINLAND

Outi Salminen
Helsinki University of Technology
Otakaari 1
02150 Espoo, FINLAND

Simo Salo
Finnish Environment Agency
P.O. Box 140
FIN-00251 Helsinki, FINLAND

Eila Salomaa
Helsinki University of Technology
Otakaari 1
02150 Espoo, FINLAND

Eija Saski
University of Helsinki
Dept. of Applied Chemistry and Microbiology
P.O. Box 27
Viikki, Building B
00014 Helsinki, FINLAND

Kaija Savelainen
Water and Environment District of Helsinki
P.O. Box 58
00421 Helsinki, FINLAND

Vladimir Savinov
Murmansk Marine Biological Inst.
Kola Science Centre
Russian Academy of Sciences
17, Vladimirskaya St.
183023 Murmansk, RUSSIA

List of participants

Tatyana Savinova
Murmansk Marine
Biological Inst.
Kola Science Centre
Russian Academy of Sciences
17, Vladimirskaya St.
183023 Murmansk, RUSSIA

Jussi Seitsonen
University of Helsinki
Dept. of Applied Chemistry and
Microbiology
P.O. Box 27
Viikki, Building B
00014 Helsinki, FINLAND

Riitta Silvo
Finnish Environment Agency
Chemicals Control Unit
P.O. Box 140
FIN-00251 Helsinki, FINLAND

Kaarina Sivonen
University of Helsinki
Dept. of Applied Chemistry and
Microbiology
P.O. Box 27
Viikki, Building B
00014 Helsinki, FINLAND

Ulla Sonck
Finnish Environment Agency
P.O. Box 140
FIN-00251 Helsinki, FINLAND

Eiliv Steinnes
University of Trondheim
Dept. of Chemistry
7055 Dragvoll, NORWAY

Anna-Mari Suortti
Finnish Environment Agency
P.O. Box 140
FIN-00251 Helsinki, FINLAND

Aino Tamsi-Joensuu
Water and Environment District of
Helsinki
P.O. Box 58
00421 Helsinki, FINLAND

Per Morten Tessem
ALLFORSK/AVH
Dept. of ecotoxicology
Overfossveien 50
7000 Trondheim, NORWAY

Lennart Torstensson
Swedish Academy of Agricultural
Sciences
Dept. of Microbiology
Box 7025
S–750 07 Uppsala, SWEDEN

Kristiina Tuominen
Helsinki University of Technology
Otakaari 1
02150 Espoo, FINLAND

Henrik Tyle
National Agency of
Environmental Protection
Strandgade 29
1401 Copenhagen, DENMARK

Lilian Törnqvist
Pesticides Approval Division
National Chemicals Inspectorate
P.O. Box 1384
S-171 27 Solna, SWEDEN

Pekka Vanhala
Finnish Environment Agency
P.O. Box 140
FIN-00251 Helsinki, FINLAND

Matti Verta
Finnish Environment Agency
P.O. Box 140
FIN-00251 Helsinki, FINLAND

Anna-Mari Virkkala
Helsinki University of Technology
Otakaari 1
02150 Espoo, FINLAND

Merja Vuorinen
Finnish Game and Fisheries
Re–search Institute
Fisheries Division
P.O. Box 202
00151 Helsinki, FINLAND

Pekka Vuorinen
Finnish Game and Fisheries
Research Institute,
Fisheries Division
P.O. Box 202
SF-00151 Helsinki, FINLAND

Heidi Vuoristo
Finnish Environment Agency
P.O. Box 140
FIN-00251 Helsinki, FINLAND

Anna-Mari Walls
University of Turku
Dept. of Biology
20500 Turku, FINLAND

Frank Wania
University of Toronto
Dept. of Chemical Engineering and
Applied Chemistry
200 College Street
Toronto, Ontario,
CANADA M5S 1A4

Karl-Erik Zachariassen
Trondheim Universitet
ALLFORSK
Dept. of Ecotoxicology
7055 Dragvoll, NORWAY

Gennadi Zaitsev
University of Helsinki
Dept. of Applied Chemistry and
Microbiology
P.O. Box 27
Viikki, Building B
00014 Helsinki, FINLAND

Xuejiao Zhou
Helsinki University of Technology
Otakaari 1
02150 Espoo, FINLAND

Leena Ylä-Mononen
Finnish Environment Agency
Chemicals Control Unit
P.O. Box 140
FIN-00251 Helsinki, FINLAND

Virpi Äystö
Helsinki University of Technology
Otakaari 1
02150 Espoo, FINLAND

Frode Ødegaard
Allforsk AVH
Dept. of Ecotoxicology
N-7055 Dragvoll, NORWAY